Converging Nanochemistry: From Disease Prevention to Diagnosis and Treatment

Converging Nanochemistry: From Disease Prevention to Diagnosis and Treatment

Editor

Jahir Orozco Holguín

Basel • Beijing • Wuhan • Barcelona • Belgrade • Novi Sad • Cluj • Manchester

Editor
Jahir Orozco Holguín
Max Plack Tandem Group in
Nanobioengineering, Institute
of Chemistry, Faculty of
Natural and Exacts Sciences
University of Antioquia
Medellín
Colombia

Editorial Office
MDPI
St. Alban-Anlage 66
4052 Basel, Switzerland

This is a reprint of articles from the Special Issue published online in the open access journal *Molecules* (ISSN 1420-3049) (available at: www.mdpi.com/journal/molecules/special_issues/nano_bioengineering).

For citation purposes, cite each article independently as indicated on the article page online and as indicated below:

Lastname, A.A.; Lastname, B.B. Article Title. *Journal Name* **Year**, *Volume Number*, Page Range.

ISBN 978-3-0365-8651-9 (Hbk)
ISBN 978-3-0365-8650-2 (PDF)
doi.org/10.3390/books978-3-0365-8650-2

Contents

About the Editor

Jahir Orozco Holguín

He is the Group Leader of the Max Planck Tandem Group in Nanobioengineering (MPTG-N) (COL0202529) at the University of Antioquia, Colombia. He is a chemist from the same University and has a Ph.D. in chemistry from the University of Barcelona and the Institute of Microelectronics of Barcelona (Spain), 2008. In addition, he completed three postdoctoral stays for seven years at the Banyuls sur Mer Oceanological Observatory (France), at the Department of Nanoengineering at University of California San Diego (USA), and at the Catalan Institute of Nanoscience and Nanotechnology (Spain). He has been awarded Beatriu de Pinós Spanish Scholarships (2010 and 2012) and the Severo Ochoa Excellence Fellow (2015), and recognised for his outstanding professional activity at the Faculty of Exact and Natural Sciences (University of Antioquia), Colombia (2017). He holds the title of Senior Researcher in Minciencias and is an Editorial Board Member of the *Molecules* journal. His areas of interest include nano(bio)technology, nano(micro)carriers, nanomotors, nano(micro)devices, nanomaterials, (bio)sensors, electrochemistry, and the development of analytical tools for environmental and biomedical applications.

Preface

Nanochemistry converges at the frontier of basic sciences and, in a joint effort with nanotechnology and nanobioengineering, searches for new or existing (bio)molecules and materials, thus explaining their interactions and providing novel functionalities for new products and unexpected applications. This exciting marriage offers outstanding opportunities for developing innovative platforms with potential use in prevention, diagnostics, and therapeutics within the healthcare field, among many others. For example, nanobioprobes and nanobioconjugates may provide alternatives for the prevention and convenient real-time diagnosis of diseases closer to the patient and opportunities for more efficient drug delivery and targeted therapeutics compared with conventional technologies. In this context, the rational design of biological molecules and structures and/or their assembly with nanomaterials allows the on-demand development of nanobioengineered platforms with improved properties and outstanding performance, which, when either acting alone or in synergy, are paving the way toward a new paradigm of personalized medicine.

This Special Issue covers recent research in nanochemistry convergence from disease prevention to diagnosis and treatment. It includes studies on new (bio)molecules and hybrids with diagnostic, chemopreventive, or therapeutical properties; novel nanobioengineered biosensors for disease diagnostics; and nanoparticles and nanobioconjugates for targeted drug delivery. This Special Issue collects a benchmark of state-of-the-art works on the computational simulation, structural modeling, and construction of new libraries of molecules to identify potential targets, synthesize molecules, and study their potential in developing diagnostic and therapeutic tools. It illustrates other topics that go from directed evolution techniques to create mutants of proteins with enhanced activity concerning the native structures to reviews on nanobioprobes conjugation approaches and biosensors as promising tools for disease detection. The issue also shows examples of drug delivery systems based on biocompatible and biodegradable materials for chemoprevention.

Therefore, this Special Issue expects to be of interest to a broad audience focused on (but not limited to) researchers in cancer and infectious diseases but with an impact on health care. Nine original research papers and two reviews were written. Finally, the Guest Editor wants to express gratitude to everyone who participated in the issue, committing to the quality of the work and the deadlines. It is difficult to mention everyone, but we would like to thank Nanobiocancer, a scientific ecosystem program supported by Minciencias Colombia, to which most of them belonged.

Jahir Orozco Holguín
Editor

 molecules

MDPI

Article

Selecting Nanobodies Specific for the Epidermal Growth Factor from a Synthetic Nanobody Library

Yunier Serrano-Rivero [1,†], Julieta Salazar-Uribe [1,†], Marcela Rubio-Carrasquilla [1], Frank Camacho-Casanova [2], Oliberto Sánchez-Ramos [2], Alaín González-Pose [1,*] and Ernesto Moreno [1,*]

[1] Faculty of Basic Sciences, University of Medellin, Medellin 050026, Colombia; 0905yunierserrano@gmail.com (Y.S.-R.); julieta.salazaru@gmail.com (J.S.-U.); marcelaru@yahoo.com (M.R.-C.)
[2] Pharmacology Department, School of Biological Sciences, University of Concepcion, Concepcion 4070386, Chile; fcamacho@udec.cl (F.C.-C.); osanchez@udec.cl (O.S.-R.)
*   Correspondence: agonzalezp@udemedellin.edu.co (A.G.-P.); emoreno@udemedellin.edu.co (E.M.)
†   These authors contributed equally to this work.

Abstract: The epidermal growth factor (EGF) is one of the most critical ligands of the EGF receptor (EGFR), a well-known oncogene frequently overexpressed in cancerous cells and an important therapeutic target in cancer. The EGF is the target of a therapeutic vaccine aimed at inducing an anti-EGF antibody response to sequester this molecule from serum. However, strikingly, very few investigations have focused on EGF immunotargeting. Since the use of nanobodies (Nbs) for EGF neutralization may be an effective therapeutic strategy in several types of cancer, in this study, we decided to generate anti-EGF Nbs from a recently constructed, phage-displaying synthetic nanobody library. To our knowledge, this is the first attempt to obtain anti-EGF Nbs from a synthetic library. By applying a selection strategy that uses four different sequential elution steps along with three rounds of selection, we obtained four different EGF-specific Nb clones, and also tested their binding capabilities as recombinant proteins. The obtained results are very encouraging and demonstrate the feasibility of selecting nanobodies against small antigens, such as the EGF, from synthetic libraries.

Keywords: epidermal growth factor; nanobody; synthetic library; phage display

Citation: Serrano-Rivero, Y.; Salazar-Uribe, J.; Rubio-Carrasquilla, M.; Camacho-Casanova, F.; Sánchez-Ramos, O.; González-Pose, A.; Moreno, E. Selecting Nanobodies Specific for the Epidermal Growth Factor from a Synthetic Nanobody Library. *Molecules* **2023**, *28*, 4043. https://doi.org/10.3390/molecules28104043

Academic Editor: Jahir Orozco Holguín

Received: 24 March 2023
Revised: 8 May 2023
Accepted: 9 May 2023
Published: 129 May 2023

1. Introduction

The human epidermal growth factor (EGF) is a small protein of 53 amino acids, whose fold is structured by three disulfide bonds. It demonstrates potent biological activity in vitro and in vivo by stimulating cell and organ proliferation [1]. Furthermore, the EGF is one of the most critical ligands of the EGF receptor (EGFR) [2], a well-known oncogene frequently overexpressed in cancerous cells, causing cell-cycle deregulation, exacerbated angiogenesis, apoptosis blockade, and tumoral cell migration [3]. It was found very early that EGF detection in carcinomas is closely related to high levels of malignancy [4]. Indeed, EGF signaling is essential for tumor-cell growth for several types of cancer [5,6], and it is linked to the epithelial–mesenchymal transition, in which epithelial cells are transformed into fibroblast-like phenotypes with high motility and invasive properties, contributing to cancer metastasis [7,8]. Therefore, along with the EGFR, the EGF is also considered a potential target for cancer therapy.

The prevention of cancer using our own immune system against self-antigens is possible due to the observed natural self-reactivity of immune cells against autologous antigens [9]. In the case of the EGF, a self-antibody response against this molecule would diminish the serum availability of EGF, thus reducing EGFR activation and, in consequence, cancer-cell proliferation. Following this active therapeutic approach, a novel anti-EGF vaccine named CIMAvax-EGF was developed. This vaccine is based on a chemical conjugate of EGF and the protein P64k from Neisseria meningitidis, adjuvanted with Montanide ISA

51 VG [10]. The main function of this immunogen is to break the EGF's self-tolerance, inducing an anti-EGF antibody response, which dramatically reduces the EGF concentration in serum [11]. More than 10 clinical trials have been completed with the therapeutic cancer vaccine CIMAvax-EGF, including phase II, III, and IV trials, demonstrating safety, long-term immunogenicity and a significant effect on survival, with several patients achieving long-term survival after vaccination [10].

An alternative, passive-therapy-based approach for sequestering EGF would involve different antibody formats, particularly, nanobodies (Nbs), which are single-domain antibody fragments derived from the heavy chain antibodies found in camelids [12]. Nbs have several advantageous properties over classical antibodies: small size, high stability and solubility, and easy tailoring for multiple applications. Furthermore, they can achieve high affinities despite the fact that their monomeric binding region displays only three hypervariable loops. Nbs have become a relevant class of biomolecules with multiple applications, especially as diagnostic tools and promising therapeutic agents in cancer and other diseases [13]. They are obtained mostly from immune libraries, constructed from animal immunization with the target antigen [14]. In recent years, however, synthetic nanobody libraries have gained ground as alternative Nb sources, offering several advantages, such as lower costs and faster results [15]. As these libraries are not generated for a specific antigen, they can be used for the selection of nanobodies against numerous antigens, including those with high toxicity or low immunogenicity [15].

A successful example of the use of nanobodies to sequester a circulating cytokine is ozoralizumab—a trivalent anti-TNF nanobody, which was recently approved in Japan for the treatment of rheumatoid arthritis [16]. Strikingly, despite the proven potential of EGF as a cancer target, there are very few investigations focused on EGF immunotargeting. A recent work by Guardiola et al. reported the generation of anti-EGF Nbs using an immune library from alpaca, and showed their ability to block EGFR phosphorylation and produce antitumor effects in vitro [17,18]. To our knowledge, these were the only anti-EGF nanobodies reported before the present study.

Since the use of Nbs for EGF neutralization may be an effective therapeutic strategy in several types of cancer, in this study, we decided to generate anti-EGF nanobodies from our recently constructed synthetic nanobody library, based on the phage-display platform [19]. After three selection rounds, each of which used four different serial elution methods, we obtained about forty positive recombinant phage clones. Twenty of these clones were sequenced, resulting in four different amino-acid sequences. These three distinct nanobodies were produced in BL21 (DE3) and purified by ion exchange (IEC) and immobilized metal affinity (IMAC) chromatography, and their ability to bind to EGF was demonstrated.

2. Results and Discussion

2.1. Design and Production of a Recombinant EGF Protein

Firstly, we designed and produced, in-house, a recombinant EGF (rEGF) protein with two tags (SV5 and 6xHis) that confer a more versatile functionality to this molecule (Figure 1a). The rEGF gene was cloned into the bacterial expression vector pET22b (Figure 1b) under the control of strong bacteriophage T7 transcriptional and translational signals, which allow the production of large amounts of recombinant proteins upon induction. The inclusion of the PelB-signal peptide before the rEGF gene allows secretion into the periplasmic compartment, in order to obtain a protein of interest that is soluble after bacterial lysis. The bacterial system (*E. coli* BL21 (DE3)) successfully produced the rEGF, as expected (Figure 1c). After cell lysis, a rEGF fraction remained soluble (Figure 1d), and it was successfully purified by IMAC and IEC (Figure 1e). Although part of the recombinant protein was lost in the different chromatographic steps, we were able to obtain a sufficient amount, with more than 90% purity. The biological activity of the rEGF was assessed in vitro using the human cell line A431, characterized by high EGFR overexpression [20]. The cells were treated with different rEGF concentrations, ranging from 100 pm to 100 nM. Similarly as found in early studies with EGF [21], the lowest applied rEGF concentration

(100 pM) stimulated cell growth in cultures with a low concentration (0.5%) of fetal bovine serum (FBS). This effect vanished with higher rEGF concentrations. On the other hand, the expected inhibitory effect of high rEGF concentrations (10 and 100 nM) [21] was markedly observed in A431 cells cultured with 5% FBS (Figure 1f). Having proven the functionality of the in-house-produced rEGF (hereafter called EGF), we proceeded to select EGF binders from our synthetic nanobody library.

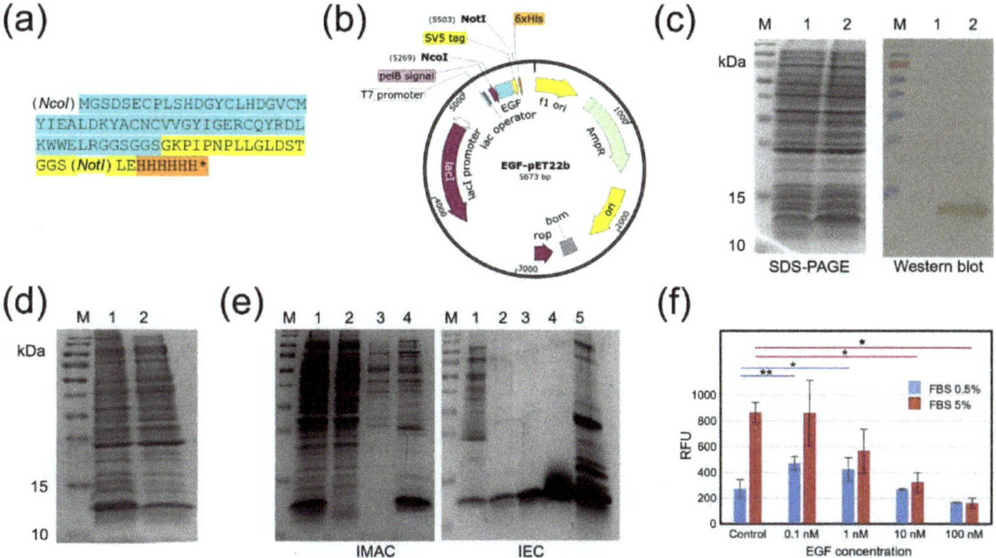

Figure 1. Production of (rEGF) in *E. coli.* (**a**) Amino-acid sequence of rEGF with C-terminal tags: cyan—EGF, yellow—SV5 tag, orange—6xHis tag. (**b**) Scheme of the pET22b-rEGF expression vector. (**c**) SDS-PAGE and Western blot of the production of rEGF in BL21 (DE3). The rEGF was identified using an anti-His tag monoclonal antibody conjugated to HRP. M: molecular-weight marker, Trident Prestained Protein Ladder GTX50875 (GeneTex, USA), 1: untransformed BL21, 2: BL21 transformed with pET22b-rEGF. (**d**) SDS-PAGE of the rEGF-soluble fraction obtained by cell lysis using several freezing/thawing rounds. M: molecular-weight marker, 1: total fraction of BL21 transformed with pET22b-hEGF, 2: soluble fraction of BL21 transformed with pET22b-hEGF. (**e**) SDS-PAGE of the rEGF-purification process. Left panel: first purification stage by IMAC. M: molecular-weight marker, 1: initial sample, 2: unbound proteins, 3: wash (40 mM imidazole), 4: elution (250 mM imidazole). Right panel: second purification stage with IEC. M: molecular-weight marker, 1: initial sample from IMAC elutions, 2, 3, and 4: unbound proteins, 5: elution from the anionic exchanger Bio-Scale Mini Macro-Prep High Q Cartridge (BioRad, Hercules, CA, USA). (**f**) In vitro evaluation of rEGF biological activity in A431 cells. Two FBS concentrations were used: (0.5%—blue bars; and 5%—red bars). Measurements were performed in triplicate. Stars indicate statistically significant differences between experimental conditions (* $p < 0.05$, ** $p < 0.005$).

2.2. Selection of EGF-Binding Phages

As a small protein, EGF presents a relatively small surface area for antibody/nanobody recognition, which presents a challenge in obtaining nanobodies from a synthetic library. We decided to implement a screening procedure based on three selection rounds, each of which was composed of four different types of elution, performed sequentially, using the following: (i) triethylamine (TEA—1), (ii) glycine-HCl (Gly—2), (iii) ultrasound (UltS—3), and (iv) the addition of the TG1 strain to previously washed wells (TG1—4). The aim was to recover binders with different physico-chemical properties that would not be eluted with TEA (a commonly used elution buffer), followed by binders not eluted either with Gly, and so on. After each round, the phages collected from each elution were pulled together for amplification, except for the final round, in which the phages from the different elutions were amplified separately.

The use of successive rounds of selection against the antigen of interest is intended to increase the proportion of antigen-specific phages [14,22]. As in this study, Guardiola et al. performed three selection rounds to obtain EGF-specific phages from an alpaca immune library [17]. In most cases, the phages bound to the target molecule were collected using a single method, which can either be triethylamine (e.g., [17,23–25]), glycine [26–28], ultrasound (albeit to a lesser extent, such as in [29,30]) and, rarely, infection by the direct addition of cells to the wells [29]. The use of a single elution method may limit phage collection, losing a fraction of phages that remain strongly bound to the immobilized target molecule in the wells. In contrast, in our study, we used four sequential elutions, recovering significant amounts of phages in each of them.

With the aim of optimizing the screening of clones by ELISA for binding to EGF, we first tested groups of selected clones, each of which consisted of four individual clones (all belonging to the same elution). We evaluated 17, 15, 16, and 16 groups from the TEA, Gly, UltS, and TG1 elutions, respectively. A plate blocked with skimmed milk was included to identify mock-phages. Thus, this assay allowed us to select groups with a high EGF binding signal and low background. Figure 2a shows the ELISA results from this screening. Following the criterion defined in the Methods, a total of 24 positive groups were obtained, which were distributed as follows: TEA-14, Gly-5, UltS-1, and TG1-4. Subsequently, the individual clones from these 24 groups were tested in the same way (Figure 2b), with one important addition. As the recombinant EGF molecule has a C-terminal tail containing SV5 and His tags, we added a third plate coated with a non-related protein with the same C-terminal tail, to discard anti-tag phages. Of the 96 individual clones, 38 met the positivity criteria, and they were distributed by elution, as follows: TEA-16, Gly-1, UltS-1, and TG1-11. A few of these clones, mostly from the TG1 elution, also showed strong anti-tag binding (Figure 2b).

2.3. Sequences of Selected Clones

Most of the EGF-positive phage clones (21 clones, discarding only a few weaker binders), as well as several of the C-tail tag binders were sent for sequencing. Figure 3 shows the sequences obtained for the CDRs (the framework is common to all the nanobodies derived from the library). The 21 EGF-positive clones corresponded to only four unique Nb sequences, and 17 of the clones were identical. These results are comparable to those obtained by Guardiola et al., who recovered six different anti-EGF clones from an immune library [17].

The use of four different elutions for binder recovery yielded intriguing results. The group of 17 identical clones was composed of phages from three elutions (TEA-8; Gly-2; TG1-7); the identical A8-1/B3-1 pair of clones and the single C10-1 clone were obtained from the TEA elution, whereas the single clone, D6-4, was recovered from the TG1 elution. The mixed origin of the repeated clones obtained after the third selection round might have been associated with the pulling of the four elutions from rounds 1 and 2 for amplification and subsequent inputs for rounds 2 and 3, respectively. On the other hand, the unique D6-4 clone was the result of the sequential application of the four types of elution; otherwise, it would not have been recovered. The five anti-tag phage clones in fact corresponded to two unique Nb sequences. In future screenings, it would be interesting, and possibly more productive, to keep each elution separately through all the selection rounds.

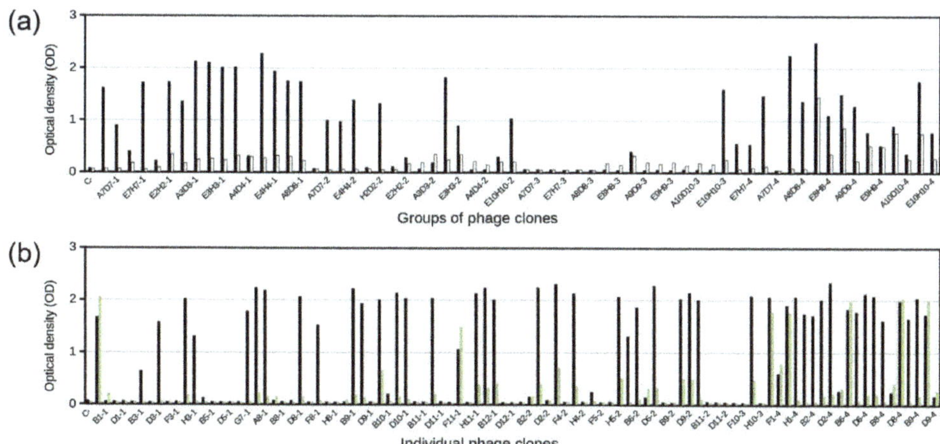

Figure 2. ELISA screening of selected phage clones for binding to EGF. (**a**) Identification of phage groups that bind to EGF (black bars). White bars correspond to background (milk) signals. (**b**) Positive phage groups were individualized and re-screened. Green bars correspond to binding to an unrelated nanobody with the same C-terminal tail as the recombinant EGF. Positivity was defined as $OD_{450nm} \geq 0.8$ for EGF binding, and an OD_{450nm} for milk or an unrelated tagged protein three or more times lower than for EGF. The last number in clone names refers to the type of elution: 1—TEA, 2—Gly, 3—UltS—3, 4—TG1.

Clone		CDR1	CDR2	CDR3
A11-2		SGDHNNQYNKFALG	AISSNGGK	ARRYQYETHRS
A2-4	★	. . . .	. . . .	. . . .
C9-1		. . . .	. . . .	. . . .
C10-1	★	. . . RQYH . SV	. . . TG . . .	. TQSGSGY . . Y
A8-1	★	. . NS . Y . . PH . D . .	. . . . G . . R	. VGSRTRNY . Y
B3-1	★	. . NS . Y . . PH . D . .	. . . . G . . R	. VGSRTRNY . Y
D6-4	★	. . TRRYK . . V . D . .	. . . TG . . R	. . . TNNANF . Y
D8-4	Tag	. . NNSYR . . R . D . .	. . . LR . . R	. MGNR . RN . D .
F11-1		. . NNSYR . . R . D . .	. . . LR . . R	. MGNR . RN . D .
A1-1		. . NRKFN . YN . D . .	. . D . S . . I	. T . TGSAYFGY
B6-4	Tag	. . NRKFN . YN . D . .	. . D . S . . I	. T . TGSAYFGY
C9-4		. . NRKFN . YN . D . .	. . D . S . . I	. T . TGSAYFGY

Framework sequence:

QVQLVESGGGSVQAGGSLRLSCTAS<u>XXXXXXXXXXXXXX</u>WFRQAPGQEREAVA<u>XXXXXXXX</u>T
YYADSVKGRFTISRDNAKNTVTLQMNNLKPEDTAIYYCAA<u>XXXXXXXXXXX</u>WGQGTQVTVS

Figure 3. CDR amino-acid sequences of selected clones. Red stars are for EGF binders, while "Tag" in blue is for tag-positive clones. Dots indicate the presence of the same amino acid as in the first sequence in the alignment. The first group of three clones (A11-2, A2-4, and C9-1), from different elutions, are representative of the group of 17 identical clones. The common framework sequence supporting the library is from the camelid nanobody cAbBCII10 [31]). CDRs are underlined, with their amino acids represented with "X".

2.4. Production of the Four Recombinant Anti-EGF Nanobodies

The four anti-EGF Nbs with different sequences (clones A8-1, A11-2, C10-1, and D6-4) were expressed as recombinant proteins by cloning their genes into pET22b and producing them in *E. coli* BL21 (DE3), which is a commonly used system for nanobody production, with a high level of success [32–34]. The recombinant-protein production behaved similarly for the four nanobodies. Upon the induction with IPTG, a majority protein band of approximately 16 kDa was observed in the four induced cultures, while it was absent in the negative control. A Western blot corroborated the observation that all four bands corresponded to recombinant nanobodies (Figure 4a). The design incorporating the signal-peptide PelB allowed the secretion of the produced proteins into the bacterial periplasm, producing a large amount of soluble protein. After production, the four nanobodies were purified by combining two chromatographic methods: ion-exchange chromatography (IEC) followed by immobilized metal-affinity chromatography (IMAC). Usually, a single IMAC purification step is not sufficient to obtain Nbs with a high degree of purity; therefore, at least one additional molecular exclusion or IEC is also used [35], being the combination of IMAC and molecular exclusion chromatography the most commonly used method for Nb purification [25,34,36]. Here, we decided to first carry out an IEC step, which removed more than 50% of the contaminating proteins, followed by IMAC, which yielded recombinant Nbs with about 90% purity (Figure 4b).

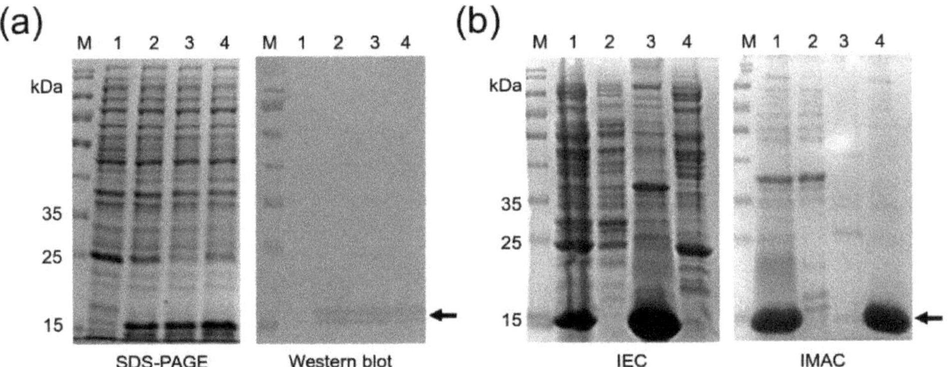

Figure 4. Production and purification of anti-EGF nanobodies. (**a**) SDS-PAGE and Western blot of nanobody production in BL21 (DE3). M: molecular-weight marker, PageRuler Prestained Protein Ladder (Thermo Fisher Scientific, USA), 1: untransformed BL21; BL21 transformed with (2) pET22b-NbA8-1, (3) pET22b-NbA11-2, and (4) pET22b-NbD6-4. The nanobodies were identified using an anti-His tag monoclonal antibody conjugated to HRP. (**b**) SDS-PAGE of the nanobody-purification process. Left panel: first purification stage by IEC. M: molecular-weight marker, 1: initial sample, 2: unbound proteins, 3: elution from the anionic exchanger, Bio-Scale Mini Macro-Prep High Q Cartridge (BioRad, USA), 4: elution from the cationic exchanger CM Sepharose Fast Flow (GE Healthcare, USA). Right panel: second purification stage by IMAC. M: molecular-weight marker, 1: initial sample from IEC, 2: unbound proteins, 3: wash (40 mM imidazole), 4: elution (250 mM Imidazole). Arrow heads indicate the protein bands corresponding to nanobodies A8-1, A11-2, and D6-4.

2.5. Binding Assays for Selected Recombinant Anti-EGF Nanobodies

The recognition of the EGF by the recombinant nanobodies was assessed by indirect ELISA. In a first attempt, we intended to use the SV5 tag in the recombinant EGF molecule for detection, by coating the wells with the recombinant nanobodies, adding the recombinant EGF and then an HRP-conjugated anti-SV5 antibody. This approach, however, yielded very weak signals. We speculate that with such a small molecule as a nanobody, adsorption on a plastic surface would reduce the binding-site availability or affect the conformation of the flexible CDR3 loop.

Since no other tag was available for detection (the His tag is common in both nanobodies and the EGF), we decided to biotinylate the recombinant nanobodies, taking advantage of the three lysine residues found in the framework region. With this method, we successfully detected the expected nanobody–antigen-binding signals, and assessed the dissociation constant (KD) for the two best binders from a titration ELISA [37] (Figure 5). The nanobodies A8-1 and D6-4 showed very similar binding strengths, with KD in the order of 10^{-7} M, while the Nbs A11-2 y C10-1 showed a weaker binding (their KD values could not be assessed).

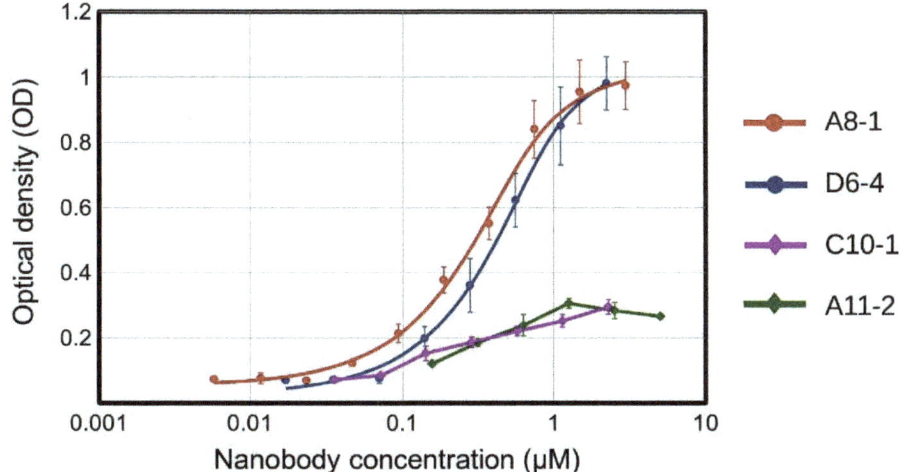

Figure 5. Binding of recombinant nanobodies to EGF by indirect ELISA. Biotinylated nanobodies were detected with streptavidin conjugated to HRP. The red and blue fitting curves were used for KD estimation, for clones A8-1 (KD = $(2.4 \pm 0.7) \times 10^{-7}$ M) and D6-4 (KD = $(3.7 \pm 0.8) \times 10^{-7}$ M), respectively. Negative controls (not shown)—BSA coating: OD = 0.11. For all Nbs, we used the maximum tested concentration for the negative controls. Experiments were performed in triplicates.

It is worth noting that while the clone A8-1 has no lysine in its CDRs, D6-4 shows a lysine in CDR1. This lysine could have been affected in the biotinylation procedure, thus abolishing or diminishing the binding capabilities of a fraction of the nanobody molecules. Therefore, the KD for the clone D6-4 might, in fact, be better. The other two clones also displayed lysine residues in their CDRs (two in A11-2, one in CDRs 1 and 2, and one in C10-1 in CDR2), which may have partially affected their binding to EGF. The impairment of the binding capacity of antibodies after biotinylation has been described, and it might be counteracted by establishing an adequate biotin:antibody ratio [38]. The incorporation of biotin into histidine, serine, threonine, and tyrosine residues has also been observed [39]. At least two of these residues are present in the CDRs of the four anti-EGF Nbs, and their possible biotinylation might have exerted a negative effect on the EGF recognition.

Despite these possible drawbacks, the affinities estimated for the clones A8-1 and D6-4 were similar to those measured by Guardiola et al. for five out of their six anti-EGF nanobodies, which showed KD values in the 10^{-7} molar order, while only one clone yielded a KD in the 10^{-8} M order [17]. That is, the anti-EGF nanobodies selected in this work from a synthetic library were comparable in terms of affinity to those obtained from an immunized alpaca. It is worth noting that these KD values are in the same order as the KD of the EGF–EGFR complex, which, in different studies using surface plasmon resonance (SPR) has been found to be within the range of 100–400 nM [40–42]. Notably, two of the anti-EGF Nbs obtained by Guardiola et al. were capable of inhibiting the binding of EGF to its receptor and EGFR phosphorylation, with IC50 constants in the micromolar order [17]. They also inhibited cell viability in tumor cells resistant to the EGFR-tyrosine-kinase inhibitor, osimertinib [18].

In this study, we applied a sequential astringent elution method for selecting specific anti-EGF Nb clones from the library, in which the triethylamine solution (suffix 1) was the first elution applied and the addition of TG1 cells (suffix 4) was the last. Therefore, it is reasonable to expect that the nanobodies obtained from the last elution should have a higher binding capacity. Paradoxically, our best binder was obtained from the first elution. This interesting outcome suggests that the nature of the molecular interactions governing nanobody–antigen binding determine the type of elution that detaches the nanobody from the immobilized antigen.

2.6. Concluding Remarks

In this study, we obtained four EGF-binding nanobodies from a synthetic library, applying a selection strategy that used four different sequential elution steps along three rounds of selection. To our knowledge, this was the first attempt to obtain anti-EGF nanobodies from a synthetic library. Notably, two of the obtained nanobodies showed KD values similar to those obtained for the Nbs derived from an immune library. It has been shown that it is possible to obtain high-affinity binders from synthetic libraries [15,43]. However, the fact that EGF is a small protein, offering a relatively small surface area for antibody/nanobody recognition, represents a major challenge. Thus, the obtained results are very encouraging. The next step will be to test the neutralizing capabilities of the obtained nanobodies. In future anti-EGF biopannings, as well as in those for other small proteins, we plan to use a different antigen-presentation system, in which the EGF is completely exposed for nanobody recognition, by, for example, using a biotinylated tag for anchoring to a streptavidin base [44]. In this way, we expect to increase the probability of obtaining high-affinity binders from the synthetic library.

3. Materials and Methods

3.1. Synthetic Library Production

The synthetic nanobody library was designed and constructed by our group as described in [19]. In this study, the library previously cloned into the phagemid pMAC and transformed in the *E. coli* strain SS320 was amplified into the amber suppressor, *E. coli* strain TG1. For this purpose, 300 mL of 2xYT medium containing 100 µg/mL ampicillin was inoculated with the TG1 strain and grown at 37 °C, and at 250 rpm, until an OD_{600nm} of 1.5. The phage library previously obtained in SS320 bacteria was used to transduce TG1 cells at a multiplicity of infection (MOI) of 58 at 37 °C for one hour (30 min static and 30 min at 50 rpm). After spinning at $2000 \times g$ for 15 min at 4 °C and discarding supernatant, the pellet was resuspended in 300 mL of 2xYT medium with 100 µg/mL ampicillin and incubated at 28 °C, 100 rpm, overnight. Next, 200 mL of the previous culture was added to a final volume of 800 mL of 2xYT medium with 100 µg/mL ampicillin and cultured at 37 °C, 250 rpm until an OD_{600nm} of 2. The culture was transduced with the helper phage M13K07 at a MOI of 25 at 37 °C for one hour (30 min static and 30 min at 50 rpm), and spined at $2000 \times g$ for 15 min at 4 °C. The pellet was diluted in 500 mL of 2xYT medium containing 100 µg/mL ampicillin, 50 µg/mL kanamycin, and 1 mM isopropyl β-D-1-thiogalactopyranoside (IPTG) for library amplification. After calculating library diversity, recombinant phages were concentrated by precipitation with PEG/NaCl (20% polyethylene glycol 8000 and 2.5 M NaCl) and aliquoted in 10% glycerol.

3.2. Production and Purification of a Recombinant EGF Protein

The gene coding the EGF was synthesized and cloned into the pET22b-expression vector (pET22b-EGF) by GenScript (Piscataway, NJ, USA). After transforming chemically competent BL21 (DE3), two inocula of 5 mL, each grown at 37 °C and at 250 rpm, overnight, in 2xYT medium with 100 µg/mL ampicillin were added to two Erlenmeyers of 1 L containing 500 mL each of Saline Minimal Medium M9 (SMM9) (45 mM Na_2HPO_4 (Sigma-Aldrich, Burlington, MA, USA), 22 mM KH_2PO_4 (Sigma, Burlington, MA, USA), 19 mM NH_4Cl (Merck, Rahway, NJ, USA), 8.4 mM NaCl (Sigma-Aldrich, Burlington, MA, USA)),

0.05% yeast extract (Oxoid, Basingstoke, UK), and 100 μg/mL ampicillin. After reaching an OD_{600nm} of 0.5, the gene expression was induced with 25 μM IPTG, and cultures were incubated at 28 °C and at 100 rpm, overnight. Next, cultures were centrifuged at 10,000× g for 15 min at 4 °C, and the pellet was resuspended in buffer A (150 mM NaCl, 15 mM Na_2HPO_4, pH 7.4) and subjected to five freeze/thaw rounds for cell lysis. Soluble rEGF was obtained in the supernatant after spinning at 10,000× g for 15 min at 4 °C. The presence of soluble EGF was verified by sodium dodecyl sulphate–polyacrylamide gel electrophoresis (SDS-PAGE) and Western blot.

For SDS-PAGE, protein samples were diluted in a buffer with beta-mercaptoethanol and run in 15% polyacrylamide and 3% stacking gels. Western blot assay was performed in a 0.2-μm PVDF-transfer membrane (Thermo Fisher Scientific, Waltham, MA, USA) and a semi-dry transfer system Trans-Blot® Turbo™ (Bio-Rad, Hercules, CA, USA) at 0.3 A and 25 V for 30 min. After blocking with 5% skimmed milk in PBS, the membrane was incubated with the HRP anti-6X His tag rabbit polyclonal antibody (ab1187, Abcam, Boston, MA, USA) diluted 1:5000 in the blocking buffer. The reaction was visualized using a DAB substrate kit (Thermo Fisher Scientific, Waltham, MA, USA).

The rEGF was purified in two chromatographic stages. (i) The lysis supernatant containing rEGF was purified by immobilized metal-affinity chromatography (IMAC) by adding 5 mM imidazole to buffer A as equilibrium buffer and the initial sample. After loading the sample, the HisPur Ni-NTA Spin Column (Thermo Fisher Scientific, Waltham, MA, USA) was washed with 25 mM imidazole, and protein was eluted with 250 mM imidazole. Samples, washing, and elution were prepared in buffer A. (ii) For ion-exchange chromatography (IEC), a serial connection of two columns was used. One column featured the weak cation-exchanger CM Sepharose Fast Flow (GE Healthcare, Chicago, IL, USA), and the other was a pre-packed column with the anion exchanger, Bio-Scale Mini Macro-Prep High Q Cartridge (BioRad, Hercules, CA, USA). Both were equilibrated with buffer B (50 mM NaCl, 7 mM Na_2HPO_4 (Sigma-Aldrich, Burlington, MA, USA), 83 mM imidazole, pH 8). Samples from IMAC were diluted three times in water and loaded into these columns. After column equilibration with the same buffer B, the anion exchanger was eluted with buffer C (500 mM NaCl, 7 mM Na_2HPO_4, pH 5), and the cation exchanger was eluted with the same buffer C at pH 10. The pH of buffers for IEC was carefully adjusted according to the theoretical isoelectric point of rEGF, calculated using Prot-Pi (https://www.protpi.ch/Calculator/ProteinTool (accessed on 10 March 2023)). All fractions were monitored using the purification system, BioLogic LP (BioRad, Hercules, CA, USA). Samples containing the rEGF were diafiltered against PBS (Sigma, USA) in Spin-X® UF concentrators of 5 kDa (Corning, Corning, NY, USA), and analyzed by SDS-PAGE and Western blot, as described above. The rEGF purity was estimated using the analytical software of the iBright 750 Imaging System (Thermo Fisher Scientific, Waltham, MA, USA), and its concentration was determined using the Pierce BCA Protein Assay Kit (Thermo Fisher Scientific, Waltham, MA, USA).

3.3. Biological Activity of Recombinant EGF

The EGF activity was determined by an in vitro assay using the A431 cell line. Ninety-six-well plates (2592, Corning, Corning, NY, USA) were used for seeding 2500 cells/well in DMEM plus 0.5% or 5% fetal bovine serum (FBS), which were treated with different EGF concentrations (0.1 nM, 1 nM, 10 nM, and 100 nM) for four days. No EGF was added to negative control. Cell viability was measured with AlamarBlue Cell Viability Reagent (Invitrogen, Waltham, MA, USA) in a Varioskan LUX Multimode Microplate Reader (Thermo Fisher Scientific, Waltham, MA, USA).

3.4. Selection of EGF-Binding Phages

Polystyrene high-binding microtiter plates (Costar) were coated with 100 μL of the EGF antigen at 10 μg/mL (eight wells) and incubated overnight at 4 °C. After three washes with phosphate-buffered saline (PBS) plus 0.1% Tween 20 (PBST), wells were blocked with

3% skimmed milk (Sigma-Aldrich, Burlington, MA, USA) in PBS (300 μL/well) for one hour. Wells were washed three times with PBST and incubated at room temperature (RT) for two hours with 100 μL of five-hundred times the library diversity diluted in 0.7% skimmed milk. The PBST was used to perform twenty washes (300 μL/well). Two additional washes were conducted with PBS. Recombinant phage collection was undertaken through sequential steps of four types of elution following neutralization. (i) Triethylamine elution: 0.1 M triethylamine, pH 12.0 (100 μL/well), for 10 min at RT, followed by neutralization with Tris-HCl 1 M pH 7.5 (100 μL/well) for 5 min at RT. (ii) Glycine elution: 0.2 M glycine, pH 2.0 (100 μL/well), for 10 min at RT, followed by neutralization with Tris-HCl 1 M pH 9.1 (100 μL/well) for 5 min at RT. (iii) Ultrasound elution: after adding 100 μL/well of PBS, wells were subjected to ultrasound (40 MHz of potency for 30 min). During the first three elutions, wells were washed twice with PBST and once with PBS (300 μL/well each). (iv) TG1 strain elution: addition of the *E. coli* strain, TG1, in exponential phase of growing (100 μL/well). This elution was performed after the three elutions described above. Simultaneously, the *E. coli* strain, TG1, was transduced with the three elutions and incubated at 37 °C for 30 min. Elutions were seeded in 2xYT plates plus 100 μg/mL ampicillin and incubated at 37 °C, and at 250 rpm, overnight.

3.5. Enrichment of Positive Phages

Three biopanning rounds were carried out to enrich the phage pull in EGF binders. For the first round, all colonies from plates obtained from the four elutions were gathered in 15 mL of 2xYT medium with 100 μg/mL ampicillin, and 5 mL was used to inoculate 50 mL of the same medium, which was incubated at 37 °C, and at 250 rpm, overnight. Next, the culture was diluted 10 times (500 mL) and grown until a OD_{600nm} of 2 was reached. Transduction with the helper phage M13KO7 was performed at a multiplicity of infection (MOI) of 20 at 37 °C for one hour (30 min static and 30 min at 50 rpm). Transduced bacteria were spun at $2000\times g$ for 15 min and diluted in 500 mL of 2xYT medium with 100 μg/mL ampicillin, 50 μg/mL kanamycin, and 1 mM IPTG for the amplification of recombinant phages at 28 °C, and at 250 rpm, overnight. Amplified phages were concentrated as previously described, for further testing against EGF. The second and third biopanning rounds were conducted in the same way, with a small variation in the final culture volume, from 500 mL to 200 mL.

3.6. Indirect ELISA for Detecting Positive Phage Clones

Polystyrene high-binding microtiter plates (Corning, Corning, NY, USA) were coated with 100 μL of EGF at 5 μg/mL and incubated overnight at 4 °C. After washing three times with PBS-0.1% Tween 20 (PBST), wells were blocked with 3% skimmed milk in PBS (300 μL/well) for one hour at RT. Supernatants of individual clones, previously amplified, were added to the plate (200 μL of supernatant) for two hours at RT. The M13KO7 phage was used as negative control. Three washes with PBST were performed, and the anti-M13 antibody (G8P) diluted 1:5000 in 3% skimmed milk was added for one hour at RT, followed by an anti-mouse antibody conjugated with horseradish peroxidase at the same dilution, time, and temperature. Plates were washed three times with PBST, and the reaction was visualized with a TMB kit solution (Thermo Fisher Scientific, Waltham, MA, USA). The reaction was stopped with 2.5 M sulphuric acid, and the absorbance was measured in a microplate reader (ES-20/80, BOECO, Hamburg, Germany) at 450 nm. Plates coated with a recombinant anti-EGFR antibody [45] (carrying the same SV5 and 6xHis tags as the recombinant EGF) at 10 μg/mL and with 3% skimmed milk were used to elude the backgrounds of recombinant phages against the SV5 and histidine tags. Positivity was considered for an OD_{450nm} signal ≥ 0.8 for EGF, and for a signal from milk or the non-related protein that was three or more times lower than for EGF.

3.7. Sequencing

The DNA from clones in TG1 were purified using the GenElute Plasmid Miniprep Kit (Sigma-Aldrich, Burlington, MA, USA) and sequenced by Macrogen (Seoul, Korea). Sequences were analyzed using the CLC Genomics Workbench v.21 (QIAGEN Aarhus, Aarhus, Denmark).

3.8. Production of Selected Recombinant Anti-EGF Nanobodies

The genes coding the anti-EGF nanobodies were extracted from the pMAC phagemid with the restriction enzymes *NcoI/NotI* (New England Biolabs, Ipswich, MA, USA) and cloned in the expression vector pET22b with the same enzymes. Nanobody production was carried out as described above, for EGF.

3.9. Nanobody Purification

The anti-EGF nanobodies were also purified in two chromatographic stages. The first was with IEC, using the same serial connection of two columns described above. Both were equilibrated with half-diluted SMM9 medium. After sample loading in the same equilibrium buffer, each column was eluted separately. as explained previously. Buffer pH was also adjusted based on the theoretical isoelectric point of nanobodies, calculated as described above. The elution from the anion exchanger was diluted three times in water and further purified by IMAC, by adding the same amounts of imidazole as previously described. Fraction monitoring, diafiltering, sample analysis, nanobody purity, and concentration were performed as described above.

3.10. Nanobody Biotinylation

Nanobodies were diluted in 25-mM carbonate/bicarbonate buffer at a concentration around 1 mg/mL. Fifty microliters of biotin (H1759, Sigma-Aldrich, Burlington, MA, USA) at 10 mg/mL in DMSO was slowly added to nanobodies. The final molar ratio biotin:nanobody was 40:1. Reaction was stirred for six hours at RT, under protection from light. Free biotin was removed by dialyzing against PBS.

3.11. ELISA for Biotinylated Anti-EGF Nanobodies

Polystyrene high-binding microtiter plates (Corning, Corning, NY, USA) were coated with 100 μL of rEGF at 5 μg/mL and incubated overnight at 4 °C. After washing three times with PBS, 0.1% Tween 20 (PBST), wells were blocked with 3% skimmed milk in PBS (300 μL/well) for 1 h at RT. Biotinylated nanobodies were serially diluted and incubated for one hour at RT. After washing, streptavidin conjugated to HRP (Biotechne R&D Systems, Minneapolis, MN, USA) diluted 1:200 was added for one hour at RT. The reaction was visualized, stopped, and measured, as described above. The KD estimation was carried out by following the method and fitting function described in [37]. A linear regression analysis using this function was performed using the MyCurveFit web server (https://mycurvefit.com/, last accessed on 4 May 2023).

Author Contributions: Conceptualization, A.G.-P. and E.M.; methodology, Y.S.-R., F.C.-C., O.S.-R., A.G.-P. and E.M.; investigation, Y.S.-R., J.S.-U., M.R.-C., F.C.-C., O.S.-R., A.G.-P. and E.M.; writing—original draft preparation, Y.S.-R., J.S.-U., M.R.-C., A.G.-P. and E.M.; writing—review and editing, Y.S.-R., F.C.-C., O.S.-R., A.G.-P. and E.M.; supervision, project administration and funding acquisition, E.M. All authors have read and agreed to the published version of the manuscript.

Funding: This research was funded by MINCIENCIAS, MINEDUCACIÓN, MINCIT and ICETEX through the Program NanoBioCancer (cod. FP44842-211-2018, project number 58676). A.G.-P. and E.M. thank the University of Medellin for its support.

Institutional Review Board Statement: Not applicable.

Informed Consent Statement: Not applicable.

Data Availability Statement: The data presented in this study are contained in the article tables.

Conflicts of Interest: The authors declare no conflict of interest.

Sample Availability: Not applicable. The phagemids and molecules used in this work were either purchased or produced in limited amounts to perform the reported experiments only.

References

1. Cohen, S.; Carpenter, G. Human Epidermal Growth Factor: Isolation and Chemical and Biological Properties. *Proc. Natl. Acad. Sci. USA* **1975**, *72*, 1317–1321. [CrossRef]
2. Singh, B.; Carpenter, G.; Coffey, R.J. EGF Receptor Ligands: Recent Advances. *F1000Research* **2016**, *5*, 2270. [CrossRef]
3. Mitsudomi, T. Molecular Epidemiology of Lung Cancer and Geographic Variations with Special Reference to EGFR Mutations. *Transl. Lung Cancer Res.* **2014**, *3*, 205–211. [PubMed]
4. Tahara, E.; Sumiyoshi, H.; Hata, J.; Yasui, W.; Taniyama, K.; Hayashi, T.; Nagae, S.; Sakamoto, S. Human Epidermal Growth Factor in Gastric Carcinoma as a Biologic Marker of High Malignancy. *Jpn. J. Cancer Res.* **1986**, *77*, 145–152. [PubMed]
5. Liu, T.-C.; Jin, X.; Wang, Y.; Wang, K. Role of Epidermal Growth Factor Receptor in Lung Cancer and Targeted Therapies. *Am. J. Cancer Res.* **2017**, *7*, 187–202. [PubMed]
6. Cheaito, K.; Bahmad, H.F.; Jalloul, H.; Hadadeh, O.; Msheik, H.; El-Hajj, A.; Mukherji, D.; Al-Sayegh, M.; Abou-Kheir, W. Epidermal Growth Factor Is Essential for the Maintenance of Novel Prostate Epithelial Cells Isolated from Patient-Derived Organoids. *Front. Cell Dev. Biol.* **2020**, *8*, 571677. [CrossRef]
7. Kalluri, R.; Weinberg, R.A. The Basics of Epithelial-Mesenchymal Transition. *J. Clin. Investig.* **2009**, *119*, 1420–1428. [CrossRef]
8. Heerboth, S.; Housman, G.; Leary, M.; Longacre, M.; Byler, S.; Lapinska, K.; Willbanks, A.; Sarkar, S. EMT and Tumor Metastasis. *Clin. Transl. Med.* **2015**, *4*, 6. [CrossRef]
9. Gonzalez, G.; Montero, E.; Leon, K.; Cohen, I.R.; Lage, A. Autoimmunization to Epidermal Growth Factor, a Component of the Immunological Homunculus. *Autoimmun. Rev.* **2002**, *1*, 89–95. [CrossRef]
10. Saavedra, D.; Crombet, T. CIMAvax-EGF: A New Therapeutic Vaccine for Advanced Non-Small Cell Lung Cancer Patients. *Front. Immunol.* **2017**, *8*, 269. [CrossRef]
11. García, B.; Neninger, E.; de la Torre, A.; Leonard, I.; Martínez, R.; Viada, C.; González, G.; Mazorra, Z.; Lage, A.; Crombet, T. Effective Inhibition of the Epidermal Growth Factor/Epidermal Growth Factor Receptor Binding by Anti–Epidermal Growth Factor Antibodies Is Related to Better Survival in Advanced Non–Small-Cell Lung Cancer Patients Treated with the Epidermal Growth Factor Cancer Vaccine. *Clin. Cancer Res.* **2008**, *14*, 840–846. [CrossRef] [PubMed]
12. Muyldermans, S. Nanobodies: Natural Single-Domain Antibodies. *Annu. Rev. Biochem.* **2013**, *82*, 775–797. [CrossRef] [PubMed]
13. Morrison, C. Nanobody Approval Gives Domain Antibodies a Boost. *Nat. Rev. Drug Discov.* **2019**, *18*, 485–487. [CrossRef]
14. Muyldermans, S. A Guide to: Generation and Design of Nanobodies. *FEBS J.* **2021**, *288*, 2084–2102. [CrossRef] [PubMed]
15. Valdés-Tresanco, M.S.; Molina-Zapata, A.; Pose, A.G.; Moreno, E. Structural Insights into the Design of Synthetic Nanobody Libraries. *Molecules* **2022**, *27*, 2198. [CrossRef] [PubMed]
16. Keam, S.J. Ozoralizumab: First Approval. *Drugs* **2023**, *83*, 87–92. [CrossRef] [PubMed]
17. Guardiola, S.; Varese, M.; Sánchez-Navarro, M.; Vincke, C.; Teixidó, M.; García, J.; Muyldermans, S.; Giralt, E. Blocking EGFR Activation with Anti-EGF Nanobodies via Two Distinct Molecular Recognition Mechanisms. *Angew. Chem. Int. Ed.* **2018**, *57*, 13843–13847. [CrossRef]
18. Guardiola, S.; Sánchez-Navarro, M.; Rosell, R.; Giralt, E.; Codony-Servat, J. Anti-EGF Nanobodies Enhance the Antitumoral Effect of Osimertinib and Overcome Resistance in Non-Small Cell Lung Cancer (NSCLC) Cellular Models. *Med. Oncol.* **2022**, *39*, 195. [CrossRef]
19. Contreras, M.A.; Serrano-Rivero, Y.; González-Pose, A.; Salazar-Uribe, J.; Rubio-Carrasquilla, M.; Soares-Alves, M.; Parra, N.C.; Camacho-Casanova, F.; Sánchez-Ramos, O.; Moreno, E. Design and Construction of a Synthetic Nanobody Library: Testing Its Potential with a Single Selection Round Strategy. *Molecules* **2023**, *28*, 3708. [CrossRef]
20. Haigler, H.; Ash, J.F.; Singer, S.J.; Cohen, S. Visualization by Fluorescence of the Binding and Internalization of Epidermal Growth Factor in Human Carcinoma Cells A-431. *Proc. Natl. Acad. Sci. USA* **1978**, *75*, 3317–3321. [CrossRef] [PubMed]
21. Kawamoto, T.; Sato, J.D.; Le, A.; Polikoff, J.; Sato, G.H.; Mendelsohn, J. Growth Stimulation of A431 Cells by Epidermal Growth Factor: Identification of High-Affinity Receptors for Epidermal Growth Factor by an Anti-Receptor Monoclonal Antibody. *Proc. Natl. Acad. Sci. USA* **1983**, *80*, 1337–1341. [CrossRef] [PubMed]
22. Anand, T.; Virmani, N.; Bera, B.C.; Vaid, R.K.; Vashisth, M.; Bardajatya, P.; Kumar, A.; Tripathi, B.N. Phage Display Technique as a Tool for Diagnosis and Antibody Selection for Coronaviruses. *Curr. Microbiol.* **2021**, *78*, 1124–1134. [CrossRef] [PubMed]
23. Rojas, G.; Pupo, A.; Gómez, S.; Krengel, U.; Moreno, E. Engineering the Binding Site of an Antibody against *N*-Glycolyl GM3: From Functional Mapping to Novel Anti-Ganglioside Specificities. *ACS Chem. Biol.* **2013**, *8*, 376–386. [CrossRef]
24. Hu, Y.; Lin, J.; Wang, Y.; Wu, S.; Wu, J.; Lv, H.; Ji, X.; Muyldermans, S.; Zhang, Y.; Wang, S. Identification of Serum Ferritin-Specific Nanobodies and Development towards a Diagnostic Immunoassay. *Biomolecules* **2022**, *12*, 1080. [CrossRef] [PubMed]
25. Hu, Y.; Zhang, C.; Lin, J.; Wang, Y.; Wu, S.; Sun, Y.; Zhang, B.; Lv, H.; Ji, X.; Lu, Y.; et al. Selection of Specific Nanobodies against Peanut Allergen through Unbiased Immunization Strategy and the Developed Immuno-Assay. *Food Sci. Hum. Wellness* **2023**, *12*, 745–754. [CrossRef]

26. Kaczmarek, J.Z.; Skottrup, P.D. Selection and Characterization of Camelid Nanobodies towards Urokinase-Type Plasminogen Activator. *Mol. Immunol.* **2015**, *65*, 384–390. [CrossRef]
27. Lakzaei, M.; Rasaee, M.J.; Fazaeli, A.A.; Aminian, M. A Comparison of Three Strategies for Biopanning of Phage-scFv Library against Diphtheria Toxin. *J. Cell Physiol.* **2019**, *234*, 9486–9494. [CrossRef]
28. Kulpakko, J.; Juusti, V.; Rannikko, A.; Hänninen, P.E. Detecting Disease Associated Biomarkers by Luminescence Modulating Phages. *Sci. Rep.* **2022**, *12*, 2433. [CrossRef]
29. Lunder, M.; Bratkovič, T.; Urleb, U.; Kreft, S.; Štrukelj, B. Ultrasound in Phage Display: A New Approach to Nonspecific Elution. *Biotechniques* **2008**, *44*, 893–900. [CrossRef]
30. Donatan, S.; Yazici, H.; Bermek, H.; Sarikaya, M.; Tamerler, C.; Urgen, M. Physical Elution in Phage Display Selection of Inorganic-Binding Peptides. *Mater. Sci. Eng. C* **2009**, *29*, 14–19. [CrossRef]
31. Conrath, K.E.; Lauwereys, M.; Galleni, M.; Matagne, A.; Frère, J.-M.; Kinne, J.; Wyns, L.; Muyldermans, S. β-Lactamase Inhibitors Derived from Single-Domain Antibody Fragments Elicited in the *Camelidae*. *Antimicrob. Agents Chemother.* **2001**, *45*, 2807–2812. [CrossRef] [PubMed]
32. de Marco, A. Recombinant Expression of Nanobodies and Nanobody-Derived Immunoreagents. *Protein Expr. Purif.* **2020**, *172*, 105645. [CrossRef] [PubMed]
33. Amcheslavsky, A.; Wallace, A.L.; Ejemel, M.; Li, Q.; McMahon, C.T.; Stoppato, M.; Giuntini, S.; Schiller, Z.A.; Pondish, J.R.; Toomey, J.R.; et al. Anti-CfaE Nanobodies Provide Broad Cross-Protection against Major Pathogenic Enterotoxigenic Escherichia Coli Strains, with Implications for Vaccine Design. *Sci. Rep.* **2021**, *11*, 2751. [CrossRef] [PubMed]
34. Nagy-Fazekas, D.; Stráner, P.; Ecsédi, P.; Taricska, N.; Borbély, A.; Nyitray, L.; Perczel, A. A Novel Fusion Protein System for the Production of Nanobodies and the SARS-CoV-2 Spike RBD in a Bacterial System. *Bioengineering* **2023**, *10*, 389. [CrossRef] [PubMed]
35. Salema, V.; Fernández, L.Á. High Yield Purification of Nanobodies from the Periplasm of *E. coli* as Fusions with the Maltose Binding Protein. *Protein Expr. Purif.* **2013**, *91*, 42–48. [CrossRef]
36. Kariuki, C.K.; Magez, S. Improving the Yield of Recalcitrant Nanobodies®by Simple Modifications to the Standard Protocol. *Protein Expr. Purif.* **2021**, *185*, 105906. [CrossRef]
37. Eble, J.A. Titration ELISA as a Method to Determine the Dissociation Constant of Receptor Ligand Interaction. *J. Vis. Exp.* **2018**, *2018*, 57334. [CrossRef]
38. Cohen, L.; Walt, D.R. Evaluation of Antibody Biotinylation Approaches for Enhanced Sensitivity of Single Molecule Array (Simoa) Immunoassays. *Bioconjug Chem.* **2018**, *29*, 3452–3458. [CrossRef]
39. Haque, M.; Forte, N.; Baker, J.R. Site-Selective Lysine Conjugation Methods and Applications towards Antibody–Drug Conjugates. *Chem. Commun.* **2021**, *57*, 10689–10702. [CrossRef]
40. Wade, J.D.; Domagala, T.; Rothacker, J.; Catimel, B.; Nice, E. Use of Thiazolidine-Mediated Ligation for Site Specific Biotinylation of Mouse EGF for Biosensor Immobilisation. *Lett. Pept. Sci.* **2001**, *8*, 211–220. [CrossRef]
41. Ferguson, K.M.; Berger, M.B.; Mendrola, J.M.; Cho, H.-S.; Leahy, D.J.; Lemmon, M.A. EGF Activates Its Receptor by Removing Interactions That Autoinhibit Ectodomain Dimerization. *Mol. Cell* **2003**, *11*, 507–517. [CrossRef] [PubMed]
42. Kuo, W.-T.; Lin, W.-C.; Chang, K.-C.; Huang, J.-Y.; Yen, K.-C.; Young, I.-C.; Sun, Y.-J.; Lin, F.-H. Quantitative Analysis of Ligand-EGFR Interactions: A Platform for Screening Targeting Molecules. *PLoS ONE* **2015**, *10*, e0116610. [CrossRef] [PubMed]
43. Moutel, S.; Bery, N.; Bernard, V.; Keller, L.; Lemesre, E.; De Marco, A.; Ligat, L.; Rain, J.C.; Favre, G.; Olichon, A.; et al. NaLi-H1: A Universal Synthetic Library of Humanized Nanobodies Providing Highly Functional Antibodies and Intrabodies. *eLife* **2016**, *5*, e16228. [CrossRef] [PubMed]
44. Predonzani, A.; Arnoldi, F.; López-Requena, A.; Burrone, O.R. In Vivosite-Specific Biotinylation of Proteins within the Secretory Pathway Using a Single Vector System. *BMC Biotechnol.* **2008**, *8*, 41. [CrossRef]
45. Cruz-Pacheco, A.F.; Monsalve, Y.; Serrano-Rivero, Y.; Salazar-Uribe, J.; Moreno, E.; Orozco, J. Engineered Synthetic Nanobody-Based Biosensors for Electrochemical Detection of Epidermal Growth Factor Receptor. *Chem. Eng. J.* **2023**, *465*, 142941. [CrossRef]

Article

Structural Modeling of Nanobodies: A Benchmark of State-of-the-Art Artificial Intelligence Programs

Mario S. Valdés-Tresanco [1,*,†], Mario E. Valdés-Tresanco [2,†], Daiver E. Jiménez-Gutiérrez [1] and Ernesto Moreno [1,*]

1 Faculty of Basic Sciences, University of Medellin, Medellin 050026, Colombia; daiverjimenez05@gmail.com
2 Centre for Molecular Simulations and Department of Biological Sciences, University of Calgary, Calgary, AB T2N 1N4, Canada; mario.valdeztresanco@ucalgary.ca
* Correspondence: mariosergiovaldes145@gmail.com (M.S.V.-T.); emoreno@udemedellin.edu.co (E.M.)
† These authors contributed equally to this work.

Abstract: The number of applications for nanobodies is steadily expanding, positioning these molecules as fast-growing biologic products in the biotechnology market. Several of their applications require protein engineering, which in turn would greatly benefit from having a reliable structural model of the nanobody of interest. However, as with antibodies, the structural modeling of nanobodies is still a challenge. With the rise of artificial intelligence (AI), several methods have been developed in recent years that attempt to solve the problem of protein modeling. In this study, we have compared the performance in nanobody modeling of several state-of-the-art AI-based programs, either designed for general protein modeling, such as AlphaFold2, OmegaFold, ESMFold, and Yang-Server, or specifically designed for antibody modeling, such as IgFold, and Nanonet. While all these programs performed rather well in constructing the nanobody framework and CDRs 1 and 2, modeling CDR3 still represents a big challenge. Interestingly, tailoring an AI method for antibody modeling does not necessarily translate into better results for nanobodies.

Keywords: artificial intelligence; protein structure; protein modeling; nanobody; antibody

Citation: Valdés-Tresanco, M.S.; Valdés-Tresanco, M.E.; Jiménez-Gutiérrez, D.E.; Moreno, E. Structural Modeling of Nanobodies: A Benchmark of State-of-the-Art Artificial Intelligence Programs. *Molecules* **2023**, *28*, 3991. https://doi.org/10.3390/molecules28103991

Academic Editor: Jahir Orozco Holguín

Received: 8 April 2023
Revised: 1 May 2023
Accepted: 5 May 2023
Published: 9 May 2023

1. Introduction

Nanobodies (Nbs) are the single binding domains of camelid heavy chain antibodies. Structurally, they share similarities with the variable heavy chain domain (VH) of traditional antibodies, consisting of a highly conserved region called framework and the antigen recognition region formed by three hypervariable loops, also called complementarity determining regions (CDRs) [1]. Nbs are much smaller (only 15 kDa) than human antibodies and their derivatives, but nonetheless can achieve similar affinities. Furthermore, they are highly stable and easy to produce [2,3]. These characteristics have positioned them as fast-growing biologic products in the biotechnology market.

The number of applications for Nbs is expanding steadily [3–8]. Several of these applications require protein engineering, which in turn would greatly benefit from having a reliable three-dimensional (3D) model of the Nb being modified [9–11]. However, as with antibodies, structural modeling of Nbs is still a challenge [12,13]. There are several hundreds of Nb crystallographic structures deposited in the Protein Data Bank (PDB) [14,15]; however, this is still insufficient to represent the huge structural and sequence variability found in Nb hypervariable loops. Furthermore, the CDR3 in Nbs shows a spectrum of conformations, lengths, and sequence variability greater than that of antibodies [16], which increases the difficulty for modeling their 3D structure. Nonetheless, homology modeling of nanobodies have been attempted for practical purposes, as in [17,18], and the recent developments in artificial intelligence (AI) methods for protein modeling, which have outperformed conventional methods [19–21], have been applied also to the modeling of antibodies and nanobodies [12,13,22–27].

Several AI methods have been developed in recent years to tackle the problem of protein modeling [28–30]. In this scenario, the development of AlphaFold represented a revolution in high-accuracy 3D protein modeling [31]. Since then, several methods have come to light, improving aspects, such as speed, computational resource consumption, and modeling accuracy [32]. AI programs especially designed to model complete antibodies and their fragments have been generated, including an AI model—Nanonet [12]—designed for modeling Nb structures. Given the similarity between Nbs and antibody VH domains, all AI models developed for antibodies can in principle be used for Nb modeling [13,26,27]. So far, however, AI methods have not been extensively tested for nanobody modeling. In this regard, comparing their modeling efficiency for this important class of proteins is important to establish to which extent the constructed models can be considered reliable for different practical applications, and as a guide for researchers in selecting the most appropriate modeling programs.

In this study, we have compared the performance in Nb modeling of six state-of-the-art AI-based programs, either designed for general protein modeling, such as AlphaFold2 [31], OmegaFold [33], ESMFold [34], and trRosetta (Yang-Server in the most recent Critical Assessment of Structure Prediction competition—CASP15) [28] or specifically designed for antibody modeling, such as IgFold [13] and Nanonet [12]. Interestingly, tailoring an AI program for antibody modeling does not necessarily translate into better results for nanobodies.

2. Results and Discussion

2.1. Dataset Selection and Validation

For this study, we built a curated, non-redundant dataset of Nbs, none of which had been included in any of the training sets of the benchmarked programs. Following the procedure described above, we obtained a dataset of 75 unique Nbs with a median resolution of 2.59 Å (Figure 1, Table S1 [Supplementary Information (SI)], Figure S1 [SI]).

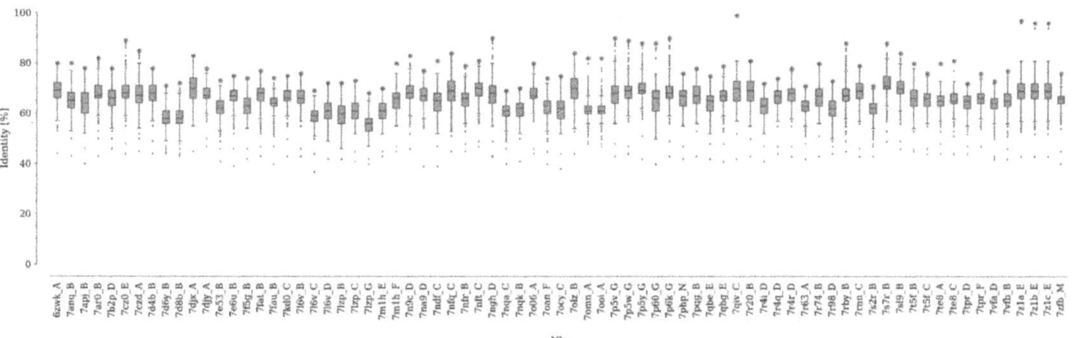

Figure 1. Sequence identity between each Nb in the dataset and the rest of the Nbs in the SAbDaB database. The sequence identity distributions are represented with boxplots. The lower and upper edges of the box represent the first (Q1) and third quartile (Q3), respectively. The difference Q3–Q1 is known as the interquartile range (IQR). Whiskers extend to the minimum and maximum points within ±1.5 × IQR, respectively. The maximum value of sequence identity for each distribution is represented as an orange dot.

The median sequence identity between the Nbs and the rest of the structures not contained in our dataset was between 56 and 71% (Figure 1). On the other hand, the maximum value of sequence identity within our dataset is below 90% in 91% of the cases, with only four pairs of Nbs showing a sequence identity higher than 95% (Figure S2 [SI]). Here, it is worth noting that for Nbs, as well as for antibodies, even point mutations can induce important structural changes in CDR3 [35].

2.2. Structure Prediction Accuracy

We compared the performance of six AI models for 3D structure prediction of Nbs: OmegaFold (OF), AlphaFold2 (AF2), IgFold (IF), NanoNet (NN), ESMFold (ESM), and tr-Rosetta (referred to as Yang-Server in the latest CASP15 and herein) (YS). Modeling accuracy was initially evaluated using global superposition structural similarity metrics—TM-score, GDT_TS, and GDT_HA—traditionally used in CASP competitions. Figure 2 shows the distribution of values by program for each metric.

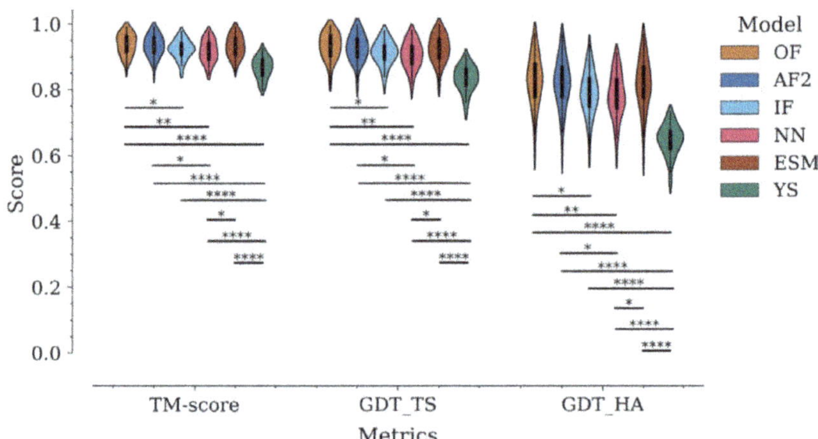

Figure 2. Assessment of the modeling accuracy of the six AI programs using global superposition metrics (TM-score, GDT-TS, and GDT-HA). The distributions of metric values are represented with violin plots, which combine a kernel density plot (outer) to show the distribution of values and a boxplot (inner) that summarizes the distribution statistics. In the boxplot, a white dot represents the median, the thick gray bar in the center represents the interquartile range, and the thin gray line accounts for the rest of the distribution. Statistical significances are represented with asterisks according to the following convention: * $p \leq 0.05$, ** $p \leq 0.01$, *** $p \leq 0.001$, and **** $p \leq 0.0001$.

In general, all tested programs performed well according to these global metrics. The Yang-Server showed the most discrete performance with medians of 0.87, 0.84, and 0.65 for TM-score, GDT_TS, and GDT_HA, respectively. On the other hand, OmegaFold, AlphaFold2, ESMFold, IgFold, and NanoNet, in decreasing order, showed medians above 0.91 for TM-score and GDT_TS, respectively, and above 0.78 for GDT_HA (Figure 2, Table S2 [SI]).

TM-score and GDT_TS estimate the percent structural similarity between the model and the experimental structure. Values above 0.5 indicate that both structures have the same folding, while values above 0.9 indicate that they are structurally identical [36]. However, unlike other protein families, antibodies and Nbs present a major challenge for modeling techniques due to their CDRs. The framework is modeled correctly in most cases due to the high conservation of this region, whereas most of the modeling errors are concentrated in the CDRs, especially in CDR3. This fact generates an important bias in the metrics. This can be reflected in the variation of the global RMSD compared to per-region RMSDs (Figure 3, Table S2 [SI]). To objectively evaluate the modeling accuracy of each program, we divided the Nbs into four regions—Framework (Fw), CDR1, CDR2, and CDR3—and calculated the RMSD for each of them (Figure 3, Table S2 [SI]).

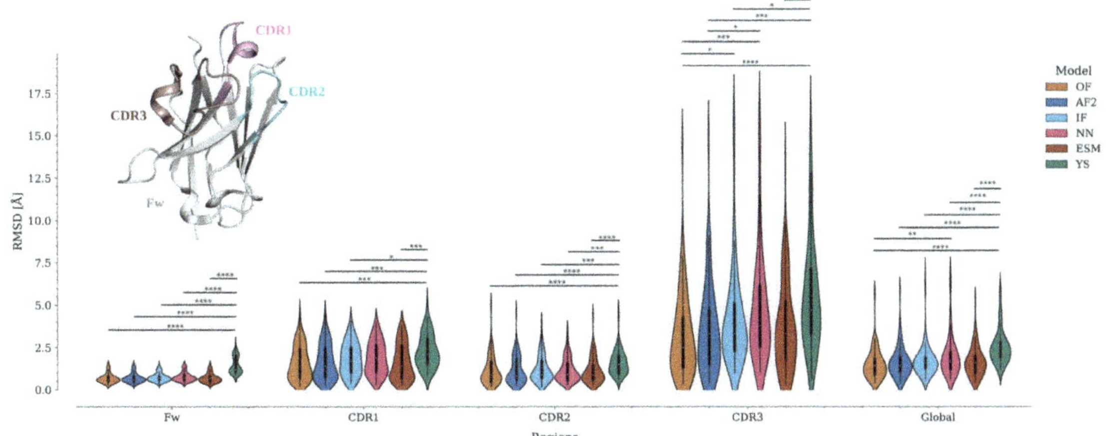

Figure 3. Assessment of modeling accuracy by RMSD for the Fw and CDR regions, for OmegaFold, AlphaFold2, IgFold, Nanonet, ESMFold and Yang-Server. RMSD distributions are represented using violin plots. Nb regions are colored as follows: Framework (Fw) as gray; CDR1—pink, CDR2—cyan, and CDR3—brown. Statistical significances: * $p \leq 0.05$, ** $p \leq 0.01$, *** $p \leq 0.001$, and **** $p \leq 0.0001$.

Because of the high conservation of the immunoglobulin domain framework, it is expected that all programs should correctly predict the structure of this region. Interestingly, while OmegaFold, AlphaFold2, IgFold, Nanonet, and ESMFold predicted the Fw structure with high accuracy ($0.6 \leq$ RMSD median ≤ 0.7), the Yang-Server yielded more discrete results (RMSD median = 1.2). In fact, only the Yang-Server shows significant statistical differences with respect to the other programs (Figure 3, Table S2 [SI]). Modeling of CDRs, in contrast, poses a challenge for all programs. CDR2 has predicted more accurately ($0.8 \leq$ median RMSD ≤ 1.5) than the other CDRs, with only a few structures showing RMSD values above 2.5 Å (Figure 3, Table S3 [SI]) and significant differences only for the Yang-Server (Figure 3, Table S2 [SI]). CDR1 predictions remain in an acceptable range ($1.4 \leq$ median RMSD ≤ 2.1) with an increase in the number of structures with RMSD > 2.5 Å (Figure 3, Table S3 [SI]), but without considerable significant statistical differences among them, except for the Yang-Server with respect to all but Nanonet.

CDR3 predictions, on the other hand, are the most inaccurate ($2.5 \leq$ median RMSD ≤ 4.7), with about or more than 50% of the structures showing RMSD > 2.5 Å (Figure 3, Table S3 [SI]). Significant statistical differences among several programs (Yang-Server, IgFold, and Nanonet) are observed for this region. OmegaFold, with a median RMSD of 2.5 Å, performs well in both overall value and RMSD per region, followed by AlphaFold2, IgFold, and ESMFold (median RMSD = 3.3 Å for CDR3), Nanonet (median RMSD = 3.8 Å) and finally Yang-Server (median RMSD = 4.7 Å)

2.3. Structure Prediction Accuracy by Sequence Position

As shown above, global superposition metrics are not suitable for estimating the accuracy of Nb modeling due to their structural characteristics. At the sequence region level, we observed a considerable variation in the accuracy of CDR modeling, especially for CDR3. We then analyzed the structures generated by the tested programs at the sequence position level to identify the regions that mark the differences in modeling. For each sequence position, we compared the RMSD values for Cα atoms and the whole amino acids between the predicted and experimental structures (Figure 4).

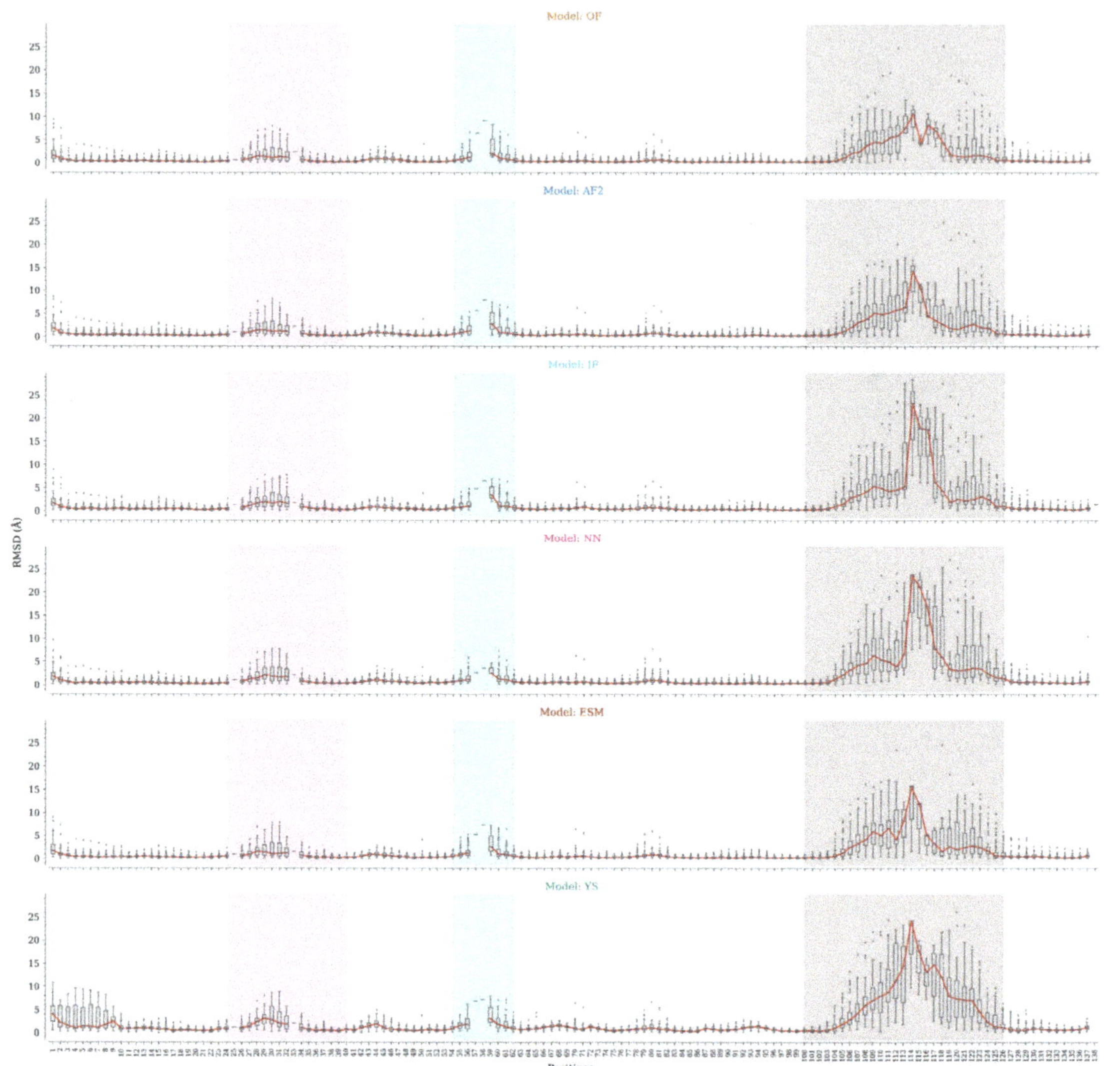

Figure 4. Distribution of Cα RMSD values by position for the OmegaFold, AlphaFold2, IgFold, Nanonet, ESMFold, and Yang-Server models. CDR1, CDR2, and CDR3 regions are colored pink, cyan, and brown, respectively. The RMSD distributions are represented by boxplots.

All programs, except the Yang-Server, are consistent regarding framework modeling, with slight variations in the N-terminal region and non-CDR loops. The Yang-Server shows slight structural variations in the whole framework as compared to the rest of the programs, while a greater variation is observed in the N-terminal segment. NanoNet uses Modeller, while IgFold and the Yang-Server use Rosetta for side-chain modeling. NanoNet shows considerable RMSD variations when all heavy atoms are considered, followed by IgFold and Yang-Server with fewer variations (Figure S3 [SI]). The side chains in the framework region are consistently well-modeled by OmegaFold, AlphaFold2, and ESMFold (Figure S3). The results for CDR1 are similar in all cases, with minor differences and slightly higher medians for Yang-Server. On the other hand, CDR2 shows appreciable variations. Positions

57, 58, and 59 are poorly represented in the dataset, with less than five structures having amino acids at these positions (Figure S4 [SI]). NanoNet slightly outperformed the rest of the programs.

Finally, the main differences were found for CDR3 modeling. The lowest RMSD distributions by position were achieved by OmegaFold, followed by AlphaFold2, IgFold, ESMFold, Nanonet, and Yang-Server. Except for Yang-Server, the differences are relatively small for short CDR3s, becoming more accentuated for Nbs with the longest loops. The C-terminal segment of CDR3 shows the lowest variations, probably associated with the frequent formation of secondary structure elements in this region, while the N-terminal part shows more discrete results. Nonetheless, in both cases the structural variations are considerable.

The observed differences in modeling performance can be due to the intrinsic characteristics of each AI model and the representation and structural variability of the Nbs with different CDR3 lengths in their training sets. OmegaFold does not require a multiple sequence alignment (MSA), instead using a new combination of a large pre-trained language model for sequence modeling and a geometry-inspired transformer model for structure prediction. According to its authors, this allows the modeling of orphan proteins and antibodies from their amino acid sequences [33]. Similarly, ESMFold is based on ESM-2 (Evolutionary Scale Model), which is a language model that internalizes evolutionary patterns linked to structure, eliminating the need for external evolutionary databases, MSAs, and templates [34]. IgFold and NanoNet do not require either a multiple sequence alignment. IgFold and NanoNet were trained to reproduce antibody and Nb structures, which limits the generation of atypical structures and therefore unrepresented in their training sets [12,13].

AlphaFold2, on the other hand, predicts the structure from neural networks and training procedures based on evolutionary, physical, and geometrical constraints of protein structures. To do so, it requires the protein primary sequence and a multiple sequence alignment, therefore, sequence identity and coverage of the different regions are crucial to obtain an accurate model [31]. Yang-Server also requires an MSA and, in most cases, including a homologous template yields better modeling results [37]. Given the number of available structures and the spectrum of lengths, composition, and conformations of CDR3, it is difficult to generate an MSA for Nbs with full coverage of their sequences. However, general protein modeling programs such as OmegaFold, AlphaFold2, and ESMFold, have been exposed to a wide and diverse set of protein structures, which may explain their better results in modeling CDRs, especially CDR3. (Figure 4).

2.4. CDR3 Structure Prediction Accuracy

The accuracy of CDR3 modeling depends mainly on its length (Figure 5). Several CDR3 lengths are poorly represented in our dataset, where the number of Nbs varies from one (for lengths 3, 7, 11, and 20) to a maximum of nine (for length 16).

Depending on the AI model, the median RMSD of the predictions varies along the CDR3 length range. In most cases, OmegaFold achieved the best predictions, followed by AlphaFold2, ESMFold, IgFold, Nanonet, and lastly, Yang-Server. Although no direct correlation between CDR3 length and RMSD values is observed among the experimental structures, the structural variability might influence the predictions. For example, for length 15, where the structural variation in CDR3 is considerable, the predictions are relatively consistent, especially for OmegaFold, which yields RMSD values all below 2 Å (Figure 5). This is probably because this length is the most represented in the PDB and, therefore, in the training sets of the tested programs (Figure S5 [SI]). On the other hand, CDR3s with lengths 17 and 18 adopt a similar conformation, hence the RMSD between the structures is relatively small and their modeling is consistently good for all the tested programs, except for the Yang-Server.

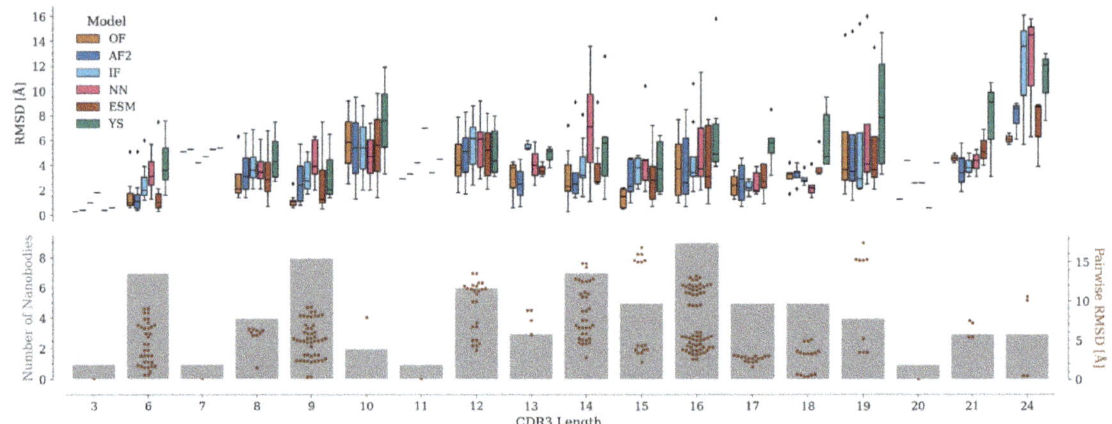

Figure 5. RMSD distributions per CDR3 length. The upper panel shows the RMSD distributions (as boxplots, with outliers as black diamonds) per CDR3 length for OmegaFold, AlphaFold2, IgFold, Nanonet, ESMFold, and Yang-Server. The lower panel shows the number of Nbs in the dataset per CDR3 length (bars in grey) and the pairwise RMSD values among CDR3s of the same length (swarm plot in brown).

For lengths 19 and 24, a few models with high RMSD are generated. For these particular cases, the experimental structure has marked differences from the rest of the Nbs with the same CDR3 length. In 7tpr_D [38] (length 19), the antigen is positioned in-between CDR3 and the framework, thus altering the common CDR3 conformation (Figure S6 [SI]). For length 24, the 7d8b_B and 7d6y_B structures [39] correspond to an engineered human variable heavy chain domain. These Nbs do not have a canonical disulfide bond and show two alpha helix segments in CDR3, which causes the N-terminal portion of this region to be displaced with respect to the rest of the structures of the same length (Figure S6 [SI]). Interestingly, although they differ in only two amino acids and have similar structures, for 7d6y_B, unlike 7d8b_B, a significant improvement was obtained when modeled with its antigen (see Section 2.6.2) (Figure S10 [SI]). However, in both cases, it was not possible to correctly reproduce the secondary structure motifs present in CDR3, probably because of the poor representation of this CDR3 length in the available structures (Figure S5 [SI]).

2.5. Nanobody Modeling Confidence

The confidence value is an important metric in protein structure modeling that allows to estimate how reliable a model can be considered. NanoNet does not produce any metric to estimate its modeling confidence. OmegaFold, AlphaFold2, ESMFold, and Yang-Server do offer a measure of confidence called pLDDT (predicted local distance difference test) on a 0–100 scale, which corresponds to the predicted model score of the lDDT-Cα metric [31,33]. IgFold, on the other hand, offers an error estimate based on per-residue Cα deviations [13]. These metrics differ both conceptually and in scale. Typically, a pLDDT above 90 indicates a highly reliable model, 70 < pLDDT < 90 is considered reliable, while a model with pLDDT below 70 should be carefully reviewed. In contrast, there is not an established RMSD value below which a model is defined as reliable, although in practice, protein models with global RMSD below 4 Å are considered good.

AlphaFold2 and OmegaFold report values of pLDDT below 70 for predicted CDR3s, which correlates with the RMSD values obtained for these loops between the models and their crystallographic structures (Tables 6 and S7 [SI], Table S4 [SI]). Yang-Server shows the lowest correlation, while OmegaFold achieves the highest.

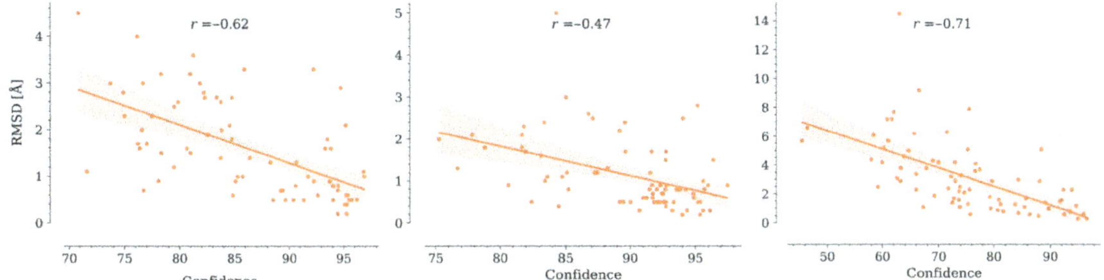

Figure 6. Correlation between the RMSD values and the average predicted confidences by OmegaFold for the CDR regions in the 75 Nbs conforming to our dataset. Regression lines are shown in orange. Translucent bands around the regression lines indicate the 95% confidence interval for the regression estimates. Spearman correlation coefficients (r) are shown in the graphs. In all cases, the *p*-value < 0.05.

Although the obtained correlation coefficients are significant, it is not possible to establish a priori whether a model is reliable or not. Since CDR3 is the region that interacts more frequently with the antigen, further studies are required to estimate whether the generated model can be used for bioinformatics approaches that demand high structural accuracy, such as protein–protein docking.

2.6. Structure Prediction Accuracy Varying Modeling Parameters

2.6.1. Number of Recycles

Among the tested programs, only AlphaFold2 and OmegaFold allow parameter modification, specifically the number of recycles, which controls the degree of structural model refinement. In several cases, AlphaFold2 has been shown to improve the prediction of disordered structures or de novo proteins by increasing the number of recycles [40]. OmegaFold has an equivalent tunable parameter, although its functionality has not been extensively assessed yet. Here, we tested several values for the number of recycles to assess their effect on the modeling of different Nb regions.

The models generated with AlphaFold2 using ten recycles slightly improved the predictions for CDR1 and CDR3, while slightly worsening those for CDR2. No considerable variations were observed for the framework and global modeling. In all cases, there were no statistically significant differences (*p*-value > 0.05). On the other hand, using 20 recycles with OmegaFold does not translate into any considerable variation for any Nb region. Interestingly, using four recycles slightly improves CDR1 and CDR2 predictions, while losing accuracy in CDR3 modeling. However, statistically, there are no significant differences in any case (Figure S8 [SI]). Based on these results, using four recycles instead of the default value (number of recycles = 10) might be preferable since it decreases the computational time (see below).

2.6.2. Modeling Nanobodies in Complex with Their Antigens with AlphaFold-Multimer

Currently, we are lacking enough Nb structures to estimate the effect of antigen binding on CDR conformations. Interestingly, there are a few cases where the same Nb shows several conformations, even in the free state (Figure S9 [SI]). In other cases, the structural variations between Nbs with the same CDR3 might be attributed to the formation of an Nb-antigen complex. Most of the Nbs used in the parameterization of AI models are complex with their antigens, thus making it difficult to determine whether the observed conformations would remain the same in their free states.

Alphafold2 can model single chains with high reliability; however, it may fail in predicting protein structures in the context of certain complexes [41]. Using AlphaFold-multimer, we tested whether there is an improvement in CDR3 modeling for Nbs complexed with their antigens. To perform this analysis, we selected 41 structures considering the size and complexity of the antigen (Table S5 [SI]). The results from these calculations were mixed. In several cases (7nfr_B, 7t5f_B, 7m1h_E, 7olz_B, 7rby_B, and 7d6y_B) significant improvements were achieved, while in other cases (7php_N, 7zfb_M, 7pqg_B, and 7e53_B) the program produced significantly worse results. In all other cases, regardless of CDR3 length, the results are similar to those obtained for the free Nb (Figure S10 [SI]).

2.6.3. Energy Minimization

Commonly, energy minimization is used to remove clashes among atoms in the structure. However, this does not imply a significant improvement in the models since such geometry optimization does not significantly change the overall conformation of loops and other regions [31]. Here, we applied energy minimization to all the generated models. The results show that, indeed, there are no significant improvements (Figure S11 [SI]).

2.7. Computation Time

Nb libraries may contain billions of sequences, with many possible different structures. In recent approaches, library design seeks to favor structures with certain CDR3 geometries (e.g., concave, or convex) that will presumably bind to specific antigens [42,43]. With the increasing development of synthetic libraries [44], methods for reliable estimation of the CDR structural diversity would be of great value for in silico design of Nb libraries with desired conformational properties. Along with accuracy, computational time becomes an important factor to be considered when modeling such a high number of structures. In this context, NanoNet takes the lead, followed by IgFold, OmegaFold, and lastly, AlphaFold2 (Figure 7). ESMFold was used in this study through the ESM Metagenomic Atlas API (application programming interface), while the Yang-Server was used through its dedicated server (https://yanglab.nankai.edu.cn/trRosetta/, accessed on 1 May 2023). ESMFold is extremely fast, obtaining results in approximately one second. However, this may depend on the demand on the server, so it might have limitations in the number of requests. Yang-Server modeling can take approximately one hour due to the algorithm and server capacity (only 30 active jobs at a time). In both cases, however, it is possible to install a standalone version for local use.

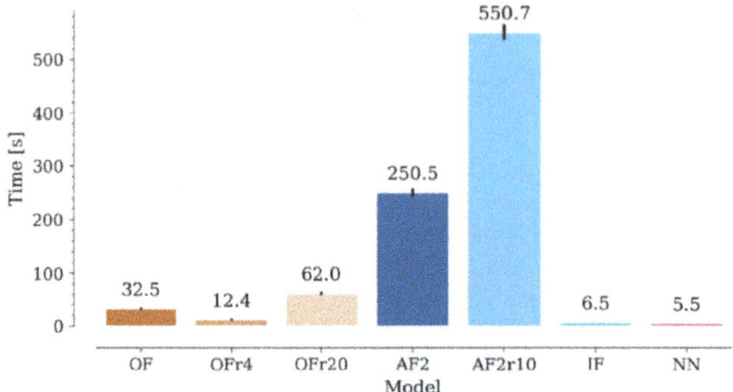

Figure 7. Computation time for the generation of a structural Nb model with OmegaFold, AlphaFold2, IgFold, and Nanonet. Computation times for OmegaFold and AlphaFold with different recycle numbers are also included.

For OmegaFold, the computation time improves when decreasing the number of recycles from ten (default) to four, without affecting its accuracy. On the other hand, we found that increasing the number of recycles beyond the default value drastically increases the computational time without any noticeable benefit in modeling accuracy for both OmegaFold and AlphaFold2. It is worth noting that NanoNet may include sidechain modeling with Modeller, which would increase the computational time by a factor of

170–900 approximately, depending on the number of sequences being simultaneously processed. Finally, energy minimization not only does not improve modeling results, but it also adds computation time. The extra time required varied between 10 and 50 s per structure using our hardware configuration.

3. Materials and Methods

3.1. Benchmark Dataset

We started from the SAbDaB database [45], containing a total of 981 structures as of 15 June 2022. Firstly, we removed the PDB structures used for the parameterization of the AI programs to be compared. Next, incomplete structures and duplicated Nbs, identified from a pairwise comparative analysis of their amino acid sequences using Blastp [46–48], were withdrawn. For the subsequent analyses, the sequences were numbered according to Aho's scheme using ANARCI [49]. All modeling was carried out from the primary structure of the Nb, without using templates, except for the modeling of Nbs in complex with their antigens, where the crystallographic structure of the antigen was used as a template.

3.2. Artificial Intelligence Models

Currently, AI methods have reached a high level of precision in protein modeling, as evidenced in the latest CASP competitions, where the first positions have been occupied by robust AI-based models (https://predictioncenter.org/index.cgi, accessed on 1 May 2023). The number of these AI protein modeling programs is rapidly increasing, making it difficult to perform comprehensive benchmarking. For this study, we selected six AI modeling programs that have stood out for their performance in general protein modeling and/or antibody modeling.

The first choice was AlphaFold2 [31], which has become a gold standard in protein modeling, inspiring the development of other AI methods. Further, we selected OmegaFold [33] and ESMFold [34], which are based on protein language models and therefore, by difference with AlphaFold2, do not involve the generation of multiple sequence alignments. As reported by its authors, OmegaFold's results are comparable to those of AlphaFold2 for proteins in general and are better for orphan proteins and antibodies [33]. ESMFold is based on ESM-2, which in a study conducted by its authors outperformed all single-sequence protein language models tested in a variety of structure prediction tasks [34]. ESMFold has gained popularity with the recent release of the ESM Metagenomic Atlas (https://esmatlas.com, accessed on 1 May 2023) that incorporates an application programming interface (API) to perform protein modeling easily and quickly. The fourth program chosen for our study is the Yang-Server [37,50], which finished as the top-ranked program in the most recent CASP competition (CASP15, https://predictioncenter.org/casp15/zscores_final.cgi, accessed on 1 May 2023). Finally, we included two programs—IgFold [13] and Nanonet [12]—that were specifically designed for antibody modeling and have proven to be considerably better than conventional homology modeling methods [13,26,33]. Below we provide a brief description of each of these programs and their use in this study.

3.2.1. AlphaFold2

AlphaFold2 is an AI model developed by DeepMind that incorporates a neural network architecture and training procedures based on evolutionary, physical, and geometrical constraints of protein structures [31]. The AlphaFold network directly predicts the 3D coordinates of all heavy atoms for a given protein using as input the primary amino acid sequence and aligned sequences of homologs.

AlphaFold2 is composed mainly of two blocks: (1) the sequence information module, and (2) the structure module, both based on transformers. The first module, called Evoformer, processes the input and generates a multiple sequence alignment (MSA) and a residue pair matrix. The main innovation in the Evoformer block is the mechanisms to exchange information within the MSA and pair representations, that enable direct reasoning on the spatial and evolutionary relationships. The second module generates the 3D structure using the pair representation and the single representation of the MSA, with a mechanism that allows simultaneous local refinement of all parts of the structure, reasoning about unrepresented side chain atoms, and weighing the correct residue orientations. After an initial structure is generated, an interactive recycling process is carried out that reuses the entire network to obtain a refined final structure [31]. AlphaFold2 was trained using the PDB and PDB70 for template search and over millions of protein families' sequences using Uniref90, BFD, Uniclust, and MGnify for the MSA construction. At CASP14, AlphaFold was the top-ranked protein structure prediction method [51].

ColabFold, on the other hand, offers a user-friendly and fast implementation of AlphaFold2 [40]. In this application, an MSA is generated with MMseqs2 [52,53], simplifying the process and reducing the computation time. For this study, we used localColabFold v1.4.0 to run the calculations on our computers (https://github.com/YoshitakaMo/localcolabfold, accessed on 1 May 2023). For Nbs in complex with an antigen, we employed AlphaFold-multimer [41] as implemented in ColabFold using the free Google Colab service. In these calculations, we kept the default AlphaFold-multimer parameters, while for modeling Nbs in the free state we used the AlphaFold2 default configuration (three recycles), as well as ten recycles.

3.2.2. OmegaFold

OmegaFold was the first computational method to successfully predict high-resolution protein structure from its single primary sequence alone [33]. It uses OmegaPLM, a deep transformer-based protein language model, to learn single- and pairwise-residue embeddings (or representations) as powerful features that model the distribution of sequences. These embeddings are fed into Geoformer, a geometry-inspired transformer neural network, to distill the structural and physical pairwise relationships between amino acids. Lastly, a structural module predicts the 3D coordinates of all heavy atoms. OmegaFold shares similarities with AlphaFold2, in their first stage of extracting per-residue pair representation information and a second stage of generating the three-dimensional structure from this representation. However, they also have notable differences. OmegaFold, by incorporating OmegaPLM, which captures structural and functional information encoded in the amino-acid sequences through the embeddings, does not require multiple sequence alignment, and the Geoformer has a focus primarily on vector geometry as opposed to the evolutionary variation of the AlphaFold2's Evoformer. The full model was jointly trained on ~110,000 single-chain structures from the PDB and all single domains from the SCOP v1.75 database with at most 40% sequence identity. According to the authors, in several cases, OmegaFold achieves the same or better precision than RoseTTaFold and AlphaFold2, particularly for orphan proteins and antibodies [33]. Here, we used the default configuration (number of recycles = 10) and tried two other values for this parameter: 4 and 20.

3.2.3. ESMFold

ESMFold [34] is an AI model for protein structure prediction that shares characteristics with AlphaFold2 [31] and OmegaFold [33]. Like OmegaFold, it uses a powerful protein language model called ESM-2 (evolutionary scale model) to process the input [34]. This model is the improved version of ESM-1b [54] (used as a reference by OmegaPLM as well), with a large number of parameters, which internalizes evolutionary patterns linked to the structure from sequences, eliminating the need for external evolutionary databases, multiple sequence alignments and templates. ESMFold uses a simple architecture that takes advantage of evolutionary information captured by the ESM-2 language model. The architecture is divided into two parts, similar to AlphaFold2. The first part is a folding module that takes the features of the language model as input and produces representations using a simplified version of AlphaFold's Evoformer. The second part is the structure module similar to AlphaFold, which generates 3D atomic coordinates from those representations. About 60 million Uniref50 [55] protein sequences were used for ESM-2 training. For its part, ESMFold was trained with selected PDB structures using the same procedure described for AlphaFold. Additionally, they incorporated around 13 million structures generated by AlphaFold with mean pLDDT > 70 [34]. For this work, we used the ESMFold API available in the ESM Metagenomic Atlas (https://esmatlas.com/, last accessed on 23 April 2023).

3.2.4. Yang-Server

The Yang-Server is a recent implementation, with several improvements, of trRosetta (transform-restrained Rosetta) [37]. Initially, the trRosetta method was inspired by other algorithms, such as RaptorX [24] and the first version of AlphaFold [56], for distance and contact prediction. Similar to these methods, it uses an MSA as input in the first step, and using a deep residual convolutional network, predicts the distance, contact, and orientation matrices of all pairs of residues in the protein. In the second step, a constrained minimization-based fastRosetta model construction protocol with distance and orientation constraints derived from the network outputs is carried out. The predicted geometries are then transformed into restraints to guide the structure prediction by direct energy minimization, which is implemented under the Rosetta framework [50]. The trRosetta version implemented in the Yang-Server has several improvements, including the MSA generation and selection improvements, a new neural network architecture for the distance and orientation prediction between residues, and the inclusion of template-based constraints [37]. Unlike the previously discussed methods, the Yang-Server was trained with just over 16,000 high-quality structures ($\leq$2.5 Å) but using a robust MSA selection method from five alternatives based on different sources, ensuring sufficient sequence representation in the MSA [50]. As mentioned above, Yang-Server was the top-ranked program in the recent CASP15 competition.

3.2.5. IgFold

IgFold uses a principle similar to ESMFold but is specifically applied to antibodies. The input is processed by AntiBERTy [23], a transformer language model pre-trained on natural antibody sequences, similar to ESM-2 or OmegaPLM. This model extracts representations of all residues from the protein sequence without requiring an MSA. These representations are then processed by the structure module, which uses a modified version of the one implemented in AlphaFold2, to generate the 3D coordinates of the model's backbone. Finally, PyRosetta [57] is used to generate the side chains of all residues and obtain the final model [13]. The AntiBERTy model was trained using 558 million antibody sequences [23]. IgFold was trained using about 4300 and 37,000 structures from SAbDab and those modeled by AlphaFold, respectively, consisting of paired and unpaired sequences, including nanobodies. This AI model showed better results than other previously proposed models [13] such as AbodyBuilder [58], DeepAb [26], and Ablooper [22]; therefore, we did not include these other programs in our study. It is worth noting that the authors of this study concluded that modeling nanobodies remains a big challenge [13].

3.2.6. Nanonet

Nanonet is a deep learning model based on a convolutional neural network [12]. This model uses an algorithm similar to trRosetta, but whose input is the one-hot encoded sequence instead of residue representations, so it does not require an MSA. Nanonet generates only the 3D structure of the protein backbone, so it requires an external tool for modeling the side chains (for this study, we used Modeller [59]). Unlike the rest of the models analyzed in this study, Nanonet seems to be the simplest model theoretically. In this sense, the model does not use representations of sequence residues, either from a comprehensive protein language model, such as ESM-2, OmegaPLM, or AntiBERTy, or extracted from MSA, likeAlphaFold2, and Yang-Server. Because of its simplicity, this model provides a great advantage in terms of time and computational resources. Nanonet was trained with about 1800 non-redundant Nbs and mAb heavy chain structures. Interestingly, Nanonet achieves good accuracy even with the simplest architecture and training data set.

3.3. Performance Evaluation Metrics

3.3.1. Structural Similarity Metrics

We used TM-score (template modeling score) [60], GDT_TS (global distance test—total score) [61], and GDT_HA (global distance test—high accuracy) [61] to evaluate the overall modeling accuracy of the different AI models. Both TM-score and GDT measure the structural similarity between two protein structures. GDT is commonly used to compare models with their corresponding crystallographic structures, being the major assessment criterion in the CASP event [51]. The Zhang group's TM-score program was used to compute the structural alignment, TM, and GDT scores [36,60]. For region-level analysis, we used RMSD (root-mean-square deviation). The RMSD for $C\alpha$ and all heavy atoms were computed using a ParmEd-based script [62].

3.3.2. Statistics

To estimate the differences between the metrics used in this study, we performed a Kruskal–Wallis one-way analysis of variance. In significant cases, we used Dunn's test with the Benjamini-Hochberg correction as a *post hoc* test. Dunn's test is the appropriate nonparametric pairwise multiple comparison procedure when a Kruskal–Wallis test is rejected. Calculations were performed using the bioinfokit tool (v 1.0.5) [63].

3.3.3. Execution Environment

Calculations were performed using low-end and mid-range hardware (AMD Ryzen 7 3700 and a GPU Nvidia 1660 Super 6GB VRAM and 16 GB RAM). All programs were installed in a standalone Miniconda environment with Python 3.8.13, following the instructions given by their developers.

3.3.4. Energy Minimization

For energy minimization, we used OpenMM v7.7.0 [64] with the Amber99SB [65] force field as described in the AlphaFold2 study [31]. The simulation was carried out for a maximum of 50,000 steps and a tolerance of 1000.0 kJ/mol/nm.

4. Conclusions

Nowadays, computer-assisted methods have become essential components in protein engineering, especially for antibodies and nanobodies. Whether for structure characterization, antigen interaction, or affinity enhancement through an in silico affinity maturation process, having reliable structural models is extremely important. Using a poor-quality model can lead to erroneous conclusions and low efficiency in further experimental development.

Multiple studies have shown the superiority of AI programs over conventional homology modeling approaches for modeling protein structures. In this study, we have evaluated the performance of six state-of-the-art AI programs in modeling Nb structures. To this aim, we generated a test dataset containing 75 unique Nbs not included in the training sets of the evaluated programs. The performance of different models was assessed using global metrics, as well as metrics for different regions within the structure. The results show that global metrics such as TM-score, GDT-TS, and GDT-HA are unsuitable for Nb structural model evaluation since the modeling errors of highly variable, but functionally important regions such as CDR3 get diluted when using these metrics. We then evaluated the modeling accuracy separately for the framework and CDR regions. OmegaFold achieved the best results, followed by AlphaFold2, ESMFold, IgFold, Nanonet, and Yang-Server.

Although the evaluated AI models represent a leap forward in Nb modeling, they are still far from providing completely reliable structural models. This study confirms that, while modeling of the framework region is consistently good in all cases, CDR modeling remains a challenge, especially for CDR3. For this loop, the RMSDs of the generated models are in most cases considerably high compared to the crystallographic structures. Although the median RMSD is relatively low for all AI models, only 52, 44, 35, 29, 25, and 15% of the CDR3 structures generated with OmegaFold, AlphaFold2, ESMFold, IgFold, Nanonet, and Yang-Server, respectively, were modeled with less than 2.5 Å difference compared to the crystallographic structures. Energy minimization did not improve the results. Since CDR3 is extremely important for antigen interaction, the obtained models may not be suitable for applications that require high accuracy, such as protein–protein docking.

Modeling with these AI programs can be performed using hardware in the low- to mid-range, which facilitates their use in common bioinformatics laboratories. In these conditions, the calculation times vary from a few to hundreds of seconds. Nanonet is the fastest model, followed by IgFold, OmegaFold, and lastly, AlphaFold2, while ESMFold and Yang-Server can be used on their dedicated servers. According to our results, OmegaFold is the most efficient AI program for Nb modeling, being relatively fast and achieving the best results. Similarly, both ESMFold and AlphaFold2 may be used as an alternative, yielding quite similar results compared to OmegaFold.

The Inherent limitations of this type of study must also be considered. The rapid development of artificial intelligence methods for protein modeling makes it almost impossible to keep the evaluation of their performance up to date in real time. In addition, the time lag between the selection of training sets and the release of the programs can lead to biases in a benchmarking study. In this study, we selected a set of nanobodies not included in the training sets of the evaluated programs. This restriction, also associated with the limited number of crystallographic structures available, reduces the representativeness of the evaluation set. Currently, several dozens of AI programs have been developed for protein modeling, as evidenced by the recent CASP15 competition (2022) that included more than twenty AI methods. For this study, we chose to evaluate six of these programs, based on their performance in CASP15, their reported applications (popularity), and accessibility. Therefore, the Nb modeling capabilities of the other AI methods have not yet been evaluated.

So far, although there have been substantial advances, the accuracy of the generated models is still limited and Nb modeling remains a challenge. However, the fast development and improvement of AI models, along with the increase of available crystallographic structures, augur significant advancements in Nb modeling in the near future.

Supplementary Materials: The following supporting information can be downloaded at: https://www.mdpi.com/article/10.3390/molecules28103991/s1, Figure S1. Resolution distribution of crystallographic structures of Nbs in the dataset. Figure S2. Sequence identity among Nbs in the dataset. Figure S3. Distribution of heavy atoms RMSD values by position for the OmegaFold, AlphaFold2, IgFold, and Nanonet models. Figure S4. Representation by positions of the Nbs in the dataset using the Aho numbering scheme. Figure S5. CDR3 length distribution of non-redundant Nbs structures in the Protein Data Bank. Figure S6. Particular cases of structural variations in the crystallographic structures are used as references. Figure S7. Correlation between the RMSD values and the average predicted confidences by AI models for the CDR regions in the 75 Nbs conforming the dataset. Figure S8. Comparison of the RMSD distribution for the models obtained varying the number of recycles of AlphaFold2 and OmegaFold. Figure S9. Structural variation of CDR3 of the same Nb in the asymmetric unit. Figure S10. Comparison between the RMSD of models generated with AlphaFold2 (AF2) and AlphaFold-multimer (AF2m) clustered by CDR3 length. Figure S11. Comparison of the RMSD distribution between the default generated model and its minimized version for OmegaFold, AlphaFold2, IgFold, Nanonet, ESMFold and Yang-server. Table S1. Nbs included in the dataset. Table S2. Results from Dunn's test non-parametric pairwise multiple comparison procedure. Table S3. Number of occurrences of predicted structures with RMSD above 2.5 Å. Table S4. Correlation between the RMSD values and the average predicted confidences by AI models for the CDR regions in the 75 Nbs conforming the dataset. Table S5. Nanobodies selected for modeling Nb-antigen complexes. All PDB files can be found at https://github.com/Valdes-Tresanco-MS/NbModelingBenchmark (accessed on 1 May 2023).

Author Contributions: Conceptualization, M.S.V.-T., M.E.V.-T. and E.M.; methodology, M.S.V.-T., M.E.V.-T. and E.M.; investigation, M.S.V.-T., M.E.V.-T. and D.E.J.-G.; writing—original draft preparation, M.S.V.-T., M.E.V.-T. and D.E.J.-G.; writing—review and editing, M.S.V.-T. and E.M.; supervision, project administration and funding acquisition, E.M. All authors have read and agreed to the published version of the manuscript.

Funding: This research was funded by MINCIENCIAS, MINEDUCACIÓN, MINCIT, and ICETEX through the Program NanoBioCancer (Cod. FP44842-211-2018, project number 58676). E.M. thanks University of Medellin for the support.

Institutional Review Board Statement: Not applicable.

Informed Consent Statement: Not applicable.

Data Availability Statement: All files and tools to reproduce the results and analyses can be found at https://github.com/Valdes-Tresanco-MS/NbModelingBenchmark (accessed on 1 May 2023).

Acknowledgments: Work by M.S.V.-T., D.E.J.-G. and E.M. was supported by the University of Medellin and MINCIENCIAS, MINEDUCACIÓN, MINCIT, and ICETEX, through the Program NanoBioCáncer, Cod. FP44842-211-2018. M.E.V.-T. is an Eyes High Doctoral Recruitment Scholarship and Alberta Graduate Student Scholarship recipient at the University of Calgary.

Conflicts of Interest: The authors declare no conflict of interest.

Sample Availability: Not applicable.

References

1. Hamers-Casterman, C.; Atarhouch, T.; Muyldermans, S.; Robinson, G.; Hammers, C.; Songa, E.B.; Bendahman, N.; Hammers, R. Naturally Occurring Antibodies Devoid of Light Chains. *Nature* **1993**, *363*, 446–448. [CrossRef] [PubMed]
2. Steeland, S.; Vandenbroucke, R.E.; Libert, C. Nanobodies as Therapeutics: Big Opportunities for Small Antibodies. *Drug Discov. Today* **2016**, *21*, 1076–1113. [CrossRef] [PubMed]
3. Gonzalez-Sapienza, G.; Rossotti, M.A.; Tabares-da Rosa, S. Single-Domain Antibodies as Versatile Affinity Reagents for Analytical and Diagnostic Applications. *Front. Immunol.* **2017**, *8*, 977. [CrossRef]
4. Zare, H.; Aghamollaei, H.; Hosseindokht, M.; Heiat, M.; Razei, A.; Bakherad, H. Nanobodies, the Potent Agents to Detect and Treat the Coronavirus Infections: A Systematic Review. *Mol. Cell. Probes* **2021**, *55*, 101692. [CrossRef] [PubMed]
5. Muyldermans, S. Applications of Nanobodies. *Annu. Rev. Anim. Biosci.* **2021**, *9*, 401–421. [CrossRef] [PubMed]
6. Yang, E.Y.; Shah, K. Nanobodies: Next Generation of Cancer Diagnostics and Therapeutics. *Front. Oncol.* **2020**, *10*, 1182. [CrossRef]
7. Wang, J.; Kang, G.; Yuan, H.; Cao, X.; Huang, H.; de Marco, A. Research Progress and Applications of Multivalent, Multispecific and Modified Nanobodies for Disease Treatment. *Front. Immunol.* **2022**, *12*, 6013. [CrossRef]
8. Njeru, F.N.; Kusolwa, P.M. Nanobodies: Their Potential for Applications in Biotechnology, Diagnosis and Antiviral Properties in Africa; Focus on Application in Agriculture. *Biotechnol. Biotechnol. Equip.* **2021**, *35*, 1331–1342. [CrossRef]
9. Wang, X.; Chen, Q.; Sun, Z.; Wang, Y.; Su, B.; Zhang, C.; Cao, H.; Liu, X. Nanobody Affinity Improvement: Directed Evolution of the Anti-Ochratoxin a Single Domain Antibody. *Int. J. Biol. Macromol* **2020**, *151*, 312–321. [CrossRef]
10. Soler, M.A.; Fortuna, S.; de Marco, A.; Laio, A. Binding Affinity Prediction of Nanobody-Protein Complexes by Scoring of Molecular Dynamics Trajectories. *Phys. Chem. Chem. Phys.* **2018**, *20*, 3438–3444. [CrossRef]
11. Hacisuleyman, A.; Erman, B. ModiBodies: A Computational Method for Modifying Nanobodies to Improve Their Antigen Binding Affinity and Specificity. *bioRxiv* **2019**. [CrossRef]
12. Cohen, T.; Halfon, M.; Schneidman-Duhovny, D. NanoNet: Rapid and Accurate End-to-End Nanobody Modeling by Deep Learning. *Front. Immunol.* **2022**, *13*, 4319. [CrossRef] [PubMed]
13. Ruffolo, J.A.; Chu, L.-S.; Mahajan, S.P.; Gray, J.J. Fast, Accurate Antibody Structure Prediction from Deep Learning on Massive Set of Natural Antibodies. *bioRxiv* **2022**. [CrossRef]
14. Berman, H.M.; Westbrook, J.; Feng, Z.; Gilliland, G.; Bhat, T.N.; Weissig, H.; Shindyalov, I.N.; Bourne, P.E. The Protein Data Bank. *Nucleic Acids Res.* **2000**, *28*, 235–242. [CrossRef]
15. Burley, S.K.; Berman, H.M.; Bhikadiya, C.; Bi, C.; Chen, L.; di Costanzo, L.; Christie, C.; Dalenberg, K.; Duarte, J.M.; Dutta, S.; et al. RCSB Protein Data Bank: Biological Macromolecular Structures Enabling Research and Education in Fundamental Biology, Biomedicine, Biotechnology and Energy. *Nucleic Acids Res.* **2019**, *47*, D464–D474. [CrossRef]
16. Mitchell, L.S.; Colwell, L.J. Analysis of Nanobody Paratopes Reveals Greater Diversity than Classical Antibodies. *Protein Eng. Des. Sel.* **2018**, *31*, 267–275. [CrossRef]
17. Xi, X.; Sun, W.; Su, H.; Zhang, X.; Sun, F. Identification of a Novel Anti-EGFR Nanobody by Phage Display and Its Distinct Paratope and Epitope via Homology Modeling and Molecular Docking. *Mol. Immunol.* **2020**, *128*, 165–174. [CrossRef]
18. Cheng, X.; Wang, J.; Kang, G.; Hu, M.; Yuan, B.; Zhang, Y.; Huang, H. Homology Modeling-Based in Silico Affinity Maturation Improves the Affinity of a Nanobody. *Int. J. Mol. Sci.* **2019**, *20*, 4187. [CrossRef]

19. Zou, J. Artificial Intelligence Revolution in Structure Prediction for Entire Proteomes. *MedComm—Future Med.* **2022**, *1*, e19. [CrossRef]
20. Bertoline, L.M.F.; Lima, A.N.; Krieger, J.E.; Teixeira, S.K. Before and after AlphaFold2: An Overview of Protein Structure Prediction. *Front. Bioinform.* **2023**, *3*, 17. [CrossRef]
21. Tunyasuvunakool, K. The Prospects and Opportunities of Protein Structure Prediction with AI. *Nat. Rev. Mol. Cell Biol.* **2022**, *23*, 445–446. [CrossRef]
22. Abanades, B.; Georges, G.; Bujotzek, A.; Deane, C.M. ABlooper: Fast Accurate Antibody CDR Loop Structure Prediction with Accuracy Estimation. *Bioinformatics* **2022**, *38*, 1877–1880. [CrossRef] [PubMed]
23. Ruffolo, J.A.; Gray, J.J.; Sulam, J. Deciphering Antibody Affinity Maturation with Language Models and Weakly Supervised Learning. *arXiv* **2021**, arXiv:2112.07782.
24. Wang, S.; Sun, S.; Li, Z.; Zhang, R.; Xu, J. Accurate De Novo Prediction of Protein Contact Map by Ultra-Deep Learning Model. *PLoS Comput. Biol.* **2017**, *13*, e1005324. [CrossRef]
25. Fernández-Quintero, M.L.; Kokot, J.; Waibl, F.; Fischer, A.L.M.; Quoika, P.K.; Deane, C.M.; Liedl, K.R. Challenges in Antibody Structure Prediction. *MAbs* **2023**, *15*, 2175319. [CrossRef] [PubMed]
26. Ruffolo, J.A.; Sulam, J.; Gray, J.J. Antibody Structure Prediction Using Interpretable Deep Learning. *Patterns* **2022**, *3*, 100406. [CrossRef]
27. Ruffolo, J.A.; Guerra, C.; Mahajan, S.P.; Sulam, J.; Gray, J.J. Geometric Potentials from Deep Learning Improve Prediction of CDR H3 Loop Structures. *Bioinformatics* **2020**, *36*, i268. [CrossRef]
28. AlQuraishi, M. Machine Learning in Protein Structure Prediction. *Curr. Opin. Chem. Biol.* **2021**, *65*, 1–8. [CrossRef]
29. Eisenstein, M. Artificial Intelligence Powers Protein-Folding Predictions. *Nature* **2021**, *599*, 706–708. [CrossRef]
30. AlQuraishi, M. Protein-Structure Prediction Revolutionized. *Nature* **2021**, *596*, 487–488. [CrossRef]
31. Jumper, J.; Evans, R.; Pritzel, A.; Green, T.; Figurnov, M.; Ronneberger, O.; Tunyasuvunakool, K.; Bates, R.; Žídek, A.; Potapenko, A.; et al. Highly Accurate Protein Structure Prediction with AlphaFold. *Nature* **2021**, *596*, 583–589. [CrossRef] [PubMed]
32. Callaway, E. What's next for AlphaFold and the AI Protein-Folding Revolution. *Nature* **2022**, *604*, 234–238. [CrossRef]
33. Wu, R.; Ding, F.; Wang, R.; Shen, R.; Zhang, X.; Luo, S.; Su, C.; Wu, Z.; Xie, Q.; Berger, B.; et al. High-Resolution de Novo Structure Prediction from Primary Sequence. *bioRxiv* **2022**. [CrossRef]
34. Lin, Z.; Akin, H.; Rao, R.; Hie, B.; Zhu, Z.; Lu, W.; Smetanin, N.; Verkuil, R.; Kabeli, O.; Shmueli, Y.; et al. Evolutionary-Scale Prediction of Atomic Level Protein Structure with a Language Model. *bioRxiv* **2022**. [CrossRef]
35. Schoof, M.; Faust, B.; Saunders, R.A.; Sangwan, S.; Rezelj, V.; Hoppe, N.; Boone, M.; Billesbølle, C.B.; Puchades, C.; Azumaya, C.M.; et al. An Ultrapotent Synthetic Nanobody Neutralizes SARS-CoV-2 by Stabilizing Inactive Spike. *Science* **2020**, *370*, 1473–1479. [CrossRef] [PubMed]
36. Xu, J.; Zhang, Y. How Significant Is a Protein Structure Similarity with TM-Score = 0.5? *Bioinformatics* **2010**, *26*, 889. [CrossRef] [PubMed]
37. Du, Z.; Su, H.; Wang, W.; Ye, L.; Wei, H.; Peng, Z.; Anishchenko, I.; Baker, D.; Yang, J. The TrRosetta Server for Fast and Accurate Protein Structure Prediction. *Nat. Protoc.* **2021**, *16*, 5634–5651. [CrossRef]
38. Hong, J.; Kwon, H.J.; Cachau, R.; Chen, C.Z.; Butay, K.J.; Duan, Z.; Li, D.; Ren, H.; Liang, T.; Zhu, J.; et al. Camel Nanobodies Broadly Neutralize SARS-CoV-2 Variants. *bioRxiv* **2021**. [CrossRef]
39. Frosi, Y.; Lin, Y.C.; Shimin, J.; Ramlan, S.R.; Hew, K.; Engman, A.H.; Pillai, A.; Yeung, K.; Cheng, Y.X.; Cornvik, T.; et al. Engineering an Autonomous VH Domain to Modulate Intracellular Pathways and to Interrogate the EIF4F Complex. *Nat. Commun.* **2022**, *13*, 4854. [CrossRef]
40. Mirdita, M.; Schütze, K.; Moriwaki, Y.; Heo, L.; Ovchinnikov, S.; Steinegger, M. ColabFold: Making Protein Folding Accessible to All. *Nat. Methods* **2022**, *19*, 679–682. [CrossRef]
41. Evans, R.; O'Neill, M.; Pritzel, A.; Antropova, N.; Senior, A.; Green, T.; Žídek, A.; Bates, R.; Blackwell, S.; Yim, J.; et al. Protein Complex Prediction with AlphaFold-Multimer. *bioRxiv* **2022**. [CrossRef]
42. Zimmermann, I.; Egloff, P.; Hutter, C.A.J.; Arnold, F.M.; Stohler, P.; Bocquet, N.; Hug, M.N.; Hub er, S.; Siegrist, M.; Hetemann, L.; et al. Synthetic Single Domain Antibodies for the Conformational Trapping of Membrane Proteins. *Elife* **2018**, *7*, e34317. [CrossRef] [PubMed]
43. Moreno, E.; Valdés-Tresanco, M.S.; Molina-Zapata, A.; Sánchez-Ramos, O. Structure-Based Design and Construction of a Synthetic Phage Display Nanobody Library. *BMC Res. Notes* **2022**, *15*, 124. [CrossRef] [PubMed]
44. Valdés-Tresanco, M.S.; Molina-Zapata, A.; Pose, A.G.; Moreno, E. Structural Insights into the Design of Synthetic Nanobody Libraries. *Molecules* **2022**, *27*, 2198. [CrossRef] [PubMed]
45. Dunbar, J.; Krawczyk, K.; Leem, J.; Baker, T.; Fuchs, A.; Georges, G.; Shi, J.; Deane, C.M. SAbDab: The Structural Antibody Database. *Nucleic Acids Res.* **2014**, *42*, D1140–D1146. [CrossRef]
46. Altschul, S.F.; Gish, W.; Miller, W.; Myers, E.W.; Lipman, D.J. Basic Local Alignment Search Tool. *J. Mol. Biol.* **1990**, *215*, 403–410. [CrossRef]
47. Camacho, C.; Coulouris, G.; Avagyan, V.; Ma, N.; Papadopoulos, J.; Bealer, K.; Madden, T.L. BLAST+: Architecture and Applications. *BMC Bioinform.* **2009**, *10*, 421. [CrossRef]
48. Altschul, S.F.; Madden, T.L.; Schäffer, A.A.; Zhang, J.; Zhang, Z.; Miller, W.; Lipman, D.J. Gapped BLAST and PSI-BLAST: A New Generation of Protein Database Search Programs. *Nucleic Acids Res.* **1997**, *25*, 3389–3402. [CrossRef]
49. Dunbar, J.; Deane, C.M. ANARCI: Antigen Receptor Numbering and Receptor Classification. *Bioinformatics* **2016**, *32*, 298–300. [CrossRef]

50. Yang, J.; Anishchenko, I.; Park, H.; Peng, Z.; Ovchinnikov, S.; Baker, D. Improved Protein Structure Prediction Using Predicted Interresidue Orientations. *Proc. Natl. Acad. Sci. USA* **2020**, *117*, 1496–1503. [CrossRef]
51. Pereira, J.; Simpkin, A.J.; Hartmann, M.D.; Rigden, D.J.; Keegan, R.M.; Lupas, A.N. High-Accuracy Protein Structure Prediction in CASP14. *Proteins* **2021**, *89*, 1687–1699. [CrossRef] [PubMed]
52. Mirdita, M.; Steinegger, M.; Söding, J. MMseqs2 Desktop and Local Web Server App for Fast, Interactive Sequence Searches. *Bioinformatics* **2019**, *35*, 2856–2858. [CrossRef] [PubMed]
53. Steinegger, M.; Söding, J. MMseqs2 Enables Sensitive Protein Sequence Searching for the Analysis of Massive Data Sets. *Nat. Biotechnol.* **2017**, *35*, 1026–1028. [CrossRef] [PubMed]
54. Rives, A.; Meier, J.; Sercu, T.; Goyal, S.; Lin, Z.; Liu, J.; Guo, D.; Ott, M.; Zitnick, C.L.; Ma, J.; et al. Biol.ogical Structure and Function Emerge from Scaling Unsupervised Learning to 250 Million Protein Sequences. *Proc. Natl. Acad. Sci. USA* **2021**, *118*, e2016239118. [CrossRef] [PubMed]
55. Suzek, B.E.; Huang, H.; McGarvey, P.; Mazumder, R.; Wu, C.H. UniRef: Comprehensive and non-redundant UniProt reference clusters. *Bioinformatics* **2007**, *23*, 1282–1288. [CrossRef]
56. Evans, R.; Jumper, J.; Kirkpatrick, J.; Sifre, L.; Green, T.F.G.; Qin, C.; Zidek, A.; Nelson, A.; Bridgland, A.; Penedones, H.; et al. De Novo Structure Prediction with Deep-Learning Based Scoring. In Proceedings of the Thirteenth Critical Assessment of Techniques for Protein Structure Prediction (ProteinStructure Prediction Center), Cancun, Mexico, 1–4 December 2018; pp. 11–12.
57. Dunbar, J.; Krawczyk, K.; Leem, J.; Marks, C.; Nowak, J.; Regep, C.; Georges, G.; Kelm, S.; Popovic, B.; Deane, C.M. SAbPred: A Structure-Based Antibody Prediction Server. *Nucleic Acids Res.* **2016**, *44*, W474–W478. [CrossRef]
58. Chaudhury, S.; Lyskov, S.; Gray, J.J. PyRosetta: A Script-Based Interface for Implementing Molecular Modeling Algorithms Using Rosetta. *Bioinformatics* **2010**, *26*, 689–691. [CrossRef]
59. Šali, A.; Blundell, T.L. Comparative protein modelling by satisfaction of spatial restraints. *J. Mol. Biol.* **1993**, *234*, 779–815. [CrossRef]
60. Zhang, Y.; Skolnick, J. Scoring Function for Automated Assessment of Protein Structure Template Quality. *Proteins* **2004**, *57*, 702–710. [CrossRef]
61. Zemla, A. LGA: A Method for Finding 3D Similarities in Protein Structures. *Nucleic Acids Res.* **2003**, *31*, 3370. [CrossRef]
62. Shirts, M.R.; Klein, C.; Swails, J.M.; Yin, J.; Gilson, M.K.; Mobley, D.L.; Case, D.A.; Zhong, E.D. Lessons Learned from Comparing Molecular Dynamics Engines on the SAMPL5 Dataset. *J. Comput. Aided Mol. Des.* **2016**, *31*, 147–161. [CrossRef] [PubMed]
63. Bedre, R. *Bioinfokit: Bioinformatics Data Analysis and Visualization Toolkit, version 1.0.5*; Zenodo: Geneva, Switzerland, 2021. [CrossRef]
64. Eastman, P.; Swails, J.; Chodera, J.D.; McGibbon, R.T.; Zhao, Y.; Beauchamp, K.A.; Wang, L.P.; Simmonett, A.C.; Harrigan, M.P.; Stern, C.D.; et al. OpenMM 7: Rapid Development of High Performance Algorithms for Molecular Dynamics. *PLoS Comput. Biol.* **2017**, *13*, e1005659. [CrossRef] [PubMed]
65. Hornak, V.; Abel, R.; Okur, A.; Strockbine, B.; Roitberg, A.; Simmerling, C. Comparison of Multiple Amber Force Fields and Development of Improved Protein Backbone Parameters. *Proteins Struct. Funct. Bioinform.* **2006**, *65*, 712–725. [CrossRef] [PubMed]

Article

Design and Construction of a Synthetic Nanobody Library: Testing Its Potential with a Single Selection Round Strategy

María Angélica Contreras [1,†], Yunier Serrano-Rivero [2,†], Alaín González-Pose [2], Julieta Salazar-Uribe [2], Marcela Rubio-Carrasquilla [2], Matheus Soares-Alves [1], Natalie C. Parra [1], Frank Camacho-Casanova [1], Oliberto Sánchez-Ramos [1,*] and Ernesto Moreno [2,*]

1 Pharmacology Department, School of Biological Sciences, University of Concepcion, Concepcion 4070386, Chile; mcontrerasv@udec.cl (M.A.C.); malves@udec.cl (M.S.-A.); natparra@udec.cl (N.C.P.); fcamacho@udec.cl (F.C.-C.)
2 Faculty of Basic Sciences, University of Medellin, Medellin 050026, Colombia; 0905yunierserrano@gmail.com (Y.S.-R.); agonzalezp@udemedellin.edu.co (A.G.-P.); julieta.salazaru@gmail.com (J.S.-U.); marcelaru@yahoo.com (M.R.-C.)
* Correspondence: osanchez@udec.cl (O.S.-R.); emoreno@udemedellin.edu.co (E.M.)
† These authors contributed equally to this work.

Abstract: Nanobodies (Nbs) are single domain antibody fragments derived from heavy-chain antibodies found in members of the Camelidae family. They have become a relevant class of biomolecules for many different applications because of several important advantages such as their small size, high solubility and stability, and low production costs. On the other hand, synthetic Nb libraries are emerging as an attractive alternative to animal immunization for the selection of antigen-specific Nbs. Here, we present the design and construction of a new synthetic nanobody library using the phage display technology, following a structure-based approach in which the three hypervariable loops were subjected to position-specific randomization schemes. The constructed library has a clonal diversity of 10^8 and an amino acid variability that matches the codon distribution set by design at each randomized position. We have explored the capabilities of the new library by selecting nanobodies specific for three antigens: vascular endothelial growth factor (VEGF), tumor necrosis factor (TNF) and the glycoprotein complex (GnGc) of Andes virus. To test the potential of the library to yield a variety of antigen-specific Nbs, we introduced a biopanning strategy consisting of a single selection round using stringent conditions. Using this approach, we obtained several binders for each of the target antigens. The constructed library represents a promising nanobody source for different applications.

Keywords: nanobody; synthetic library; phage display; CDR randomization; biopanning; tumor necrosis factor; vascular endothelial growth factor; Andes virus

Citation: Contreras, M.A.; Serrano-Rivero, Y.; González-Pose, A.; Salazar-Uribe, J.; Rubio-Carrasquilla, M.; Soares-Alves, M.; Parra, N.C.; Camacho-Casanova, F.; Sánchez-Ramos, O.; Moreno, E. Design and Construction of a Synthetic Nanobody Library: Testing Its Potential with a Single Selection Round Strategy. *Molecules* **2023**, *28*, 3708. https://doi.org/10.3390/molecules28093708

Academic Editor: Jahir Orozco Holguín

Received: 20 March 2023
Revised: 18 April 2023
Accepted: 18 April 2023
Published: 25 April 2023

1. Introduction

Nanobodies (Nbs) are single domain antibody fragments derived from heavy-chain antibodies, lacking the light chain present in classical immunoglobulins [1]. These special antibodies are found in members of the Camelidae family, which includes camels, dromedaries, llamas and alpacas. Nbs have several important advantages as compared to antibodies and their fragments, such as their small size (~15 kDa) and high thermal stability (median melting temperature (Tm) ~67 °C [2]). These tiny proteins have found multiple applications in many different areas, from basic research—for example, as affinity capture reagents and crystallization chaperones [3]—to the clinics, with more than 40 clinical trials reported for different Nb-based products in the ClinicalTrials.gov web repository maintained by the National Institutes of Health (https://clinicaltrials.gov) and two Nbs approved for clinical use: one in the United States [4] and another one in Japan [5]. Such application versatility is due in large part to the single-domain structure of Nbs, which makes them easy to engineer and integrate into many different constructs. Notably, Nbs can achieve high affinities in spite of their smaller binding region displaying only three hypervariable loops [6].

Nbs are obtained mostly from immune libraries generated by animal immunization [6]. During the last few years, however, synthetic libraries with different designs are gaining ground as reliable Nb sources, offering important advantages in terms of cost and speed [7]. Two key features define a synthetic Nb library: framework selection and the design of the complementarity-determining regions (CDRs). A few recent works have relied on both sequence and structural data to define the CDR positions to be randomized, as well as the sets of amino acids (aa) to be introduced at those positions [8–14], including a recent report by our group [15].

A first, comprehensive work in designing and validating a synthetic Nb library was reported in 2016 by Moutel and coworkers [8] using as scaffold an in-house developed framework. They kept CDRs 1 and 2 with a constant length (7 aa each), randomizing each position in a way that resembles the natural diversity observed for these two CDRs. For CDR3, four different CDR3 lengths were introduced (9, 12, 15 and 18 aa) and all the positions were randomized, allowing all amino acids except cysteine. Two years later, McMahon and coworkers [9] reported the structure-based design and construction of a yeast-displayed library in which the amino acid variability in CDRs 1 (7 aa) and 2 (5 aa) recapitulates the natural diversity observed in a set of over 90 Nb crystal structures available at that time. CDR3 was constructed with different lengths (7, 11 and 15 aa), fully randomizing every position. That same year (2018), Zimmermann et al. [10] reported the design and construction of a ribosome-displayed library composed of three sub-libraries with different CDR3 lengths, using two different frameworks. Their design was based on a structure-based analysis of Nb crystal structures, finding that Nbs with a short CDR3 (6 aa) show a concave shape, those with an intermediate length (12 aa) show a protrusion, and those with a longer loop (16 aa) display a convex surface. CDR randomization focused on achieving an optimal balance between charged, polar, aromatic, and non-polar amino acids to keep a moderate hydrophobicity on the binding site surface. In a more recent (2021) report, Chen et al. [14] constructed a ribosome-displayed library using four CDR3 lengths (6, 9, 10 and 13 aa) and fully randomizing each CDR position. Several other synthetic nanobody libraries have been reported during the last few years following similar design strategies, as recently reviewed [7]. Library sizes range from 10^8–10^{10} for phage-displayed libraries, and up to 10^{12} when using ribosome display [7].

An important issue to consider in Nb library design is the length of CDR3. It has been shown that nanobodies can recognize clefts and cryptic epitopes in proteins that are less accessible to conventional antibodies [16–18]. This important capability is due to the compact prolate shape of Nbs together with their usually long CDR3 loop that folds over the framework region, generating a convex paratope. In several cases, this effect may be enhanced by a protruding loop structure. Such convex–concave Nb–antigen interface provides an interaction surface as large as that of a two-domain antibody paratope, while interacting with a smaller section of the antigen [16]. As observed by Zimmermann and coworkers [10] from the analysis of a large number of nanobody crystal structures, medium length CDR3 loops (10–12 aa) adopt an extended, protruding conformation that can be inserted into a receptor cavity.

Here, we describe the design and construction of a new synthetic nanobody library with a 10 amino acid-long CDR3, in which the three hypervariable loops were subjected to position-specific randomization schemes. The design follows a structure-based approach that seeks to maintain the high stability shown by the original framework-donor nanobody and increase the number of functional variants within the combinatorial space of mutations. As scaffold, we used the framework region from the camelid nanobody cAbBCII10 [19]. This "universal" framework has been shown to be highly stable (Tm = 68 °C [20]), capable of accepting many different CDRs [21], and has been used for the construction of several Nb libraries [15,22–24]. The capabilities of the new library were explored by selecting nanobodies specific for three therapeutically relevant antigens: tumor necrosis factor (TNF), vascular endothelial growth factor (VEGF) and the glycoprotein complex (GnGc) of Andes virus. To test the potential of the library to yield a variety of antigen-specific Nbs, we introduced a biopanning strategy consisting of a single selection round using stringent conditions, aiming to wash out the weaker binders. By applying this strategy, we obtained several binders for each of the target antigens. For one of the obtained anti-TNF clones, we constructed a recombinant fusion protein that incorporates an albumin binding domain and confirmed the functionality of the two binding modules.

2. Results

2.1. Structure-Based Library Design

The design of this library follows a rationale similar to the approach described in a previous work by our group [15]. The amino acid sequence of the framework region was taken from the camelid nanobody cAbBCII10 [19]—a universal scaffold used for the construction of several Nb libraries [15,22–24]. The design of the CDRs relied on the analysis of the crystal structure of the parent cAbBCII10 nanobody (entry 3DWT in the Protein Data Bank [25]), focusing on the structural role played by individual residues in defining CDR conformation or exposing their side chains for antigen binding. The principles followed in the design of CDRs 1 and 2 are explained in detail in [15]. Briefly, the lengths of these two CDRs were kept as in the original cAbBCII10. Furthermore, the amino acids whose sidechains are packaged against framework residues in the 3D structure, as well as those found to be highly conserved in nanobody sequences, were kept as in the parent nanobody. This way we intended to preserve as much as possible the structural stability of the library mutants. CDR residues with surface-exposed side chains were subjected to tailored randomization by introducing degenerate codons in the gene sequence [26]. The allowed codons did not include cysteines and were carefully chosen to restrict the presence of hydrophobic amino acids at these solvent-exposed positions.

For this library, we chose a 10-long CDR3, which for most of the resulting nanobody variants should create a "concave" binding site topology with an "upright"-oriented and protruding CDR3 loop. This represents an important difference as compared with our previously constructed library, carrying a 14 aa-long CDR3 that bends over the framework flank, creating a "convex" topology [15]. Codons VRN and WMY were introduced at several positions to favor the presence of polar/charged amino acids, while the relatively high probability of Gly in the VRN codon may favor a conformational diversity. The highly variable VNN codon was also used. For the C-terminal part of CDR3 (the last two residues), we took into account the amino acid frequencies observed at these positions in the crystal structures of nanobodies with short CDR3 loops, which show that Ser and Tyr are the most frequent aa at the C-terminal end (position "n"), while polar residues are frequent at position "n − 1" (our own data). For the framework region, codon usage was optimized for bacterial expression. Figure 1 shows the library design at the amino acid, nucleotide and structural levels, as well as the amino acid repertoire corresponding to each of the degenerate codons employed in the design.

A total of 22 sequence positions were randomized. The theoretical variability resulting from this tailored design (calculated by multiplying the numbers of the different amino acids coded at each randomized position) is in the order of 10^{18}. This huge number, however, is in practice drastically reduced in the next two construction steps: firstly, by the actual number of genes that are synthesized and, secondly, by the number of bacteria that become transformed in the process of library construction, as explained below.

(a)
```
QVQLVESGGGSVQAGGSLRLSCTASGXXXXXYXXFXLGWFR
QAPGQEREAVAAIXXXGGXTYYADSVKGRFTISRDNAKNTV
TLQMNNLKPEDTAIYYCAAXXXXXXXXXXWGQGTQVTVS
```

(c)
```
CAGGTGCAACTGGTCGAGAGTGGCGGCGGCTCCGTCCAA
GCTGGTGGCTCGCTTCGCTTGTCCTGCACGGCTTCGGGT
DMY VRN VRN HWY VRN TAC DMY VNN TTC KMY
TTAGGCTGGTTTCGCCAAGCACCAGGACAGGAACGTGAG
GCTGTAGCCGCAATC KMY WBG VRN GGT GGA ADM
ACTTACTACGCAGATTCCGTAAAAGGCCGCTTTACAATT
TCGCGTGATAACGCCAAGAATACCGTGACATTGCAAATG
AATAACTTGAAACCGGTAGACACTGCCATTTACTATTGC
GCGGCC VNN VRN WMY VRN WMY VNN WMY YWY
VRN TMY TGGGGACAGGGAACCCAGGTCACCGTGTCT
```

(d) Degenerate codons – encoded amino acids

```
DMY – A:2 N:2 D:2 S:2 T:2 Y:2
VRN – R:6 N:2 D:2 Q:2 E:2 G:4 H:2 K:2 S:2
HWY – N:2 H:2 I:2 L:2 F:2 Y:2
VNN – A:4 R:6 N:2 D:2 Q:2 E:2 G:4 H:2 I:3 L:4 K:2 M:1 P:4 S:2 T:4 V:4
KMY – A:2 D:2 S:2 Y:2
WBG – R:1 L:1 M:1 S:1 T:1 W:1
ADM – R:1 N:1 I:2 K:1 S:1
WMY – N:2 S:2 T:2 Y:2
YWY – H:2 L:2 F:2 Y:2
TMY – S:2 Y:2
```

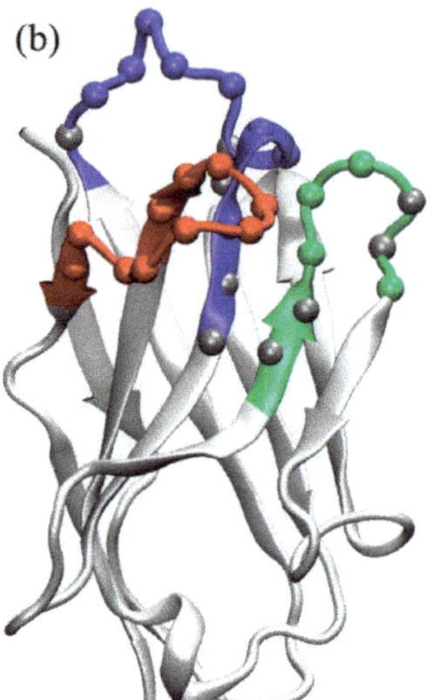

(b)

Figure 1. Library design. (**a**) Sequence design at the amino acid level. Positions chosen for randomization are shown as "X" in bold. The CDR sequences are highlighted in colors (blue, green and red for CDRs 1, 2 and 3, respectively), while the framework region is shown in light gray; (**b**) 3D model of a representative library nanobody based on the cAbBCII10 crystal structure (PDB: 3DWT). CDRs are colored following the same code as in panel (**a**). The colored spheres at the alpha carbons in CDRs represent the randomized positions, while gray spheres represent CDR positions that were kept fixed. (**c**) Nucleotide sequence with degenerate codons, colored by CDR; (**d**) degenerate codons used in library design and their encoded amino acids, showing also the numbers of resulting codons for each amino acid type.

2.2. Library Construction

The randomized genes were synthesized by GenScript (Piscataway, NJ, USA) and cloned as described in Methods into our ad hoc-designed pMAC phagemid vector [15] (see Figure S1). The amount of synthetic genes used for cloning (4 µg) corresponds roughly to 10^{13} individual molecules, that is, five orders of magnitude lower than the theoretical library variability. The pMAC vector employed for cloning includes a pelB leader containing a *NcoI* restriction site at its 3′ end, followed by three other unique restriction sites (*EcoRI*, *BamHI* and *NotI*, in this order). To avoid unnecessary N-terminal and/or C-terminal additions to the recombinant nanobodies, we used the outer *NcoI* and *NotI* sites for cloning. Then, the phagemid codes a short linker (SGGGG), a 6xHis tag, an amber stop codon and, finally, the M13 PIII protein. The amber codon allows the expression of recombinant nanobodies directly from recombinant library plasmids using a non-amber suppressor *E. coli* strain [27], and the obtained nanobodies can then be purified by affinity chromatography using the His tag. The library of recombinant phagemids was transformed by electroporation into SS320 *E. coli* cells as described in Methods.

The library size, which corresponds to its diversity, since with a very high probability each transformed bacterium acquired a unique nanobody gene, was assessed by colony-forming units (CFU) counting. The estimated size was 1.5×10^8. Phage titration by CFU counting yielded a phage concentration of 3.6×10^{10} cfu/µL.

2.3. Assessing Library Quality and Diversity

One hundred randomly picked library clones were sequenced to evaluate the quality of the constructed library and its diversity, as compared to the theoretical design. From these clones, 76 contained a correct nanobody sequence, 15 showed a reading frame shift,

5 clones contained nanobody sequences with no CDR3, 3 clones yielded arbitrary unknown sequences and 1 clone contained an empty phagemid vector. From these results we obtain an estimate of 76% correct clones in the library, which keeps its actual size in the same order of magnitude previously determined (10^8).

Figure 2 shows a sequence logo obtained from the alignment of the 76 correct nanobody sequences. All the randomized positions show an amino acid variability in correspondence with the gene library design, as illustrated in the figure for three CDR positions. Furthermore, and in spite of the relatively limited number of sequenced clones, even the highly variable positions (e.g., for codons VNN and VRN) show a large diversity, matching the expected repertoire of amino acids. For example, between 14 and 16 different residues, out of 16 possible amino acids, are found at the three positions (34, 102 and 107) coded with the VNN triplet.

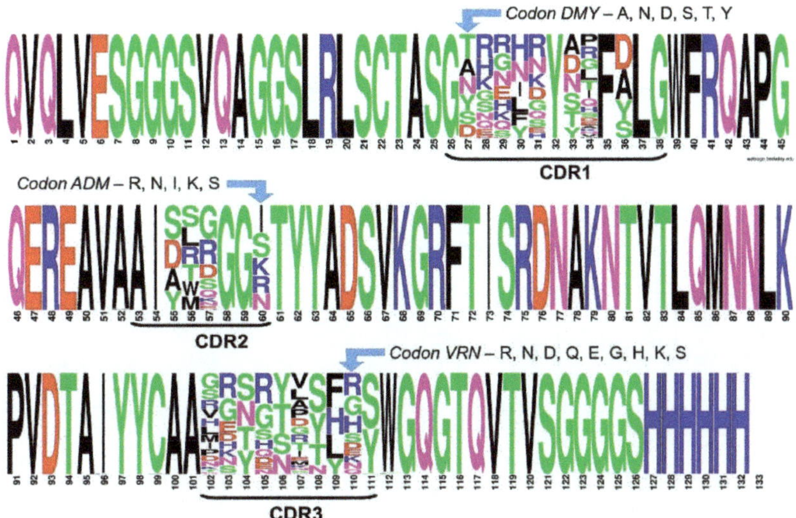

Figure 2. Amino acid distribution per sequence position for the ensemble of 76 correct nanobody clones, shown as a sequence logo. The framework and fixed CDR positions display their conserved amino acid as a single big letter. The amino acid variability found at each randomized position is represented as a stack of letters, each of them with a size that is proportional to its frequency in the multiple alignment. The close match between the theoretical design and the actual experimental diversity is illustrated for three CDR positions (one for each CDR).

2.4. Library Screening

The capability of the library to yield specific binders was tested for three protein antigens: tumor necrosis factor (TNF), vascular endothelial growth factor (VEGF) and the glycoprotein complex (GnGc) of Andes virus. Both TNF and VEGF, as well as their receptors, are relevant therapeutic targets in cancer and autoimmune diseases, and several monoclonal antibodies targeting these molecules have been used in the clinics for several years [28–30]. Furthermore, several nanobodies specific for VEGF have been reported [31], and very recently (Sept/2022) a trivalent anti-TNF nanobody called ozoralizumab was approved in Japan for the treatment of rheumatoid arthritis [5]. Regarding the viral GnGc antigen, to our knowledge no nanobodies specific for this molecule have been yet reported.

2.4.1. Selection of Antigen-Specific Binders in a Single Round

Here we decided to implement a screening procedure based on a single selection round using stringent conditions, aiming at a quick enrichment of the selected phages with the strongest binders in only one selection step, and also as a way of probing the capabilities of the newly designed library. Before elution, we applied four serial washes with glycine-HCl pH 2.2, a buffer commonly used for elution in phage display biopannings. No phage collection was carried out in this step since the aim of these stringent washes was to remove a large part of the phages that would bind with weaker affinity. In a subsequent, final step, the wells were incubated with a relatively high concentration of the antigen (10 μg/well, 10-fold the amount used for coating) to recover the bound recombinant phages by binding competition against the coated and soluble antigens.

For each antigen, the whole eluted phage sample was used to infect *E. coli* TG1 bacteria, which were seeded on 2xYT/ampicillin plates. The numbers of obtained colonies were 97, 1404 and 1656 for TNF, VEGF and GnGc, respectively. We then proceeded to select individual clones to produce recombinant phages and analyze their ability to bind to their corresponding antigens. For TNF, we tested all the 97 obtained clones, whereas for both VEGF and GnGc we randomly picked 180 clones. Figure 3 shows the results from the binding experiments. Notably, we obtained a high number of positive clones in only one selection round, for the three antigens, several of them showing high OD signals.

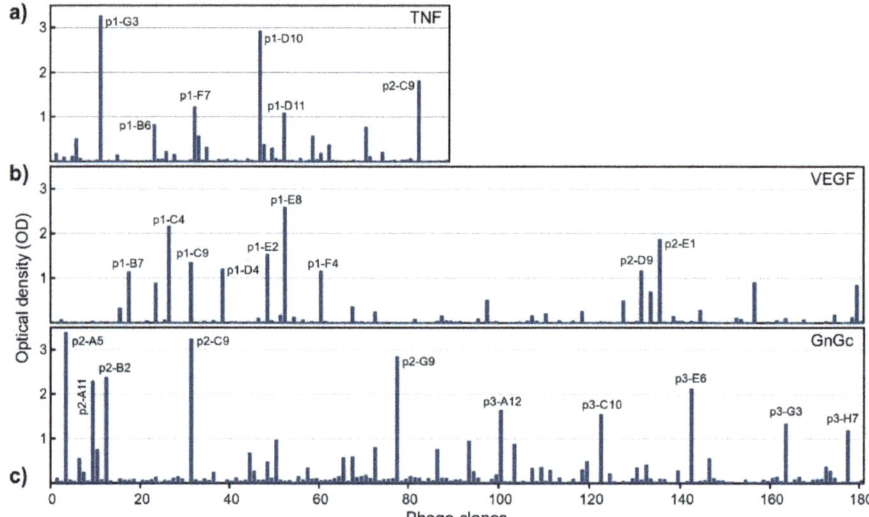

Figure 3. Binding of individual phage clones to their antigens, as measured by ELISA, (**a**) TNF, (**b**) VEGF, (**c**) GnGc. The optical density (OD) values for each clone corresponds to antigen binding with subtracted binding to BSA. For a few clones showing a negative value for this difference, the OD was set to 0 in the graph. The X-axis scale (clone numbers) is common for the three panels. Clones with high binding signal (OD > 1) are labeled, matching their IDs in Figure 4.

2.4.2. Sequencing of Selected Groups of Phage Clones

We decided to sequence all the clones showing OD values above 0.15, for the three antigens, resulting in 28, 24 and 44 clones for TNF, VEGF and GnGc, respectively. In practice, we obtained the sequences for 22, 22 and 34 clones, respectively, since a few of the samples could not be correctly sequenced. Nonetheless, we obtained the sequences for practically all of the best binders shown in Figure 3, with the exception of the anti-TNF clone p1-F7 and the anti-GnGc clone p2-C9.

As shown in Figure 4, for the three antigens we obtained sets of unique different binders (with only one identical pair of anti-TNF clones) as a consequence of performing a single selection round, without further binding clone enrichment. In the three cases, no common sequence motifs are evident from the alignment, for any of the CDRs, which suggests that these clones have different binding modes, likely recognizing different epitopes on the antigen surface.

(a) TNF

Clone		CDR1	CDR2	CDR3
p1-D11	★	GTNRFSYTRFALG	AIAMQGGR	IQSSNLTHRY
p1-E10		. . . .	. . . .	. . . .
p2-B5		.YK.ND.SP....	...WG...	.RYDYAY.DS
p2-B7		.A.KNR.AL....	...S....	RRT.TANYHS
p1-D9		..HQNR.DT....	...RG...	NRTQ.VN.QS
p1-B2		.NHDLR.AS....	..YWR..I	RGYG.RYLH.
p2-E2		..G.LR.AV.D..	..Y.R..I	VR.G.ASFN.
p2-B6		.ARNNG.NP....	..YWH..I	TR.KSRSYG.
p1-C8		..R.LQ.S.....	...SR...	PDTHTTNFG.
p2-B2		..EDYQ.S.....	..Y.H..S	PKYQ.TY.G.
p1-B3		.YR.NK.NT.Y..	...LG..S	PETNY.SFG.
p2-C3		..RENN.DG.D..	..STR...	HGYRYR.YG.
p1-G6		.DQ.I..S.....	..SSE..K	GRYH.VN...
p1-D10	★	.NQ..R.YG....	..SSS..N	LRYNSI...S
p1-G3	★	.ARK.R.YA....	..S.N..N	TRTR.RSFG.
p1-B6	★	.YGNYR.NT.S..	..SSS..I	.G.KTAS..S
p1-E6		.DKKYK.S..S..	..SLD..I	VRTRYVSLES
p1-F8		.NREHH.YA.Y..	..YSH..S	NRYNSVS..S
p2-D4		.NQK.R.YV.Y..	..YRK..S	LR..YNS.K.
p1-G2		.YQ.HK.NH.Y..	..SRK..K	QRTESDNY.S
p2-C9	★	.YR.NG..V.D..	...SK..N	NKYR.PSFK.
p1-G7		.SRNNH.YT....	..SWG..K	LGYGYVSYG.

(b) VEGF

Clone		CDR1	CDR2	CDR3
p1-C1		GARRHRYTMFSLG	AISLRGGK	RGSSYTTFRS
p1-E2	★	.N.K...YL....	...WG..S	.RTR.GN.S.
p1-F4	★	.T.D...SR....	..A.....	L.Y.S.SHK.
p1-E7		.NGKNN..T....	..A....R	L.NGSSNH.Y
p1-B5		.DHG...NP....	...RD..I	PH.RTAN.H.
p1-C4	★	.YGKLG.YN....	...R...R	ND.GSRYHHY
p2-D9	★	.YGKLG.YN....	...R...R	ND.GSRYHHY
p1-H7		.NNHLS.YV....	....G...	LR.RSSNHQY
p2-A9		.SG.L..AA....	..YSG...	.KYGSINY.Y
p2-E1	★	.S...K..G....	..D....R	SRYGS.S.G.
p2-F10		.SGGLG.AR.Y..	..AM...N	.KNR.M.HN.
p1-G4		..GNYK.SR.A..	...SG..I	VR.HSP.YG.
p2-H9		.YGQIK.SA.A..	...S...I	..TR.HYYK.
p2-D11		...GI..NK.Y..	...R...R	LR.RSPY.NY
p1-C9	★	.TN.N..AI.Y..	..A.H..S	.RNG.V..KY
p1-B7	★	.DGHLK.YK.A..	....G..R	IQNRSVN...
p1-E8	★	.S.GN..AR.D..	....H..R	QRTRSVSHKY
p2-D5		.S.KLG..R.Y..	..A.D...	G.Y..AYL.Y
p2-C8		.NKK.N.YR.Y..	..ASD..R	SRY.SIYHG.
p2-E10		.SHKNK.NR.Y..	..AWE..R	VRYRTGYLEY
p2-B10		.D..IN.AR....	..ASN..R	GRNH.SS.S.
p1-D4	★	.TS.YN.AI...C	...TQ..N	LRYR.LY.KY

(c) GnGc

Clone		CDR1	CDR2	CDR3
p2-A5	★	GYRRHNYYGFALG	AISMRGGN	PRSGSDYFGY
p3-D6		.N.GLE.......	...L....	VGTNYRT...
p2-F11		.D...R.NQ.D..	...K....	S...YV.L..
p3-C7		.SN.NS.NS.D..	...W...K	S..NNLTLRS
p3-D8		.SGHNR.SH.D..	...TS..K	T.TRNPNLRS
p2-B2	★	.S..LR.AD....	I	GKYRYTS.RS
p3-B7		..DHIH..V....	...LH..I	L.N.N.S.RS
p2-A11	★	.....R.NM.D..	..DWG..I	H..S.LS.RS
p3-C10	★	.N..NH.TK.S..	..DRG..I	H..HTSSYRS
p3-C12		.DE.LR.NH.Y..	..YWQ..I	S.TKYQS.ES
p2-A12		.TGSFR.ST.S..	..AL...S	RG.RYTNL.S
p2-E6		.T.KFR..R.Y..	...TN..S	RG.R.PTLHS
p2-G2		.S.SIQ.S..S..	..AS...R	QNTHYR.HH.
p2-H6		.S.G.R.S..D..	..ATS..R	RHY..S.HH.
p3-A11		.TNQIR.TR.D..	...R...S	TDYK.GTYS.
p2-G12		.DQGYG.SL....	...TG..S	QG.NYG.LRS
p3-H7	★	.AGHYG.NK.Y..	..DTE..S	Q..RYRSY..
p2-A9		.DSNYR..P.Y..	..YL...I	.GYHYIS.Q.
p2-F9		.NQ...DR....	..DR...R	VG.KTNSYRS
p3-B3		.TK.IG..R.D..	..ALQ...	.S.RNITLES
p3-G3	★	.TKG.G.TP....	..A....K	R...YR.HRS
p3-A5		.N.GNR.SA....	..ASG..I	.KNNYRN.R.
p3-E10		.ANHLR.TK....	..YTS..K	GGYHNP.Y.S
p2-D12		.SGHNK.SR.S..	...WG..R	RGYR.GS.KS
p3-H2		.S..N..SR.D..	...K..R	ENN..GS...
p2-D2		.S..FD..A....	..A....R	RSNR.GS.N.
p2-E1		...DYG.AT....	..AWD..R	RSNR.H..R.
p3-E3		..NGNK.TP....	..A.G..S	GGYR.IS.Q.
p2-G4		.SGQ.R.A..S..	..YLH...	R.NKYRTLKS
p3-H3		.SD.YH.DK.D..	..DW...R	LQY.YATLR.
p2-G9	★	..HHI..DR.Y..	..YRG...	K.YQTTN.RS
p3-B9		.TGQYR.TR.S..	...RG...	VNY.TQSYDS
p3-E6	★	.S.GNS..R....	...S..S	VGNRTANHR.
p3-A12	★	...HIR.SR.D..	..AS...I	GGY.YGS.DS

Figure 4. Alignments of CDR sequences obtained for groups of clones selected from the biopannings against the following: (**a**) TNF (22 clones); (**b**) VEGF (22 clones) and (**c**) GnGc (34 clones). Dots indicate the presence of the same amino acid as in the first sequence in the alignment. Red stars denote a high binding signal by ELISA (OD > 1), matching their labels in Figure 3.

2.5. Design and Expression of a Recombinant Fusion Protein with an Anti-TNF Nb

A known drawback for the therapeutic use of nanobodies is their short half-life in serum due their small size. Several strategies can be followed to prolong the Nb half-life, one of them being the genetic fusion or chemical conjugation to a molecule capable of binding to serum albumin [32]. Here, we decided to construct a fusion protein (NbB6-ABD) composed of an anti-TNF Nb and an albumin binding domain (ABD) from the *Streptococcus* sp. G protein, which shows high specificity and affinity for human serum albumin (HSA), with a dissociation constant (KD) in the nanomolar order [33,34] (Figure 5a). The anti-TNF clone p1-B6 was selected for this purpose. Although this clone is not among the strongest binders (OD = 0.9), it was chosen because of its very low background signal to BSA and skim milk (data not shown). In addition to the 46 aa constituting the ABD domain, six additional aa (AVDANS) of the protein were included at the N-terminal end since they are packed with the ABD domain in its crystal structure. A c-Myc tag was included between the nanobody and ABD, separated by short linkers. An *EcoRI* restriction site inserted right after the Nb sequence allows switching the Nb binder to target any antigen of interest. The gene coding for the fusion protein was cloned into the pET22b plasmid, which adds a C-terminal His tag, as shown in Figure 5a.

For binding assays, the fusion protein was biotinylated as described in Methods. Figure 5b shows the ELISA results for the binding to TNF and HSA. The ABD domain kept its binding capability to HSA (Figure 5b, right panel). For the binding of the nanobody domain to TNF, a titration ELISA [35] was performed in order to estimate the dissociation constant, obtaining a KD = $(1.48 \pm 0.35) \times 10^{-7}$ M. This is quite an encouraging result, taking into account that this was an initial test for this fusion protein design, using one of the obtained anti-TNF clones, which was not among the strongest binders in the phage-based ELISAs.

a) (*NcoI*) QVQLVESGGGSVQAGGSLRLSCTASGYGNYRYNTFSLGWFRQAPGQEREAVAAISSSG
GITYYADSVKGRFTISRDNAKNTVTLQMNNLKPVDTAIYYCAAIGSKTASHRSWGQGTQVTVSS
EFGGGSEQKLISEEDLGGSGGSAVDANSLAEAKVLANRELDKYGVSDYYKNLINNAKTVEGVKA
LIDEILAALPGGSGGS (*XhoI*) HHHHHH*

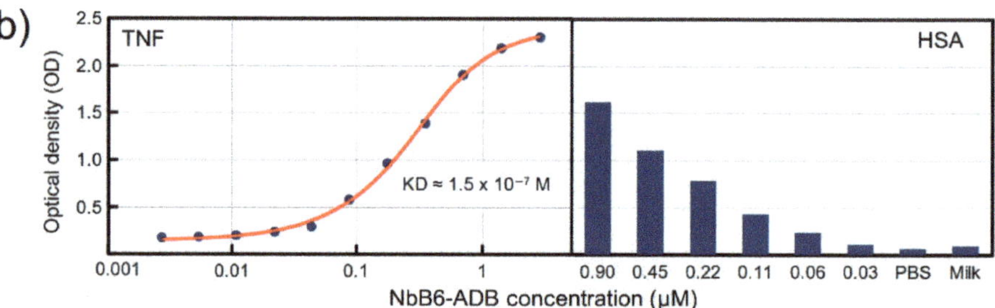

Figure 5. (**a**) Amino acid (aa) sequence of the NbB6-ABD fusion protein. Legend: White background—anti-TNF nanobody clone p1-B6; green—the two aa coded by an inserted *EcoRI* restriction site; gray—spacers (linkers); cyan—c-Myc tag; yellow—albumin binding domain; orange—6xHis tag. The *NcoI* and *XhoI* restriction sites were used for cloning into the pET22b vector, which adds the C-terminal histidine tag. (**b**) Binding of NbB6-ABD to TNF (**left chart**) and to HSA (**right chart**) as measured by ELISA. (**Left chart**): Negative control (not shown)—BSA coating: OD = 0.11. The red fitting curve for TNF binding was used for KD estimation. (**Right chart**): The Y-axis scale is the same as for the left chart; negative controls—PBS (instead of NbB6-ABD) and skim milk. For both antigens, we used the maximum tested NbB6-ABD concentration for the BSA/milk negative control. Experiments were performed in duplicates.

3. Discussion

We have constructed a new synthetic nanobody library following a tailored, structure-based design. Synthetic libraries are nonspecific and therefore seek to recreate a large clonal variability to increase the probability of obtaining good binders. For this reason, synthetic libraries must be large, at least 10^8 in size, preferably larger [6,7]. Most of the reported synthetic, phage-displayed nanobody libraries have sizes in the order of 10^9, as recently reviewed [7]. The clonal diversity of our library is in the order of 10^8, that is, at the lower limit of the accepted range. This level of diversity, however, proved to be enough to produce a high rate of specific clones against three different, relevant therapeutic targets. In this regard, we believe that the library's CDR design creates a high-quality repertoire of binding paratopes that may, to a certain extent, counteract the relatively smaller size of the library.

As scaffold for the library, we chose a well-proven framework—from the cAbBCII10 Nb—that has been shown to support CDR loops of different lengths [21]. This is an important base point in the design to ensure that most of the inserted CDR sequences yield functional nanobodies. As a basis for the CDR design, firstly we carefully analyzed the structural role played by each aa in CDRs 1 and 2 in the parental cAbBCII10 Nb. A first rule applied here was to keep fixed every aa whose sidechain is buried in the structure, as well as those aa found to be highly or relatively conserved in nanobody sequences, or thought to be important in holding CDR conformation in cAbBCII10. This approach differs from the all-position randomization strategies followed in many reported libraries [7], e.g., in [8,11,13,14].

A few recent reports, however, incorporate structure-based strategies to select the positions in CDRs 1 and 2 to be randomized. For example, McMahon and coworkers [9] selected four positions in CDR1 and one in CDR2, based on their large variability in a set of analyzed Nb sequences. All these positions were fully randomized (avoiding Cys and Met). Zimmerman et al. [10] selected five residues in CDR1 and also in CDR2 for randomization, using three different mixtures of nucleotide triplets. The most used mixture coded for 18 aa (excluding Cys and Pro). CDR residues contributing to the Nb hydrophobic core were kept fixed, as in our library. By difference with these designs, here we used 10 different degenerate codons instead of triplets, tailoring the use of these codons at a position level. The amino acid repertoires resulting from these codons vary from 2 to 16 aa, with most of the sets having only 4 or 6 aa. Even so, the theoretical diversity is huge, in the order of 10^{18}.

Although the cAbBCII10 Nb has been shown to accept CDR1 loops of different lengths [21], in this library we kept the full length of the parental CDR1, randomizing 8 out of its 13 positions. In contrast, only 4 positions were randomized in CDR2. Therefore, most of the variability in the library comes from CDR1 and CDR3, which in a modeled structure (Figure 1) form a shallow concave surface between them. We speculate that this shape would be likely fitting for binding to relatively small globular proteins and slightly concave surface patches on proteins in general. Furthermore, the protruding CDR3 might bind to protein cavities.

The new library was tested against three protein antigens of therapeutic relevance: TNF, VEGF and GnGc (a viral antigen). To evaluate its capabilities, we decided to apply a selection strategy consisting in applying stringent washing conditions, followed by competitive elution, aiming to retrieve mostly strong binders in a single screening step. For stringent washing (repeated four times) we used glycine-HCl pH 2.2—a commonly used elution buffer in phage display biopannings [36]. This way, many phages that otherwise would be collected for a second biopanning round were discarded. Subsequently, the wells were incubated with a relatively high concentration of antigen (100 µg/mL) to recover bound recombinant phages by binding competition against the coated antigens. Such competitive phage elution is also a common procedure used to collect phages with high affinity for their target molecule [6,37].

Stringent washes are very often used before the elution step, but in general, such stringency consists in increasing the washing time, number of washes and/or Tween 20 concentration [6,37–39], as well as decreasing the antigen concentration in each subsequent selection round [6,40,41]. There are few reports, however, in which a glycine-HCl solution was used as a wash buffer. Lunder et al. [42], for example, implemented several protocols that included four glycine-HCl (0.2 M, pH 2.2) washings and then eluted the phages that remained bound to the antigen by direct infection with *E. coli*, ultrasound or competition. On the other hand, although using several selection-amplification rounds enriches the library in clones specific for the target molecule, it may also have a negative effect by reducing the diversity of the finally obtained clones [37].

Here, applying stringent washes and without further enrichment rounds, we were able to obtain a significant number of clones with high, specific binding signals by ELISA, with a positivity of 13–29%. For TNF, we obtained 97 clones, all of which were tested individually. For VEGF and GnGc the number of clones was much higher—around 1400 and 1600, respectively—of which we tested only 180 in each case. Notably, all the positive clones, for the three antigens, corresponded to unique sequences (with the only exception of a pair of clones for TNF), with no evident common motifs. Since for VEGF and GnGc we tested only about 12% of the total number of clones, we would expect about a 10-fold higher number of positive clones for these two antigens, most of them most likely with unique sequences.

Finally, we tested the functionality of one of the obtained anti-TNF nanobodies, in a format of a recombinant fusion protein that incorporates an albumin binding domain—a strategy used to prolong the half-life in serum of therapeutic Nbs [32]. A similar solution was employed in the design of the anti-TNF nanobody trimer ozoralizumab (approved for clinical use in Japan), which consists of two anti-human TNF Nbs and an anti-human serum albumin Nb [5]. The estimated dissociation constant for TNF was in the order of 10^7 M, which is an encouraging result considering that the anti-TNF clone chosen for this construction showed a moderate OD signal in the phage ELISA. This design can be also extended to multimeric Nb constructions, using Nbs targeting the same antigen in a non-competitive manner to synergically increase the affinity, as with ozoralizumab.

For future development of this library, we plan to include other CDR3 lengths to enrich its conformational variability. We are also exploring other selection strategies involving different stringent conditions and numbers of selection rounds.

4. Materials and Methods

4.1. In Silico Design and Analyses

Several bioinformatics tools were used along the design process and sequence analyses. The program VMD [43] was employed for visualization and analyses of nanobody structures. The Degenerate Codon Designer online tool (https://www.novoprolabs.com/tools/degenerate-codon-designer, NovoPro, Shanghai, China, last accessed on 20 January 2023) was used for codon analyses. The CLC Genomics Workbench v.21 (QIAGEN Aarhus, Aarhus, Denmark) was employed for sequence analyses.

4.2. Library Construction

The nanobody gene library was synthesized by GenScript (NJ, USA) following the theoretical design. The genes were flanked with the restriction sites *NotI* and *NcoI* for cloning in the pMAC phagemid vector [15]. After cloning (using 4 µg of both the gene library and pMAC), the recombinant plasmids were transformed by electroporation (voltage 2.5 kV, resistance 200 Ω, capacitance 25 µF) in the *E. coli* strain SS320, previously transduced with the helper phage M13KO7 (New England Biolabs, Ipswich, MA, USA).

Transformed bacteria were recovered in SOC medium for (i) determining library diversity by seeding serial dilutions in plates containing solid 2xYT medium supplemented with 100 µg/mL ampicillin, and (ii) amplifying the recombinant phage library in 2xYT medium containing 100 µg/mL ampicillin, 50 µg/mL kanamycin, 1 mM isopropyl β-D-1-thiogalactopyranoside (IPTG) by incubating 20 h at 30 °C and 185 rpm. Phage library was precipitated from the supernatant with 0.2 volumes of a solution containing PEG/NaCl (20% polyethylene glycol 8000 and 2.5 M NaCl) at 4 °C for two hours, and aliquoted in 10% glycerol until further use [44,45].

4.3. Library Screening

Antigens. Recombinant TNF [46], VEGF [47] and GnGc [48] antigens were produced and purified in-house, at the Pharmacology Department, University of Concepcion, as previously described.

Polystyrene high-binding microtiter plates (Costar) were coated with 100 µL of the antigen (TNF, VEGF or GnGc) at 10 µg/mL (24 wells per antigen for TNF and GnGc, 12 wells for VEGF), and incubated overnight at 4 °C. After two washes with phosphate buffered saline (PBS), wells were blocked with 5% skim milk (Sigma-Aldrich, Burlington, MA, USA) in PBS (300 µL/well) overnight at 4 °C. Wells were washed twice with PBS plus 0.05% Tween 20 and incubated at room temperature (RT) for two hours with 100 µL of library phages (in a quantity 500 times bigger than the library diversity) diluted in 5% skim milk. PBS plus 0.1% Tween 20 was used to perform twenty washes (250 µL/well) of five minutes each. Four additional 5 min washes were made with glycine-HCl (0.2 M, pH 2.2), and subsequently neutralized with PBS pH 7.2 for five minutes. Afterwards, recombinant phages were obtained by competitive elution with 100 µg/mL (100 µL/well) of the antigen of interest (TNF, VEGF or GnGc) for one hour at RT and 300 rpm. The E. coli strain TG1 in exponential phase of growing was transduced with the elution and incubated at 37 °C overnight in 2xYT plates supplemented with 100 µg/mL ampicillin and 2% glucose. Deep well plates were used to amplify individual clones in a final volume of 0.5 mL. Individual phage-infected colonies were picked and used to produce phagemid particles in a 96-well plate scale to test their target recognition [49].

4.4. Binding Assays to Detect Positive Phage Clones

Polystyrene high-binding microtiter plates (Costar) were coated with 100 µL of the antigens (TNF, VEGF or GnGc) at 5 µg/mL and incubated overnight at 4 °C. After washing with PBS, wells were blocked with 3% BSA in PBS (250 µL/well) for two hours at 37 °C. Supernatants of individual clones, previously amplified, were added to the plate (50 µL of supernatant plus 50 µL BSA 3%) for one hour at 37 °C. After three washes with PBS-0.1% Tween 20, the anti-M13 antibody conjugated to horseradish peroxidase (GE Healthcare, Chicago, IL, USA) diluted 1:5000 in BSA 1% plus PBS-0.05% Tween 20 was added for one hour at 37 °C. Plates were washed with PBS 0.1% Tween 20 and the reaction was developed with a solution of o-phenylenediamine dihydrochloride (Sigma-Aldrich) and hydrogen peroxide as substrate, and stopped with 2.5 M sulfuric acid. The absorbance was measured in a Synergy/HTX multi-mode reader (BioTek Instruments, Winooski, VT, USA) at 492 nm.

4.5. Sanger Sequencing

Recombinant phagemids from selected TG1 clones were purified using the GenElute Plasmid Miniprep Kit (Sigma-Aldrich) and sequenced by Macrogen (Seoul, Korea) using the standard M13R primer. Sequences were analyzed using the CLC Genomics Workbench v. 21 (QIAGEN Aarhus, Aarhus, Denmark).

4.6. Production of Recombinant Fusion Protein

The sequence of the ABD from the *Streptococcus* sp. G protein was taken from the PDB structure 1GJS [34]. The gene coding the chimeric protein Nb-TNFB6-ABD was synthesized and cloned into the plasmid pET22b, using the *NcoI* and *XhoI* restriction sites, by GenScript (USA). The production of NbB6-ABD was carried out in two 1L Erlenmeyers containing 500 mL each of SMM9 medium, 0.05% yeast extract (Oxoid, Basingstoke, UK), and 100 µg/mL ampicillin. After inducing the gene expression with 25 µM IPTG, the culture was stirred at 100–120 rpm and incubated at 28 °C for 18 h in a shaker-incubator (ES-20/80, BOECO, Hamburg, Germany). Next, the culture was centrifuged at 10,000× g for 15 min at 4 °C, the pellet was re-suspended in half-diluted SMM9, and then subjected to five freeze/thaw rounds.

Soluble NbB6-ABD was obtained in the supernatant after centrifugation at 10,000× g for 15 min at 4 °C. The presence of the soluble fusion protein was verified by sodium dodecyl sulphate-polyacrylamide gel electrophoresis (SDS-PAGE) and Western blot.

For SDS-PAGE, protein samples were diluted in a buffer with beta-mercaptoethanol and run in 15% polyacrylamide and 3% stacking gels. Western blot assay was performed using a 0.2 µm PVDF transfer membrane (Thermo Fisher Scientific, Waltham, MA, USA) in a semi-dry transfer system Trans-Blot® Turbo™ (Bio-Rad, USA) at 0.3 A and 25 V for 30 min. After blocking with 5% skim milk in PBS, the membrane was incubated with the HRP anti-6xHis tag rabbit polyclonal antibody (ab1187, Abcam, Boston, MA, USA) diluted 1:5000 in the blocking buffer. The reaction was visualized using a DAB substrate kit (Thermo Fisher Scientific, USA).

Protein purification was performed by immobilized metal affinity chromatography (IMAC) by adding 5 mM imidazole to the equilibrium buffer (150 mM NaCl, 10 mM Na2HPO4, pH 7.7) and the initial sample diluted in the same EB. Wash and elution was done in EB by adding 25 mM and 250 mM imidazole, respectively. All fractions were monitored using the purification system BioLogic LP (BioRad, Hercules, CA, USA). Imidazole from the elution sample was removed by diafiltering against PBS (Sigma-Aldrich, Burlington, MA, USA) in 5 kDa Spin-X® UF concentrators (Corning, Corning, NY, USA). Samples were analyzed by SDS-PAGE and Western blot as described above. NbB6-ABD purity was estimated using the analytical tool of the iBright 750 Imaging System (Thermo Fisher Scientific, USA), and its concentration was determined using a Pierce BCA Protein Assay Kit (Thermo Fisher Scientific, USA).

For biotinylation of NbB6-ABD, 50 µL of Na_2CO_3/$NaHCO_3$ buffer (500 mM, pH 9.6) were mixed with 900 µL of the fusion protein (1.1 mg/mL). Next, 50 µL of biotin (H1759,

Sigma-Aldrich, USA) prepared at 10 mg/mL in dimethyl sulfoxide (Merck, Rahway, NJ, USA) was slowly added at a rate of 10 µL/min and mixed. The amounts used correspond to an 80:1 biotin/NbB6-ABD molar ratio. The reaction was incubated for 6 h at room temperature (RT) under stirring. Free biotin was removed by dialysis (88244, Thermo Fisher Scientific, Waltham, MA, USA) against 4 L of 1X PBS overnight at RT.

4.7. Binding Assay for the Fusion Protein

Binding of biotinylated NbB6-ABD to TNF and HSA was determined by ELISA, using streptavidin-HRP (DY998, Biotechne R&D Systems, USA) for detection. Plate wells (2592, Corning, USA) were coated with 1µg of TNF or HSA in carbonate buffer pH 9.6 overnight at 4 °C, then washed three times with 0.3 mL of PBS 1X 0.1% Tween 20 (PBST) and blocked with 3% BSA or 5% milk in 1X PBS for 1 h at room temperature. Wells were then washed three times with 0.3 mL PBST and incubated for 1h at RT with 100 µL of different concentrations of the biotinylated protein. Wells were again washed three times with 0.3 mL PBST and incubated for 1 h at RT with 100 µL streptavidin-HRP (1:200, DY998, Biotechne R&D Systems, USA). They were then washed four times with 0.3 mL PBST, revealed with 100 µL of 3,3′,5,5,5′-Tetramethylbenzidine (TMB) (DY999, Biotechne R&D Systems, USA), and stopped with 50 µL of 2N H2SO4. Binding signals were read at 450 nm in a plate reader (BOECO, Germany). KD estimation was carried out followed the method and fitting function described in [35]. Linear regression analysis using this function was performed using the MyCurveFit web server (https://mycurvefit.com/, last accessed on 20 March 2023).

Supplementary Materials: The following supporting information can be downloaded at: https://www.mdpi.com/article/10.3390/molecules28093708/s1, Figure S1: Map of the designed pMAC phagemid vector.

Author Contributions: Conceptualization, O.S.-R. and E.M.; methodology, M.A.C., A.G.-P., Y.S.-R., M.R.-C., F.C.-C., O.S.-R. and E.M.; investigation, M.A.C., Y.S.-R., A.G.-P., J.S.-U., M.R.-C., M.S.-A., N.C.P., F.C.-C., O.S.-R. and E.M.; writing—original draft preparation, M.A.C., A.G.-P., Y.S-R., J.S.-U., M.S.-A. and E.M.; writing—review and editing, M.A.C., A.G.-P., M.R.-C., N.C.P., F.C.-C., O.S.-R. and E.M.; supervision, project administration and funding acquisition, O.S.-R. and E.M. All authors have read and agreed to the published version of the manuscript.

Funding: This research was funded by MINCIENCIAS, MINEDUCACIÓN, MINCIT and ICETEX through the Program NanoBioCancer (Cod. FP44842-211-2018, project number 58676). M.A.C., F.C.-C. and O.S.-R. thank the University of Concepción for its support. A.G.-P. and E.M. thank the support from the University of Medellin.

Institutional Review Board Statement: Not applicable.

Informed Consent Statement: Not applicable.

Data Availability Statement: The data presented in this study are contained in the article tables and supplementary materials.

Conflicts of Interest: The authors declare no conflict of interest.

Sample Availability: Not applicable. The phagemids and molecules used in this work were either purchased or produced in limited amount only to perform the reported experiments.

References

1. Hamers-Casterman, C.; Atarhouch, T.; Muyldermans, S.; Robinson, G.; Hammers, C.; Songa, E.B.; Bendahman, N.; Hammers, R. Naturally Occurring Antibodies Devoid of Light Chains. *Nature* **1993**, *363*, 446–448. [CrossRef] [PubMed]
2. Valdés-Tresanco, M.S.; Valdés-Tresanco, M.E.; Molina-Abad, E.; Moreno, E. NbThermo: A New Thermostability Database for Nanobodies. *Database* **2023**, baad021. [CrossRef] [PubMed]
3. Hassanzadeh-Ghassabeh, G.; Devoogdt, N.; De Pauw, P.; Vincke, C.; Muyldermans, S. Nanobodies and Their Potential Applications. *Nanomedicine* **2013**, *8*, 1013–1026. [CrossRef] [PubMed]
4. Morrison, C. Nanobody Approval Gives Domain Antibodies a Boost. *Nat. Rev. Drug. Discov.* **2019**, *18*, 485–487. [CrossRef]
5. Keam, S.J. Ozoralizumab: First Approval. *Drugs.* **2023**, *83*, 87–92. [CrossRef]

6. Muyldermans, S. A Guide to: Generation and Design of Nanobodies. *FEBS. J.* **2021**, *288*, 2084–2102. [CrossRef]
7. Valdés-Tresanco, M.S.; Molina-Zapata, A.; Pose, A.G.; Moreno, E. Structural Insights into the Design of Synthetic Nanobody Libraries. *Molecules* **2022**, *27*, 2198. [CrossRef]
8. Moutel, S.; Bery, N.; Bernard, V.; Keller, L.; Lemesre, E.; de Marco, A.; Ligat, L.; Rain, J.-C.; Favre, G.; Olichon, A.; et al. NaLi-H1: A Universal Synthetic Library of Humanized Nanobodies Providing Highly Functional Antibodies and Intrabodies. *Elife* **2016**, *5*, e16228. [CrossRef]
9. McMahon, C.; Baier, A.S.; Pascolutti, R.; Wegrecki, M.; Zheng, S.; Ong, J.X.; Erlandson, S.C.; Hilger, D.; Rasmussen, S.G.F.; Ring, A.M.; et al. Yeast Surface Display Platform for Rapid Discovery of Conformationally Selective Nanobodies. *Nat. Struct. Mol. Biol.* **2018**, *25*, 289–296. [CrossRef]
10. Zimmermann, I.; Egloff, P.; Hutter, C.A.; Arnold, F.M.; Stohler, P.; Bocquet, N.; Hug, M.N.; Huber, S.; Siegrist, M.; Hetemann, L.; et al. Synthetic Single Domain Antibodies for the Conformational Trapping of Membrane Proteins. *Elife* **2018**, *7*, e34317. [CrossRef]
11. Sevy, A.M.; Chen, M.-T.; Castor, M.; Sylvia, T.; Krishnamurthy, H.; Ishchenko, A.; Hsieh, C.-M. Structure- and Sequence-Based Design of Synthetic Single-Domain Antibody Libraries. *Protein. Eng. Des. Selection.* **2020**, *33*, gzaa028. [CrossRef]
12. Zimmermann, I.; Egloff, P.; Hutter, C.A.J.; Kuhn, B.T.; Bräuer, P.; Newstead, S.; Dawson, R.J.P.; Geertsma, E.R.; Seeger, M.A. Generation of Synthetic Nanobodies against Delicate Proteins. *Nat. Protoc.* **2020**, *15*, 1707–1741. [CrossRef]
13. Zhao, Y.; Wang, Y.; Su, W.; Li, S. Construction of Synthetic Nanobody Library in Mammalian Cells by DsDNA-Based Strategies. *Chem. BioChem.* **2021**, *22*, 2957–2965. [CrossRef]
14. Chen, X.; Gentili, M.; Hacohen, N.; Regev, A. A Cell-Free Nanobody Engineering Platform Rapidly Generates SARS-CoV-2 Neutralizing Nanobodies. *Nat. Commun.* **2021**, *12*, 5506. [CrossRef]
15. Moreno, E.; Valdés-Tresanco, M.S.; Molina-Zapata, A.; Sánchez-Ramos, O. Structure-Based Design and Construction of a Synthetic Phage Display Nanobody Library. *BMC. Res. Notes.* **2022**, *15*, 124. [CrossRef]
16. De Genst, E.; Silence, K.; Decanniere, K.; Conrath, K.; Loris, R.; Kinne, J.; Muyldermans, S.; Wyns, L. Molecular Basis for the Preferential Cleft Recognition by Dromedary Heavy-Chain Antibodies. *Proc. Natl. Acad. Sci. USA* **2006**, *103*, 4586–4591. [CrossRef]
17. Uchański, T.; Pardon, E.; Steyaert, J. Nanobodies to Study Protein Conformational States. *Curr. Opin. Struct. Biol.* **2020**, *60*, 117–123. [CrossRef]
18. Shi, Z.; Li, X.; Wang, L.; Sun, Z.; Zhang, H.; Chen, X.; Cui, Q.; Qiao, H.; Lan, Z.; Zhang, X.; et al. Structural Basis of Nanobodies Neutralizing SARS-CoV-2 Variants. *Structure* **2022**, *30*, 707–720.e5. [CrossRef]
19. Conrath, K.E.; Lauwereys, M.; Galleni, M.; Matagne, A.; Frère, J.-M.; Kinne, J.; Wyns, L.; Muyldermans, S. β-Lactamase Inhibitors Derived from Single-Domain Antibody Fragments Elicited in the *Camelidae*. *Antimicrob. Agents. Chemother.* **2001**, *45*, 2807–2812. [CrossRef]
20. Dumoulin, M.; Conrath, K.; Van Meirhaeghe, A.; Meersman, F.; Heremans, K.; Frenken, L.G.J.; Muyldermans, S.; Wyns, L.; Matagne, A. Single-Domain Antibody Fragments with High Conformational Stability. *Protein. Sci.* **2002**, *11*, 500–515. [CrossRef]
21. Saerens, D.; Pellis, M.; Loris, R.; Pardon, E.; Dumoulin, M.; Matagne, A.; Wyns, L.; Muyldermans, S.; Conrath, K. Identification of a Universal VHH Framework to Graft Non-Canonical Antigen-Binding Loops of Camel Single-Domain Antibodies. *J. Mol. Biol.* **2005**, *352*, 597–607. [CrossRef] [PubMed]
22. Wei, G.; Meng, W.; Guo, H.; Pan, W.; Liu, J.; Peng, T.; Chen, L.; Chen, C.-Y. Potent Neutralization of Influenza A Virus by a Single-Domain Antibody Blocking M2 Ion Channel Protein. *PLoS ONE* **2011**, *6*, e28309. [CrossRef] [PubMed]
23. Yan, J.; Li, G.; Hu, Y.; Ou, W.; Wan, Y. Construction of a Synthetic Phage-Displayed Nanobody Library with CDR3 Regions Randomized by Trinucleotide Cassettes for Diagnostic Applications. *J. Transl. Med.* **2014**, *12*, 343. [CrossRef] [PubMed]
24. Chi, X.; Liu, X.; Wang, C.; Zhang, X.; Li, X.; Hou, J.; Ren, L.; Jin, Q.; Wang, J.; Yang, W. Humanized Single Domain Antibodies Neutralize SARS-CoV-2 by Targeting the Spike Receptor Binding Domain. *Nat. Commun.* **2020**, *11*, 4528. [CrossRef]
25. Vincke, C.; Loris, R.; Saerens, D.; Martinez-Rodriguez, S.; Muyldermans, S.; Conrath, K. General Strategy to Humanize a Camelid Single-Domain Antibody and Identification of a Universal Humanized Nanobody Scaffold. *J. Biol. Chem.* **2009**, *284*, 3273–3284. [CrossRef]
26. Cornish-Bowden, A. Nomenclature for Incompletely Specified Bases in Nucleic Acid Sequences: Rcommendations 1984. *Nucleic Acids Res.* **1985**, *13*, 3021–3030. [CrossRef]
27. Hoogenboom, H.R.; Griffiths, A.D.; Johnson, K.S.; Chiswell, D.J.; Hudson, P.; Winter, G. Multi-Subunit Proteins on the Surface of Filamentous Phage: Methodologies for Displaying Antibody (Fab) Heavy and Light Chains. *Nucleic Acids Res.* **1991**, *19*, 4133–4137. [CrossRef]
28. van Loo, G.; Bertrand, M.J.M. Death by TNF: A Road to Inflammation. *Nat. Rev. Immunol.* **2022**, *15*, 1–15. [CrossRef]
29. Leone, G.M.; Mangano, K.; Petralia, M.C.; Nicoletti, F.; Fagone, P. Past, Present and (Foreseeable) Future of Biological Anti-TNF Alpha Therapy. *J. Clin. Med.* **2023**, *12*, 1630. [CrossRef]
30. Ghalehbandi, S.; Yuzugulen, J.; Pranjol, M.Z.I.; Pourgholami, M.H. The Role of VEGF in Cancer-Induced Angiogenesis and Research Progress of Drugs Targeting VEGF. *Eur. J. Pharmacol.* **2023**, 175586. [CrossRef]
31. Arezumand, R.; Alibakhshi, A.; Ranjbari, J.; Ramazani, A.; Muyldermans, S. Nanobodies As Novel Agents for Targeting Angiogenesis in Solid Cancers. *Front. Immunol.* **2017**, *8*, 1746. [CrossRef]
32. Dennis, M.S.; Zhang, M.; Meng, Y.G.; Kadkhodayan, M.; Kirchhofer, D.; Combs, D.; Damico, L.A. Albumin Binding as a General Strategy for Improving the Pharmacokinetics of Proteins. *J. Biolo. Chem.* **2002**, *277*, 35035–35043. [CrossRef]

33. Jonsson, A.; Dogan, J.; Herne, N.; Abrahmsen, L.; Nygren, P.-A. Engineering of a Femtomolar Affinity Binding Protein to Human Serum Albumin. *Protein Eng. Des. Sel.* **2008**, *21*, 515–527. [CrossRef]
34. Johansson, M.U.; Frick, I.-M.; Nilsson, H.; Kraulis, P.J.; Hober, S.; Jonasson, P.; Linhult, M.; Nygren, P.-Å.; Uhlén, M.; Björck, L.; et al. Structure, Specificity, and Mode of Interaction for Bacterial Albumin-Binding Modules. *J. Biol. Chem.* **2002**, *277*, 8114–8120. [CrossRef]
35. Eble, J.A. Titration ELISA as a Method to Determine the Dissociation Constant of Receptor Ligand Interaction. *J. Vis. Exp.* **2018**, *15*, 57334. [CrossRef]
36. Lakzaei, M.; Rasaee, M.J.; Fazaeli, A.A.; Aminian, M. A Comparison of Three Strategies for Biopanning of Phage-scFv Library against Diphtheria Toxin. *J. Cell. Physiol.* **2019**, *234*, 9486–9494. [CrossRef]
37. Jaroszewicz, W.; Morcinek-Orłowska, J.; Pierzynowska, K.; Gaffke, L.; Węgrzyn, G. Phage Display and Other Peptide Display Technologies. *FEMS Microbiol. Rev.* **2022**, *46*, fuab052. [CrossRef]
38. Scarrone, M.; González-Techera, A.; Alvez-Rosado, R.; Delfin-Riela, T.; Modernell, Á.; González-Sapienza, G.; Lassabe, G. Development of Anti-Human IgM Nanobodies as Universal Reagents for General Immunodiagnostics. *New Biotechnol.* **2021**, *64*, 9–16. [CrossRef]
39. Roshan, R.; Naderi, S.; Behdani, M.; Cohan, R.A.; Ghaderi, H.; Shokrgozar, M.A.; Golkar, M.; Kazemi-Lomedasht, F. Isolation and Characterization of Nanobodies against Epithelial Cell Adhesion Molecule as Novel Theranostic Agents for Cancer Therapy. *Mol. Immunol.* **2021**, *129*, 70–77. [CrossRef]
40. Kazemi-Lomedasht, F.; Behdani, M.; Bagheri, K.P.; Habibi-Anbouhi, M.; Abolhassani, M.; Arezumand, R.; Shahbazzadeh, D.; Mirzahoseini, H. Inhibition of Angiogenesis in Human Endothelial Cell Using VEGF Specific Nanobody. *Mol. Immunol.* **2015**, *65*, 58–67. [CrossRef]
41. Wu, M.; Tu, Z.; Huang, F.; He, Q.; Fu, J.; Li, Y. Panning Anti-LPS Nanobody as a Capture Target to Enrich Vibrio Fluvialis. *Biochem. Biophys. Res. Commun.* **2019**, *512*, 531–536. [CrossRef] [PubMed]
42. Lunder, M.; Bratkovič, T.; Urleb, U.; Kreft, S.; Štrukelj, B. Ultrasound in Phage Display: A New Approach to Nonspecific Elution. *Biotechniques* **2008**, *44*, 893–900. [CrossRef] [PubMed]
43. Humphrey, W.; Dalke, A.; Schulten, K. VMD: Visual Molecular Dynamics. *J. Mol. Graph.* **1996**, *14*, 33–38. [CrossRef] [PubMed]
44. Tonikian, R.; Zhang, Y.; Boone, C.; Sidhu, S.S. Identifying Specificity Profiles for Peptide Recognition Modules from Phage-Displayed Peptide Libraries. *Nat. Protoc.* **2007**, *2*, 1368–1386. [CrossRef]
45. Chen, G.; Sidhu, S.S. Design and Generation of Synthetic Antibody Libraries for Phage Display. In *Monoclonal Antibodies: Methods and Protocols*; Ossipow, V., Fischer, N., Eds.; Humana Press: Totowa, NJ, USA, 2014; pp. 113–131.
46. Contreras, M.A.; Macaya, L.; Neira, P.; Camacho, F.; González, A.; Acosta, J.; Montesino, R.; Toledo, J.R.; Sánchez, O. New Insights on the Interaction Mechanism of RhTNFα with Its Antagonists Adalimumab and Etanercept. *Biochem. J.* **2020**, *477*, 3299–3311. [CrossRef]
47. Parra, N.C.; Mansilla, R.; Aedo, G.; Vispo, N.S.; González-Horta, E.E.; González-Chavarría, I.; Castillo, C.; Camacho, F.; Sánchez, O. Expression and Characterization of Human Vascular Endothelial Growth Factor Produced in SiHa Cells Transduced with Adenoviral Vector. *Protein J.* **2019**, *38*, 693–703. [CrossRef]
48. Beltrán-Ortiz, C.E.; Starck-Mendez, M.F.; Fernández, Y.; Farnós, O.; González, E.E.; Rivas, C.I.; Camacho, F.; Zuñiga, F.A.; Toledo, J.R.; Sánchez, O. Expression and Purification of the Surface Proteins from Andes Virus. *Protein. Expr. Purif.* **2017**, *139*, 63–70. [CrossRef]
49. Baek, H.; Suk, K.; Kim, Y.; Cha, S. An Improved Helper Phage System for Efficient Isolation of Specific Antibody Molecules in Phage Display. *Nucleic. Acids. Res.* **2002**, *30*, e18. [CrossRef]

molecules

Article

Identifying Potential Molecular Targets in Fungi Based on (Dis)Similarities in Binding Site Architecture with Proteins of the Human Pharmacolome

Johann E. Bedoya-Cardona [1], Marcela Rubio-Carrasquilla [1,2], Iliana M. Ramírez-Velásquez [1,3], Mario S. Valdés-Tresanco [1] and Ernesto Moreno [1,*]

1   Facultad de Ciencias Básicas, Universidad de Medellín, Medellin 050026, Colombia
2   Corporación para Investigaciones Biológicas, Medellin 050034, Colombia
3   Instituto Tecnológico Metropolitano, Medellin 050034, Colombia
*   Correspondence: emoreno@udemedellin.edu.co

Abstract: Invasive fungal infections represent a public health problem that worsens over the years with the increasing resistance to current antimycotic agents. Therefore, there is a compelling medical need of widening the antifungal drug repertoire, following different methods such as drug repositioning, identification and validation of new molecular targets and developing new inhibitors against these targets. In this work we developed a structure-based strategy for drug repositioning and new drug design, which can be applied to infectious fungi and other pathogens. Instead of applying the commonly accepted off-target criterion to discard fungal proteins with close homologues in humans, the core of our approach consists in identifying fungal proteins with active sites that are structurally similar, but preferably not identical to binding sites of proteins from the so-called "human pharmacolome". Using structural information from thousands of human protein target-inhibitor complexes, we identified dozens of proteins in fungal species of the genera *Histoplasma*, *Candida*, *Cryptococcus*, *Aspergillus* and *Fusarium*, which might be exploited for drug repositioning and, more importantly, also for the design of new fungus-specific inhibitors. As a case study, we present the in vitro experiments performed with a set of selected inhibitors of the human mitogen-activated protein kinases 1/2 (MEK1/2), several of which showed a marked cytotoxic activity in different fungal species.

Keywords: fungal pathogens; drug repurposing; drug development; new therapeutic targets; structural bioinformatics; MEK inhibitors

Citation: Bedoya-Cardona, J.E.; Rubio-Carrasquilla, M.; Ramírez-Velásquez, I.M.; Valdés-Tresanco, M.S.; Moreno, E. Identifying Potential Molecular Targets in Fungi Based on (Dis)Similarities in Binding Site Architecture with Proteins of the Human Pharmacolome. *Molecules* **2023**, *28*, 692. https://doi.org/10.3390/molecules28020692

Academic Editors: Jahir Orozco Holguín and Wei Li

Received: 11 November 2022
Revised: 23 December 2022
Accepted: 4 January 2023
Published: 10 January 2023

1. Introduction

Invasive fungal infections (IFIs), caused by yeasts and filamentous fungi, are opportunistic infections that occur mostly in immunodepressed patients and in patients in critical conditions, causing a high morbidity and mortality [1]. IFIs may manifest with different intensities, from simple and mild infections, as is the case of external mycoses, to severe systemic and disseminated mycoses that can cause death [2]. The epidemiological landscape of invasive mycoses is in continuous change, driven by etiological variations among hospitals, countries and the influence of multiple local variables, patient risk factors and medical and surgical praxis [3].

The current repertoire of antifungal drugs includes different classes of molecules: pyrimidines, polyenes, echinocandins and azoles [4–6]. These antifungal drugs, however, present several important drawbacks, such as their adverse side effects, the increasing resistance developed by many fungal pathogens and long treatment times [7]. Therefore, there is a compelling medical need for broadening the therapeutic alternatives to treat these infections. Two main alternatives in this route are drug repositioning and the development of new drugs directed to new molecular targets in fungi. In either case, a favorable balance between clinical benefits and adverse effects is a relevant issue to take into account.

Commonly, the identification of new targets in pathogens focuses on unique proteins, not present in humans, or with low sequence similarity with human proteins. This approach intends to minimize possible cross-reactions leading to adverse secondary effects. For example, Sosa et al. (2018) developed a database (Target-Pathogen) and a search system that integrates multiple sources of information for the identification of possible targets in pathogens [8]. Among the filters applied in this search system is the so-called "off-target criterion", which discards proteins with close homologues in humans. Similarly, in more recent works, Mukherjee and coworkers applied a "subtractive genomics" approach to filter out homologous proteins in the search for targets in *Candida* species [9], while Palumbo and coworkers applied the same off-target criterion in a search for potential targets in *Listeria monocytogenes*, following a multilayer omics strategy [10].

Drug repositioning, also referred to as drug repurposing, is a different strategy consisting in finding new medical uses for approved drugs or compounds that have shown an acceptable safety profile in clinical trials, including those that have failed in later stages during the development. This strategy implies shorter development times, lower costs and fewer risks [11,12]. Successful drug repurposing cases have been reported in various therapeutic areas, prompting pharmaceutical companies to open up collaborations with biotech firms and academic communities to synergize research in this area [12–14].

The emergence of various genomic, drug and disease knowledge databases has promoted the rapid development of a variety of computational approaches to guide drug repositioning and new drug development projects. Thus, the so-called network-based methods combine and exploit various kinds of information from multiple data sources, e.g., transcriptomics, drug-induced expression profiling, disease–disease associations, drug–drug interactions, among others [13,15]. Structure-based methods, on the other hand, rely on techniques such as protein–ligand docking, molecular dynamics simulations, virtual screening and quantitative structure–activity relationship (QSAR). Additionally, in recent years, the use of artificial intelligence methods in drug development is gaining a big momentum [15,16].

Drug repurposing approaches have been applied as well to fungal infections [17]. For example, finasteride, a drug generally used for the treatment of benign prostatic hyperplasia, showed efficacy in the prevention of biofilm formation by *Candida albicans*, when used alone and in combination with fluconazole, and showed an effect also in the treatment of preformed biofilms [18]. Another example is atorvastatin, a drug used as a plasma cholesterol reducer, which showed antifungal activity by inhibiting the production of ergosterol from the cell wall in five *Candida* strains (*C. albicans*, *C. glabrata*, *C. kefyr*, *C. stellatoidea* and *C. krusei*) [19]. Furthermore, several drug libraries have been screened against different pathogenic fungi in the search of drugs with previously unknown antifungal effects [20–26].

Currently, the more than one and a half thousand FDA-approved drugs and several thousand compounds in clinical trials, together with their molecular targets, constitute a rich repertoire for drug repurposing and new drug development. According to a study published by Santos et al. (2017), by the year 2015 the Food and Drug Administration (FDA) in the United States had approved a total of 1578 drugs, targeting 893 different human and pathogen-derived biomolecules. This set of targets is defined in the article as the "pharmacolome", which is spread across the fourteen groups of the anatomical therapeutic chemical (ATC) classification system. A quick survey over the compiled pharmacolome shows the limited availability of approved drugs to treat invasive fungal infections [27].

In this work we developed a structural bioinformatics strategy to identify potential therapeutic targets in fungi, test them in vitro using known drugs and inhibitors and, in suitable cases, intend to develop new fungus-specific inhibitors. The core of this approach consists in identifying fungal proteins with active sites that are structurally similar, but preferably not identical to binding sites of proteins from the human pharmacolome. On the one hand, a high structural similarity with a human counterpart allows validation of the fungal target using cross-reactive inhibitors of the human protein (possibly leading to drug repurposing). On the other hand, a few amino acid differences in the binding pocket would produce local topological and chemical changes that might be exploited for the design of new specific inhibitors of the fungal target.

Using structural information, we have identified dozens of proteins in several fungal species of the genera *Histoplasma*, *Candida*, *Cryptococcus*, *Aspergillus* and *Fusarium*, which might be exploited for drug repurposing and for the design of new antifungal agents. As case study we analyze a few fungal proteins showing binding sites similar to the non-ATP competitive binding site of the human mitogen-activated protein kinases 1 and 2 (MEK1/2), and present the in vitro experiments performed with a set of selected (MEK1/2) inhibitors, several of which showed marked cytotoxic activity in various fungal species. Importantly, the binding sites of the MEK analogs in several fungal species show mutations that create opportunities for the design of fungus-specific inhibitors.

2. Results and Discussion

2.1. Selected Set of Human Protein Targets and Binding Site Definition

The primary data source for this work was the list compiled by Santos and coworkers (2017), which included 549 protein targets of small drugs approved by the FDA up to 2015 [27]. We complemented these data by including the small drugs (and their protein targets) approved between 2016 and 2020, which added another 90 small drugs, resulting in a total of 639 human protein targets. From this set, 433 proteins included in their UniProt records cross-referenced to PDB structures, which amounted to more than 8500 PDB entries. The automated and subsequent manual analysis of all these structures, as described in Section 3, yielded 264 different protein targets in complex with one or more ligands.

Figure 1 shows the distribution of the number of PDB complexes per protein target. Most of the targets are represented in the PDB by more than a single protein–ligand complex, which allows a more comprehensive definition of the binding site. An extreme case is the estrogen receptor (UniProt ID: P03372), with more than 500 protein–ligand structures. In spite of this disparity in the numbers of complexes per target, we found consistent binding pocket definitions for most of the proteins. For example, for the estrogen receptor the only found binding site region, located between sequence positions 342 and 544, corresponds to the estradiol binding site.

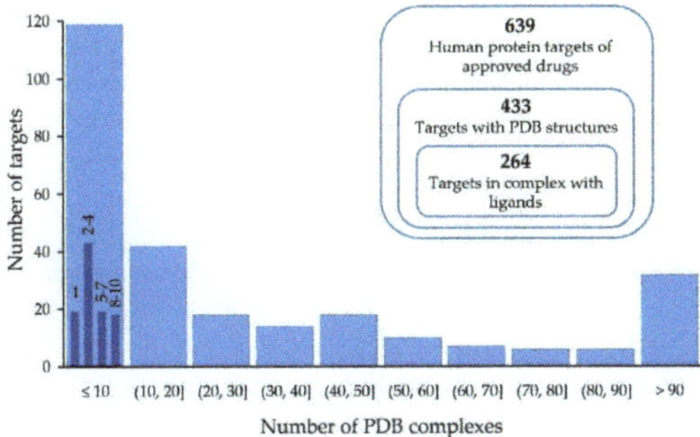

Figure 1. Distribution of the number of human targets per number of PDB protein–ligand complexes in the analyzed set of 264 proteins from the pharmacolome. For the first gross interval [1, 10), a distribution in smaller intervals (1, 2–4, 5–7 and 8–10) is shown in dark blue bars.

The obtained set of PDB entries included 86 ligands corresponding to FDA-approved small drugs (Table S1), which were distributed across > 400 complexes. The large majority of these drugs have >60% of their surface area buried in the protein upon complexation (Figure 2A), while the few cases showing a lower percent of buried area (for example, for cholic acid) corresponds to extra copies of the ligand lying on external areas of the protein surface. We decided to use this value of 60% of buried ligand surface, covering most of the complexes, as cutoff for further analysis of the binding sites. Likewise, we applied a molecular weight cutoff, allowing a maximum of 80 heavy atoms (corresponding roughly to 1.1 kDa), to discard large ligands, which were mostly peptides and oligonucleotides (see Figure 2B).

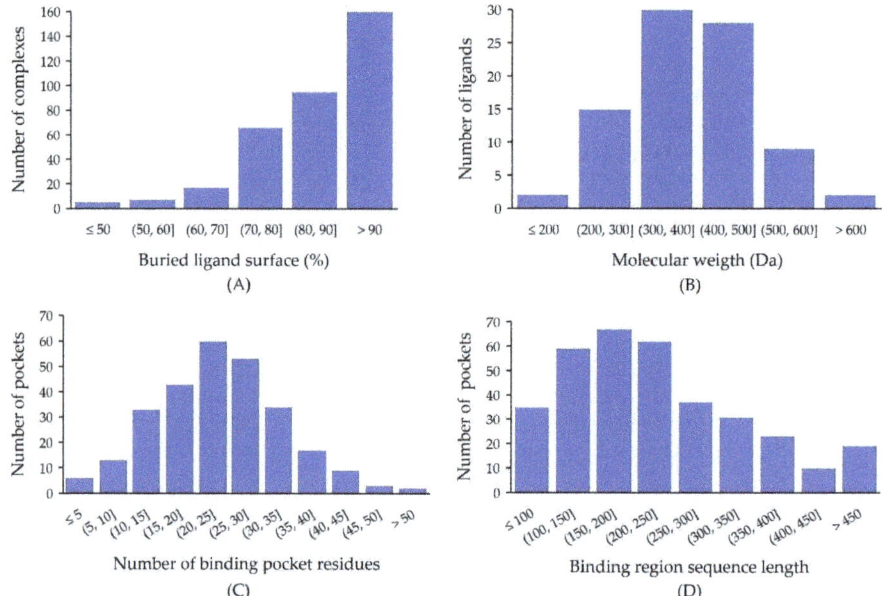

Figure 2. Statistics from the protein–ligand complexes of targets from the human pharmacolome. (**A**) Histogram of the buried ligand area upon complexation for FDA-approved drugs (data for cholic acid were omitted). (**B**) Distribution of ligand molecular weights for FDA-approved drugs. (**C**) Distribution of the number of residues per binding pocket for the 264 selected human targets. (**D**) Distribution of binding region sequence lengths for the 264 human targets.

The analysis performed to delimit the binding regions within the protein sequences yielded around 1200 clusters of sequence regions, corresponding to 272 protein targets. By manually reviewing these clusters, we selected 343 binding regions in a total of 264 proteins from the human pharmacolome. About 30% of these proteins contained more than one pocket region.

Figure 2C shows the distribution of the number of amino acid residues per binding pocket, as defined here following a contact distance criterion. This means that each amino acid belonging to a binding pocket has at least one atom within a contact distance (4.5 Å) from a ligand in at least one PDB complex. These contacts include mostly amino acid side chains, but also residues that interact only through their backbone atoms. The numbers of pocket amino acids across different targets span a wide range, having a maximum at around 20–30 residues. These residues are distributed along sequence regions of different lengths, mostly within a range of 100–250 residues (Figure 2D). The largest regions correspond to transmembrane proteins, such as the alpha units of the sodium channel proteins 2, 9 and 4 (Q99250, Q15858, P35499) and the Voltage-dependent T-type calcium channel subunit alpha-1G (O43497), where the protein chain crosses the cell membrane several times, with large sequence stretches separating the ligand-binding segments.

2.2. Searching a Fungal Proteome for Binding Sites–Case Study: Histoplasma capsulatum

Here we present the results obtained for the *Histoplasma capsulatum* proteome as example of the application of the developed strategy. Figure 3A shows the significant differences between the results obtained using the full human protein sequences for BLAST and those obtained using the defined 343 binding site regions, even though the restrictions imposed for the second type of search were stronger: $\geq$80% sequence coverage vs. $\geq$40% for the full sequences (for most of the proteins, the binding region covers around 40–50% of the full sequence). As shown in Figure 3A, BLAST with binding regions yielded a significantly higher number of hits.

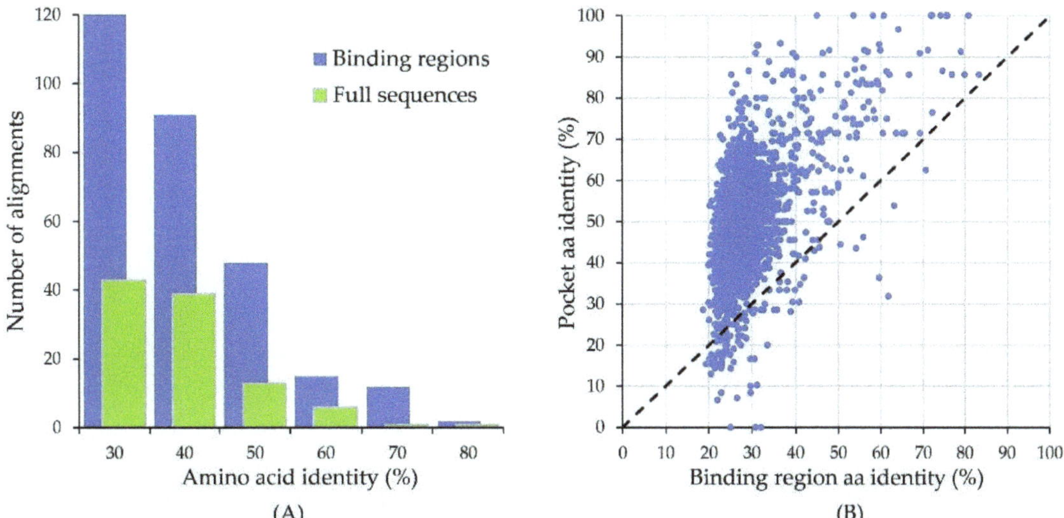

Figure 3. (**A**) Distribution of aa identity percent in the BLAST sequence alignments obtained for the *Histoplasma capsulatum* proteome, using as input either the full sequences (green bars) or the binding region sequences (blue bars) for the selected 264 proteins from the human pharmacolome. (**B**) Scatter plot showing the higher similarity between human and fungal binding pockets, as compared with the similarity between the binding region sequences that encompass the binding pocket amino acids.

The similarity further increases when comparing only the sets of amino acids forming the binding pockets (Figure 3B), which for the fungal proteins were defined from their alignments with the human binding region sequences, as explained in Methods (Section 3). Even for proteins with low similarity (<30% aa identity) in their binding region sequences, the identity between the binding pocket amino acids may be considerably high. For example, the alignment for the aromatic-L-amino-acid decarboxylase (P20711, sequence region 147–303) yields a 33% aa identity with a sequence segment of a fungal protein (UniProt identifier C0NW51; annotated as a glutamate decarboxylase-like protein), while the identity of the corresponding binding pocket residues reaches 85%. Not surprisingly, highly similar binding pockets belong to proteins with conserved roles in the cell, as is the case of polymerases and other enzymes. Several of these binding pockets correspond to binding sites for ATP and different cofactors.

2.3. Expanding the Search to Other Fungal Proteomes

The above analysis carried out for *Histoplasma capsulatum* was extended to other five fungal proteomes of microorganisms of medical relevance: *Aspergillus fumigatus*, *Candida albicans*, *Candida parapsilosis*, *Cryptococcus neoformans* and *Fusarium oxysporum*. The main results from these analyses are summarized in Table 1, while the full list of hits is presented in Table S1. The fungal proteins listed in Table 1 contain binding pockets showing $\geq$70% aa identity with their human counterparts. Interestingly, four of the human targets have orthologs with 100% conserved binding sites in all or most of the investigated fungal species.

Table 1. Proteins with similar binding pockets (≥70% aa identity) for the six fungal species.

Human Protein Target	UniProt Code	Fungal Species *											
		Histoplasma capsulatum	%	*Aspergillus fumigatus*	%	*Candida albicans*	%	*Candida parapsilosis*	%	*Cryptococcus neoformans*	%	*Fusarium oxysporum*	%
Ribonucleotide reductase	P23921	F0U515	100	Q4WNR5	100	Q5A0N3	96	G8BE28	96	Q5KGK1	100		
Ornithine decarboxylase	P11926	F0U9K9	100	Q4WBW7	95	P78599	100	G8B654	100	Q5KJY8	100	A0A0D2XUF0	95
HMG-CoA reductase	P04035	F0UKH1	97	Q4WHZ1	97	A0A1D8PD39	100	G8B666	100	Q5KEN6	93	A0A0D2XPE5	97
P-glycoprotein 1	P08183	F0UGI7	93	Q4WTT9	87					Q5KN49	87	A0A0D2XVG0	80
IMP dehydrogenase 2	P12268	F0UKI1	93	Q4WHZ9	93	Q59Q46	97	G8B7I7	97	Q5KP44	97	A0A0D2XAA8	93
FPP synthase	P14324	F0UP55	91	Q4WEB8	91	A0A1D8PH78	91	G8B7C3	91	Q5KG83	96	A0A0D2XDQ9	91
Ribonucleotide reductase	P23921	F0U515	91	Q4WNR5	87	Q5A0N3	96	G8BE28	91	Q5KGK1	91	A0A0D2XVB2	91
Glutathione reductase	P00390	F0UFI2	91	Q4WRK8	91	Q6FRV2	73			Q5KH19	91	A0A0C4DHU2	91
Xanthine dehydrogenase	P47989	F0UCF6	90	Q4WQ15	90							A0A0D2Y4E4	90
DNA topoisomerase 2-β	Q02880	F0UWA9	87.5	Q4WLF7	75	O93794	82			Q5KP97	71	A0A0C4DHU5	87
Histone deacetylase 2	Q92769	F0UKC3	87	Q4WHY0	87	Q5ADP0	95.7	G8BBB0	96	Q5KF65	95.7	A0A0D2X821	87
GSK-3 beta	P49841	F0UQX6	87	Q4WDL1	93	A0A2H0ZU47	83	G8BDX2	80	Q5KMR8	90	A0A0D2XCF2	87
DHOdehase	Q02127	F0UDX1	87	Q4X169	87	Q874I4	100	G8BA68	100	Q5KK62	100		
Histone deacetylase 7	Q8WUI4	F0UVW7	85	Q4WE71	75	Q5A960	80	G8BBK1	80	Q5KL48	80	A0A0D2YC83	72.7
Thymidylate synthase	P04818	F0URV8	84	Q4W9N9	88	A0A2H0ZCG7	88	G8BCL7	88	P0CS12	78	A0A0D2X8Z9	88
PKC-alpha	P17252	F0UE28	84	Q4WVG0	84	A0A2H1A7H5	84	G8BAI0	79	A0A0S2LIC5	84	A0A0D2XWP4	89.5
IMP dehydrogenase 2	P12268	F0UKI1	82	Q4WHZ9	86	Q59Q46	82	G8B7I7	86	Q5KP44	82	A0A0D2XAA8	82
IMPase 1	P29218	F0UHX4	82	Q4WEX3	82	Q6FSE7	87			Q5KKG2	83	A0A0D2XJP5	78
FPP synthase	P14324	F0UP55	82	Q4WEB8	82	A0A1D8PH78	77	G8B7C3	77	Q5KG83	74	A0A0D2XDQ9	82
GARS/AIRS/GART	P22102	F0UHE5	81	Q4WDH1	81	A0A1D8PE67	75	G8BD50	75	Q5K7B4	75	A0A0D2Y2N6	88
Histone deacetylase 6	Q9UBN7	F0UVW7	81.2			Q5A960	75	G8BBK1	75	Q5KL48	86.7	A0A0D2YC83	87.5
DHOdehase	Q02127	F0UDX1	80	Q4X169	88								
Histone deacetylase 8	Q9BY41	F0UKC3	80	Q4WHY0	80	A0A2H0ZKW1	80	G8BBB0	76	Q5KF65	76.2	A0A0D2X821	80
PPIase FKBP1A	P62942	F0URT3	78			P28870	74			P0CP94	78		
HMG-CoA reductase	P04035	F0UKH1	77	Q4WSY2	77	A0A1D8PD39	77	G8B666	77	Q5KEN6	85	A0A0D2XPE5	77
Histone deacetylase 1	Q13547	F0UKK7	77	Q4WI19	77	A0A1D8PSA6	82	G8BBQ5	88	Q5KF65	71	A0A0D2X821	71
Rho kinase 2	O75116	F0UBW5	76	Q4WQ81	76	Q5AP53	73	G8BKE8	71	Q5KEJ1	73		
ALDH class 2	P05091	F0UNE9	75	Q4WM26	79	Q6FPK0	74			Q5KEX3	73	A0A0D2XAL2	74
CPSase 1	P31327	F0UNF7	75										
Tubulin beta-3 chain	Q13509	F0UQK5	74	Q4WA70	93	A0A1D8PC97	96	G8B7W7	96	Q5K876	89	A0A0C4DHQ2	93
MEK2	P36507	F0UAN5	74	Q4WWH7	76	Q6FQU4	82	G8BFG7	79	Q5KKP1	87	A0A0D2XNJ1	79
MEK1	Q02750	F0UAN5	74	Q4WWH7	76	A0A2H0ZYL6	77	G8BAQ2	76	Q5KKP1	77	A0A0D2XNJ1	77
Succinate dehydrogenase	P51649	F0U4T1	72.2	Q4WPA5	78	Q6FVP8	81	G8B862	78	Q5K8N2	81	A0A0D2Y168	88
CFTR	P13569	F0ULL9	72	Q4WIK7	74	Q6FWS5	79			Q5KL35	78		
DPDE4	Q07343	F0UHN7	71.4										
FADK 2	Q14289	F0UKJ3	70.8	Q4X028	76	A0A2H0ZKT3	83					A0A0D2XB54	71
MAPK 11	Q15759	F0USW8	70,4	Q4WSF6	70	Q92207	70	G8BE43	70	Q5KC34	70	A0A0D2XQS0	74
ATP-binding cassette G2	Q9UNQ0	F0U6H3	70	Q4WXJ0	70	Q6FQ96	76.2			Q5KCK1	71.4	A0A0D2Y998	71.4
IDPc	O75874			Q4WX92	100	A0A1D8PHH7	100	G8BDJ9	100	Q5KLU0	100	A0A0D2XM79	100
ICD-M	P48735			Q4WX92	100	A0A1D8PS79	100	G8BDJ9	100	Q5KLU0	100	A0A0D2XM79	100
DNA topoisomerase 2	P11388			Q4WLF7	89	A0A1D8PMM1	96	G8BHF5	96	Q5KP97	86	A0A0C4DHU5	93
PARP-2	Q9UGN5			Q4WU62	83.9							A0A0D2XUC2	87.1
IDPc	O75874			Q4WX92	77	A0A2H0ZLU3	77					A0A0D2XM79	71
Cyclin-dependent kinase 6	Q00534			Q4WN13	77	P43063	73	G8BG79	73			A0A0D2Y7P8	71
Thioredoxin reductase 1	Q16881			Q4WRK8	73.9								

Table 1. *Cont.*

Human Protein Target	UniProt Code	Histoplasma capsulatum	%	Aspergillus fumigatus	%	Candida albicans	%	Candida parapsilosis	%	Cryptococcus neoformans	%	Fusarium oxysporum	%
						Fungal Species *							
DHOdehase	Q02127					Q874I4	84	G8BA68	84	Q5KK62	74		
Tyrosine kinase CSK	P41240					A0A1D8PR87	80	G8BKZ2	80				
ROCK-I	Q13464					Q6FP74	71			Q5KEJ1	70		
BCNG-2	Q9UL51					Q59V20	70,6						
Proto-oncogene c-Src	P12931					Q9Y7W4	70						
NTK38	P51813											A0A0C4DJR2	73
CFTR	P13569											A0A0D2XXA6	71
ALDH class 2	P05091											A0A0D2YFW3	70

* For each human target and each proteome, only the fungal protein with the highest aa identity percent is shown. The color code goes from dark to light gray following the decreasing percent of pocket aa identity (from 100% to 70%).

It is worth noting that Table 1 shows, for each human target, only the highest ranked fungal protein. However, for several human targets we found two or three fungal proteins (within the same species) having similar binding pockets, with relatively small differences in their aa identity percentages. This is the case, for example, of the DNA polymerase delta catalytic subunit, which yielded two matches in each of the six proteomes. The binding region sequences of these fungal proteins differ in aa identity (38–60%) compared to the corresponding sequence region in the human target, but all of them contain very similar binding pockets (~90% aa identity). Table S2 shows the full lists of matches; additionally, see below as an example the results for MEK1/2 in Figure 4.

Protein	Organism	%	78 79 80 · · 97 · 99 · · 115 · 118 · · 126 127 128 129 · · 141 · 143 · · 190 · · 207 208 209 210 211 212 · 215 216 · 219
Q02750	*Homo sapiens*	-	N G G - - K - I - - L - L - - I V G F - - I - M - - D - - C D F G V S - L I - M
P36507		-	N G G - - K - I - - L - L - - I V G F - - I - M - - D - - C D F G V S - L I - M
Q4WWH7	*A. fumigatus*	75.5	N G G - - K - I - - L - **G** - - I V T F - - L - M - - D - - C D F G V A - T V - I
Q4WRK9		73.6	N **Y** G - - K - I - - L - L - - I I D F - - M - V - - D - - C D F G V S - L V - I
Q4WJJ0		62.3	**A** G G - - K - I - - L - N - - I C R Y - - I - M - - D - - C D F G V S - F T - D
A0A1D8PKE2	*C. albicans*	73.6	N **Y** G - - K - V - - L - L - - I V D F - - M - I - - D - - C D F G V S - L V - L
Q5A6T5		73.6	N **S** G - - K - I - - L - L - - I I E F - - I - M - - D - - C D F G V S - L T - M
A0A1D8PHG7		66.0	N G G - - K - V - - L - **A** - - I V K Y - - I - M - - D - - C D F G V S - A V - F
G8BAQ2	*C. parapsilosis*	75.5	N **Y** G - - K - V - - L - L - - I V D F - - M - I - - D - - C D F G V S - L V - M
G8BFG7		73.6	N **S** G - - K - I - - L - L - - I I E F - - I - M - - D - - C D F G V S - L T - M
G8BDG0		67.9	N G G - - K - I - - L - **A** - - I V N Y - - I - M - - D - - C D F G V S - A V - F
Q5KN75	*C. neoformans*	77.4	N **Y** G - - K - I - - L - L - - I V E F - - Y - M - - D - - C D F G V S - L E - L
Q5KKP1		77.4	N G G - - K - I - - L - M - - I V G Y - - I - M - - D - - C D F G V S - L I - I
Q5KGY3		67.9	T G G - - K - I - - L - L - - I V E H - - I - M - - D - - C D F G V S - L V - L
A0A0D2XNJ1	*F. oxysporum*	77.4	N G G - - K - I - - L - M - - I V T F - - M - M - - D - - C D F G V S - L V - I
A0A0D2XGR3		73.1	N **Y** G - - K - M - - L - L - - I I D F - - M - I - - D - - C D F G V S - L V - I
A0A0D2XDY6		62.3	**A** G G - - K - I - - L - N - - I C K Y - - I - M - - D - - C D F G V S - F T - E
F0UAN5	*H. capsulatum*	73.6	N G G - - K - I - - L - **G** - - I V T V - - L - M - - D - - C D F G V A - T V - I
F0UFB3		73.6	N **Y** G - - K - I - - L - L - - I I D F - - I - V - - D - - C D F G V S - L V - I
F0U908		62.3	**A** G G - - K - I - - L - N - - I C R Y - - I - M - - D - - C D F G V S - F T - D

Figure 4. Alignment of the binding pocket residues for the non-ATP competitive site of MEK1/2 (Q02750 and P36507), with the corresponding residues in the identified fungal proteins having similar binding pockets, for the six analyzed species. The third column in the table shows the calculated aa identity percentages for the set of binding pocket residues. Fully conserved positions are highlighted in blue; non-conservative substitutions at positions 79 and 118 are marked in bold.

Several of the human proteins included in Table 1 are the targets of drugs and inhibitors that have been tested in fungi. For example, the cancer drug sorafenib, which targets multiple proteins, among them the P-glycoprotein 1 (P08183), was identified from a kinase inhibitor library screening as a strong inhibitor of *Histoplasma capsulatum* and *Cryptococcus neoformans* [28]. Statins such as atorvastatin and simvastatin, targeting the HMG-CoA reductase (P04035) have shown inhibitory effects in *Candida albicans*, *Candida Glabrata* and *Aspergillus fumigatus* [29]. Disulfiram, a drug inhibiting the aldehyde dehydrogenase (P05091) that is used to treat chronic alcoholism, showed strong inhibitory effects in *Candida albicans* and *Candida auris* [30]. The immunosuppressive drug tacrolimus, targeting the peptidyl-prolyl cis-trans isomerase FKBP1A (P62942) had effects in 11 fungi and 3 oomycetes of agricultural importance [31]. Finally, vorinostat, targeting histone deacetylases (Q92769, Q9UBN7) and used in the treatment of cutaneous T cell lymphomas, showed strong effects in *Aspergillus* spp. [32].

The identification in this work of fungal proteins with binding pockets similar to those of human proteins targeted by drugs that have shown inhibitory effects in fungi, not only serves as a strong support of the developed strategy, but also helps to identify the actual fungal targets and to understand the mechanisms of action of such drugs in these microorganisms. Furthermore, many of the human proteins included in Tables 1 and S2, are the targets of drugs and inhibitors that have not been tested yet in fungi, which opens up a large research space for drug repositioning and new drug development.

Since the fungal proteomes have been annotated mostly in an automated way, functional assignments for the identified proteins are not always reliable. Therefore, it would be difficult in many cases to establish direct functional relationships between the human targets and the identified fungal proteins having similar binding sites. For practical purposes, nonetheless, the obtained results lead straightforwardly to the use of known inhibitors of the human targets to test their effects in fungi. Such chemical probing of the predicted targets may be accomplished either by following a comprehensive in vitro testing of a large number of inhibitors (when available), or by following a computational modeling approach to define a more limited set of molecules to be tested, as we illustrate below with the in silico predictions and in vitro assays performed with inhibitors of the human MEK1/2 proteins.

2.4. Several MEK1/2 (MEK) Inhibitors Have Strong Inhibitory Effects in Various Pathogenic Fungi

In humans, the dual specificity mitogen-activated protein kinases 1 and 2 (MEK1 and MEK2, also known as MAP2K1 and MAP2K2), are essential components of the mitogen activated protein (MAP) kinase signal transduction pathway. Both MEK1 and MEK2 have a unique inhibitor-binding pocket adjacent to the Mg/ATP-binding site [33]. Currently, four MEK inhibitors have been approved by the FDA for cancer treatment: trametinib, binimetinib, selumetinib and cobimetinib [34] while others are in clinical trials. The web platform of Selleck Chemicals (Houston, TX, USA), for example, currently lists 33 commercially available MEK inhibitors.

In general, inhibitors of the PI3K/AKT/mTOR, RAS/RAF/MEK/ERK pathway, which are used in the treatment of malignancies and immune-mediated diseases, may predispose to fungal infections by suppressing important components of the adaptive and innate immune response [35], therefore, they would not likely be used as antifungal agents. Nonetheless, there are a few reports where MEK inhibitors have been tested in plant pathogenic fungi. For example, the MEK1/2 inhibitor U0126 was found to decrease germination and hyphae growth in *Aspergillus fumigatus* [36] and to inhibit the conidial germination and pathogenicity of *Setosphaeria turcica*, a plant pathogen [37].

The binding region sequence encompassing the non-ATP binding pocket in MEK1/2 goes from residue 78 to 219 (ca. 200 aa). In this region we identified 23 amino acids (identical in the two proteins) shaping the binding pocket inner surface. Running BLAST using the MEK1/2 binding region sequences yielded three proteins in each of the six analyzed proteomes, showing 62–77% of aa identity between their binding pocket residues and those of MEK (Figure 4).

The alignment in Figure 4 reveals a high degree of binding pocket conservation, with 10 out of 23 residues fully conserved across the human and all the fungal variants. Furthermore, in most cases the amino acid substitutions are conservative, as in positions 78, 99, 127, 141, 143, 212, 215 y 216. At positions 79 and 118, drastic substitutions (G/Y; L/G or L/A, respectively) appear in a few proteins in several fungal species. As discussed below, some of these substitutions represent interesting opportunities for the design of fungus-specific inhibitors.

We decided to test our predictions by assaying in vitro a set of reported MEK inhibitors on the six fungal species analyzed in silico. Docking simulations on the constructed models for proteins F0UAN5 and A0D2XNJ1 from *Histoplasma capsulatum* and *Fusarium oxysporum*, respectively, were performed for 25 inhibitors found in complex with MEK1 in the Protein Data Bank. As result, we selected seven inhibitors: cobimetinib [38], myricetin [39], refametinib [40], trametinib [41], GDC0623 [42], AZD6244 [43] and TAK-733 [44] for the in vitro assays.

Table 2 shows the results of the growth inhibition experiments performed for the six fungal species. The most susceptible microorganism was *Histoplasma capsulatum*, with four inhibitors (cobimetinib, GDC-0623, myricetin and refametinib) showing IC50 values in the low micromolar range. Similarly, *Aspergillus fumigatus* was strongly affected by three inhibitors (cobimetinib, GDC-0623 and TAK-733), while only one inhibitor (cobimetinib) showed a marked effect on *Fusarium oxysporum*. No inhibitor had effects on all the fungal species. The two tested Candida species were affected by two inhibitors each, but only at a high micromolar range (>100 μM). The use of a very low concentration of the SDS surfactant (0.002%), which most likely increases inhibitor solubility, improved the observed inhibitory effects in most cases. This concentration of SDS alone, or in combination with DMSO or ethanol, had only minor effects in fungal viability.

Table 2. Result of the in vitro susceptibility assays (IC50 values, (μM)).

Inhibitor	Fungal Species											
	Histoplasma capsulatum		*Cryptococcus neoformans*		*Candida albicans*		*Candida parapsilosis*		*Fusarium oxysporum*		*Aspergillus fumigatus*	
	Solvent *	+SDS **	Solvent	+SDS	Solvent	+SDS	Solvent	+SDS	Solvent	+SDS	Solvent	+SDS
cobimetinib	53	<12.5	>188	>188	>188	>188	>188	158	~100	65	>100	82
GDC-0623	114	63	194	83	>219	>219	>219	>219	>100	~100	8	7
myricetin	>251	36	>251	>251	>251	>251	>251	>251	>251	>251	>251	>251
TAK-733	>198	198	>198	198	>198	198	>198	>198	>100	>100	>100	54
AZD-6244	>328	>328	>328	>328	>328	>328	>328	>328	>328	>328	>328	>328
refametinib	25	<17.5	81	54	175	114	>175	175	ND	ND	ND	ND
trametinib	>49	>49	>49	>49	>49	>49	>49	>49	>49	>49	>49	>49

IC50 values < 100 μM are marked in bold and shadowed in gray. The "<" and ">" signs are used when the IC50 value is lower/greater than the minimum/maximum tested concentration. * Compounds were dissolved in DMSO or ethanol, and added to culture medium. ** Same as above, with the addition of 0.002% SDS.

Since for each of the investigated fungal species we found three proteins with binding sites similar to that of the human MEKs, it is not possible to attribute the observed cytotoxic effects to a particular protein. Furthermore, and although less probable, the actual target might be a different, so far unidentified fungal protein. Reliable target validation would require complementary experiments, e.g., genetic manipulations to affect protein expression. In addition, as discussed below, target validation could be supported with growth inhibition assays involving compounds predicted to be specific for a particular fungal protein.

2.5. Opportunities for the Design of Fungus-Specific Inhibitors

Several of the fungal proteins in Figure 4 show amino acid substitutions in their binding pockets, as compared with the human MEKs, that cause small local topological changes, in particular mutations L118G (*A. fumigatus*, *H. capsulatum*) and L118A in the two *Candida* species. As illustrated in Figure 5 for the *Histoplasma capsulatum* protein F0UAN5, mutation L118G creates a void space within the binding site, previously occupied by the bulky Leu sidechain. This additional small cavity could be filled up by compounds with suitable chemical structures, which, on the other hand, would not bind to human MEK1/2 because of the steric hindrances caused by the leucine sidechain. As discussed above, the actual antifungal effect of these fungus-specific inhibitors would depend on the relevance of their targets for cell vitality.

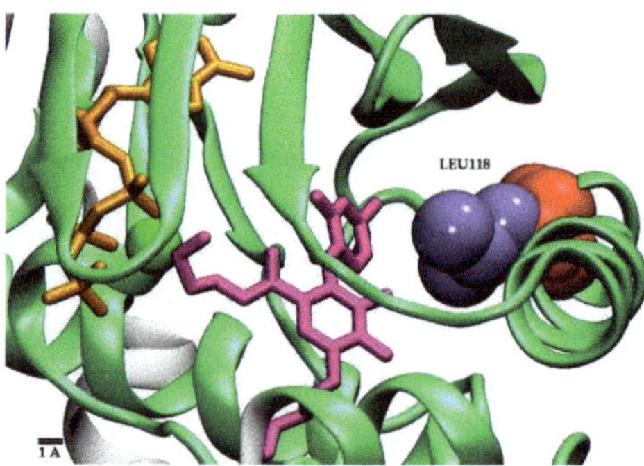

Figure 5. Structure of the MEK1/2 binding site showing the Leu 118 side chain (in blue spheres). Mutation of this residue by Glycine (in red spheres) creates a void space that can be occupied by atoms of fitting compounds.

Performing this kind of analysis on the different pairs of human and fungal proteins having similar binding pockets, as found in this study, may disclose many potential fungal targets with binding site mutations that open up a design space for fungus-specific inhibitors. The zone between 60–75% binding pocket aa identity (Figure 3B), which includes dozens of fungal proteins, looks particularly interesting in this regard.

3. Computational and Experimental Methods

3.1. Computational Strategy to Identify Potential Targets in Fungi and Other Pathogens

Our approach consists in identifying fungal proteins with active sites (meaning the set of residues lining the binding pocket) that are similar to active sites of proteins from the human pharmacolome. As mentioned in the Introduction, a high structural similarity with the binding site of a human counterpart facilitates a chemical validation of the fungal target using known inhibitors of the human protein and, ultimately, may lead to a drug repurposing strategy. We, however, are more focused on exploiting one or a few relevant amino acid differences in the binding pocket that would create a "design space" for new specific inhibitors of the fungal target.

Briefly, we employed a structural approach to identify binding site similarities, taking advantage of the thousands of available crystal structures for proteins of the human pharmacolome, many of them in complex with inhibitors. As explained in detail in the following sections, we used these bound inhibitors as anchors to define the binding site amino acids for each human target, followed by local sequence searches and analyses against the proteomes of several fungal species. The workflow is represented in Figure 6.

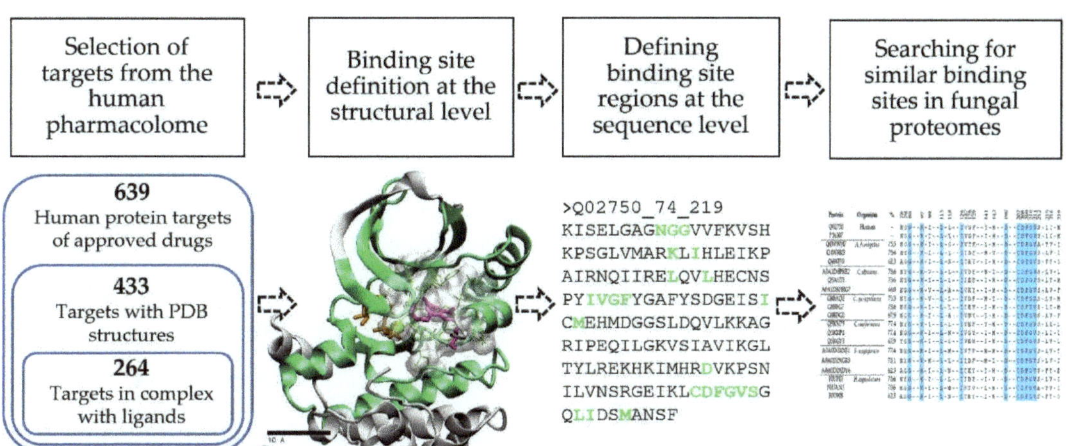

Figure 6. Computational strategy to search for fungal proteins with similar binding sites, taking as reference a selected set of proteins from the human pharmacolome.

3.1.1. Selection of the Human Protein Targets to Be Used for Fungal Proteome Searches

The list of FDA-approved small drugs and their protein targets, up to 2015 as compiled by Santos et al. (2017), was the main primary source for our work. We updated this list up to 2020 by including the small drugs approved by the FDA between 2016 and 2020, taken from the "Compilation of CDER NME and New Biologic Approvals 1985–2020" (www.fda.gov, accessed on 15 November 2021) and mapping their protein targets using the DrugBank database [45]. The compiled data included the generic drug names, their molecular weights, as well as the UniProt identifier [46] of their protein targets, which were used to retrieve the amino acid sequences and the available crystal structures that are associated with many of these proteins.

3.1.2. Binding Site Definition at the Structural Level in the Human Targets

Binding site determination for a human target relied on the existence of at least one protein–ligand complex in the Protein Data Bank (PDB) [47]. Therefore, the next step was to determine which of the thousands of PDB structures associated with hundreds of human clinical protein targets contain bound inhibitors. For this purpose, we used our own program 'complex_info' [48], which identifies bound small ligands and carries out a detailed geometric analysis of the protein–ligand interactions, providing information on ligand size (number of heavy atoms), percent of buried ligand surface area, contacting protein atoms and amino acids, among other useful data. We used a filter of 10 heavy atoms as minimum to identify bound ligands, including small peptides and small nucleic acid chains.

Next, we focused the analysis on protein–ligand complexes containing FDA-approved drugs to gather statistics on the number of heavy atoms, surface area buried in the protein upon complexation and the number of contacting protein residues. We then used these data to adjust our search parameters and define more precisely the binding pockets in the human protein targets. In this process we excluded crystallographic molecules such as buffers and polyethylene glycols, heme groups and large peptides and nucleic acid ligands. Finally, for each obtained protein–ligand complex we defined the pocket region as the set of amino acid residues found within 4.5 Å from the ligand, using the VMD program [49]. For each of the identified complexes we tabulated the protein UniProt identifier, the PDB ligand ID, the number of ligand heavy atoms and the PDB sequence number of each binding pocket residue.

3.1.3. Defining Binding Site Regions at the Sequence Level for the Human Targets

We reasoned that using the functionally conserved binding site regions of the human targets for a BLAST search would increase the chances of finding similar regions in fungal proteins. Therefore, the next step was to delimit, for each selected protein target, a continuous sequence region containing the binding site pocket, based on the list of individual binding site amino acids identified in the previous step. Commonly, these binding site residues were scattered along a large sequence segment of a few hundred amino acids. In many cases, more than one protein–ligand complex was available in the PDB for the same target, yielding slightly different binding site lists depending on the size and geometry of each ligand. In addition, the sequence numbering for the same protein may differ between PDB entries, which created an additional difficulty for mapping the binding site residues to the reference Uniprot sequence. To solve this problem, we used pentamer sequence segments, each containing at least one of the binding site amino acids, to find its position in the reference sequence by simple string search. From this mapping procedure we could define a continuous sequence region containing all the binding site residues.

For those human target proteins having several binding site lists (originated from different protein–ligand complexes), we clustered and aligned the obtained sequence regions and manually revised each cluster. From this analysis we defined a unique consensus binding region sequence for each target protein.

3.1.4. Searching for Similar Binding Sites in Fungal Proteomes

The binding region sequences for the obtained set of human targets, as defined in the previous step, were used as query sequences for BLAST searches [50] in fungal proteomes, aiming to focus the search into regions that are more likely to be conserved among evolutionary distant organisms, such as humans and fungi. For comparison purposes, we performed BLAST searches using also the full sequences of the human targets.

For the subsequent analyses, we considered as hits only those alignments covering > 80% of the query sequence (i.e., the binding region sequence). The obtained alignments were then used to establish functional relationships between the binding pocket residues of the human targets and the corresponding amino acids in the fungal sequences. This way, the fungal binding sites became also defined at the amino acid level, as illustrated in Figure 7.

The similarity (percent of amino acid identity) between a human binding site and its corresponding fungal binding pocket was evaluated taking into account only the binding pocket residues. Lastly, we analyzed the alignments showing > 70% identity for the set of binding pocket residues. From the DrugBank we retrieved the list of approved drugs for a small set of these human targets, using also web services such as Drugs.com ("Drugs.Com | Prescription Drug Information, Interactions & Side Effects," 2021, last accessed on 10 April 2022).

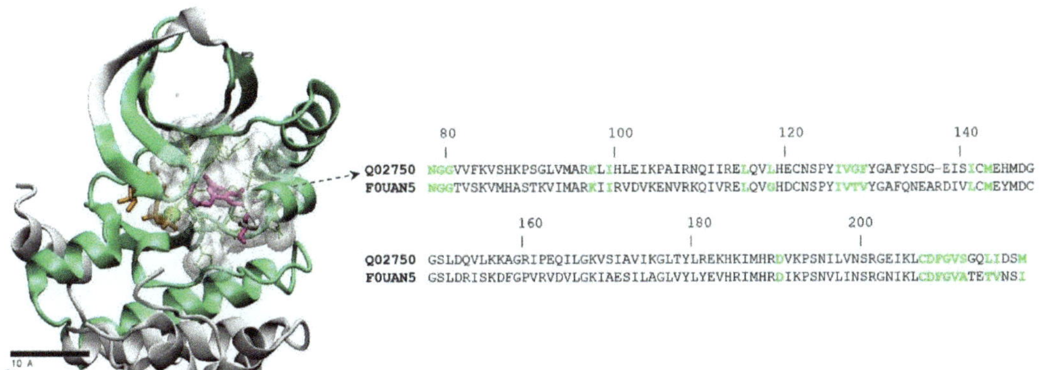

Figure 7. Definition of the binding site pocket and binding region sequence for the human MEK1 target and a fungal protein from *Histoplasma capsulatum* having a highly similar region. The continuous binding region sequence is represented as a green ribbon in the structure (PDB code 3dv3) and shown in full in one-letter code. Binding pocket amino acids are shown with their side chains (green, thin sticks) enclosed in a whitish volume, and are highlighted in green bold letters in the sequence. The MEK1 inhibitor in the 3dv3 structure is shown in thick sticks, colored in magenta. The ATP ligand is shown in orange sticks.

3.2. Fungal Proteomes Included in the Study

We analyzed the proteomes of six fungal species: *Aspergillus fumigatus* (UP000002530), *Candida albicans* (UP000000559), *Candida parapsilosis* (UP000005221), *Cryptococcus neoformans* (UP000002149), *Fusarium oxysporum* (UP000009097) and *Histoplasma capsulatum* (UP000008142), retrieved from the UniProt database.

3.3. Homology Modeling and Molecular Docking

For homology modeling of fungal proteins, we used the SwissModel server [51]. Structural models of the *Histoplasma capsulatum* protein with UniProt identifier F0UAN5 and the *Fusarium oxysporum* protein A0D2XNJ1 were constructed using as template the crystal structure of human MEK1 in complex with an inhibitor (PDB code 3dv3) [52]. AutoDock Tools [53] was employed to prepare molecules for docking simulations, which were carried out with AutoDock Vina [54] using default parameters and a box enclosing the non-ATP competitive binding site.

3.4. In Vitro Assays of MEK Inhibitors

The in vitro tests to assess the susceptibility to MEK inhibitors were carried out in 96-well microplates, seeding 300,000 cells/well for yeasts (*Histoplasma capsulatum, Cryptococcus neoformans, Candida albicans* and *Candida parapsilosis*) and 40,000 conidia/well for *Fusarium oxysporum* and *Aspergillus fumigatus*. *Histoplasma capsulatum* was cultured for 6 days in HAMF12 medium supplemented with cysteine and glutamine. The other yeasts were cultured in RPMI 1640 supplemented with 2% glucose for 24 h (for the two *Candidas*) or 72 h (*Cryptococcus*), all of them at 37 °C and stirring at 150 rpm.

MEK inhibitors were purchased from Cayman Chemicals (Ann Arbor, MI, USA). For each compound, the maximum tested concentration was determined by the solubility data reported by the manufacturer. Each inhibitor was dissolved either in DMSO or ethanol according to manufacturer's instructions. The stock solution for each compound was used at 1% as maximum, so that the DMSO concentration in the culture medium (kept at 1%) would not have toxic effects on the fungi. The compounds were tested also with the addition of 0.002% SDS, which most likely increased their solubility. Controls with 1% DMSO or ethanol, alone or combined with 0.002% SDS, were included in each microplate. To determine the half maximal inhibitory concentration (IC50), a 2-fold dilution series of 4 or 5 inhibitor concentrations was used. Fungal viability was determined using the XTT colorimetric assay.

4. Concluding Remarks

We have developed a strategy for a rational, structure-based approach to drug repositioning and new drug design, which can be applied not only to infectious fungi, but also to other pathogens. Following this methodology, we have identified fungal proteins having high binding site similarities with human targets of drugs that have shown inhibitory effects in fungi. These results not only support the developed strategy, but also contribute to identify the fungal targets responsible for these effects. Importantly, they also expose new routes to explore many drugs and inhibitors not yet tested in fungi.

Not all the identified fungal proteins, even if they are essential for the microorganism, are suitable for drug repositioning to treat fungal infections, especially in cases where the treatment produces severe side effects (as for many cancer drugs) or when it has immunosuppressive effects, which opens a door to opportunistic mycotic and bacterial infections. For a number of human targets, however, the available drugs may have only mild secondary effects, so they might be used to treat fungal infections if they show strong cytotoxic effects on these pathogens. Last but not least, the small structural differences in binding pocket architecture between some pairs of human and fungal proteins can be exploited to design specific antifungal drugs.

Supplementary Materials: The following supporting information can be downloaded at: https://www.mdpi.com/article/10.3390/molecules28020692/s1, Table S1: PDB entries of complexes including FDA-approved small drugs; Table S2: Full list of proteins with similar binding pockets.

Author Contributions: Conceptualization, E.M.; Methodology, J.E.B.-C., M.R.-C., M.S.V.-T. and E.M.; investigation, J.E.B.-C., M.R.-C., I.M.R.-V., M.S.V-T. and E.M.; writing—original draft preparation, J.E.B.-C., M.R.-C., I.M.R.-V. and E.M.; writing—review and editing, J.E.B.-C. and E.M.; supervision, project administration and funding acquisition, E.M. All authors have read and agreed to the published version of the manuscript.

Funding: This research was funded by the University of Medellin and Minciencias, Colombia (grant 795-2018).

Institutional Review Board Statement: Not applicable.

Informed Consent Statement: Not applicable.

Data Availability Statement: The data presented in this study are contained in the article tables and Supplementary Materials.

Acknowledgments: E.M. and M.R.-C. are grateful to Luz Elena Cano for her support in the early stages of the project.

Conflicts of Interest: The authors declare no conflict of interest.

Sample Availability: All the compounds used in the experiments were purchased from Cayman Chemicals.

References

1. Zeng, H.; Wu, Z.; Yu, B.; Wang, B.; Wu, C.; Wu, J.; Lai, J.; Gao, X.; Chen, J. Network Meta-Analysis of Triazole, Polyene, and Echinocandin Antifungal Agents in Invasive Fungal Infection Prophylaxis in Patients with Hematological Malignancies. *BMC Cancer* **2021**, *21*, 404. [CrossRef] [PubMed]
2. Vallabhaneni, S.; Mody, R.K.; Walker, T.; Chiller, T. The Global Burden of Fungal Diseases. *Infect. Dis. Clin. N. Am.* **2016**, *30*, 1–11. [CrossRef] [PubMed]
3. Seagle, E.E.; Williams, S.L.; Chiller, T.M. Recent Trends in the Epidemiology of Fungal Infections. *Infect. Dis. Clin. N. Am.* **2021**, *35*, 237–260. [CrossRef]
4. Campoy, S.; Adrio, J.L. Antifungals. *Biochem. Pharmacol.* **2017**, *133*, 86–96. [CrossRef]
5. Houšt', J.; Spížek, J.; Havlíček, V. Antifungal Drugs. *Metabolites* **2020**, *10*, 106. [CrossRef]
6. Nami, S.; Aghebati-Maleki, A.; Morovati, H.; Aghebati-Maleki, L. Current Antifungal Drugs and Immunotherapeutic Approaches as Promising Strategies to Treatment of Fungal Diseases. *Biomed. Pharmacother.* **2019**, *110*, 857–868. [CrossRef] [PubMed]
7. Wheat, L.J.; Azar, M.M.; Bahr, N.C.; Spec, A.; Relich, R.F.; Hage, C. Histoplasmosis. *Infect. Dis. Clin. N. Am.* **2016**, *30*, 207–227. [CrossRef]
8. Sosa, E.J.; Burguener, G.; Lanzarotti, E.; Defelipe, L.; Radusky, L.; Pardo, A.M.; Marti, M.; Turjanski, A.G.; Fernández Do Porto, D. Target-Pathogen: A Structural Bioinformatic Approach to Prioritize Drug Targets in Pathogens. *Nucl. Acids Res.* **2018**, *46*, D413–D418. [CrossRef]
9. Mukherjee, S.; Kundu, I.; Askari, M.; Barai, R.S.; Venkatesh, K.V.; Idicula-Thomas, S. Exploring the Druggable Proteome of Candida Species through Comprehensive Computational Analysis. *Genomics* **2021**, *113*, 728–739. [CrossRef]
10. Palumbo, M.; Sosa, E.; Castello, F.; Schottlender, G.; Serral, F.; Turjanski, A.; Palomino, M.M.; do Porto, D.F. Integrating Diverse Layers of Omic Data to Identify Novel Drug Targets in Listeria Monocytogenes. *Front. Drug Discov.* **2022**, *2*, 969415. [CrossRef]
11. Ashburn, T.T.; Thor, K.B. Drug Repositioning: Identifying and Developing New Uses for Existing Drugs. *Nat. Rev. Drug Discov.* **2004**, *3*, 673–683. [CrossRef] [PubMed]
12. Pushpakom, S.; Iorio, F.; Eyers, P.A.; Escott, K.J.; Hopper, S.; Wells, A.; Doig, A.; Guilliams, T.; Latimer, J.; McNamee, C.; et al. Drug Repurposing: Progress, Challenges and Recommendations. *Nat. Rev. Drug. Discov.* **2018**, *18*, 41–58. [CrossRef] [PubMed]
13. Xue, H.; Li, J.; Xie, H.; Wang, Y. Review of Drug Repositioning Approaches and Resources. *Int. J. Biol. Sci.* **2018**, *14*, 1232–1244. [CrossRef]
14. Jourdan, J.P.; Bureau, R.; Rochais, C.; Dallemagne, P. Drug Repositioning: A Brief Overview. *J. Pharm. Pharmacol.* **2020**, *72*, 1145–1151. [CrossRef] [PubMed]
15. Vanhaelen, Q. *Computational Methods for Drug Repurposing*; Humana: New York, NY, USA, 2019; Volume 1903, ISBN 978-1-4939-8954-6.
16. Paul, D.; Sanap, G.; Shenoy, S.; Kalyane, D.; Kalia, K.; Tekade, R.K. Artificial Intelligence in Drug Discovery and Development. *Drug Discov. Today* **2021**, *26*, 80–93. [CrossRef] [PubMed]
17. Peyclit, L.; Yousfi, H.; Rolain, J.M.; Bittar, F. Drug Repurposing in Medical Mycology: Identification of Compounds as Potential Antifungals to Overcome the Emergence of Multidrug-Resistant Fungi. *Pharmaceuticals* **2021**, *14*, 488. [CrossRef] [PubMed]
18. Chavez-Dozal, A.A.; Lown, L.; Jahng, M.; Walraven, C.J.; Lee, S.A. In Vitro Analysis of Finasteride Activity against Candida Albicans Urinary Biofilm Formation and Filamentation. *Antimicrob. Agents Chemother.* **2014**, *58*, 5855–5862. [CrossRef]
19. Nasr Esfahani, A.; Golestannejad, Z.; Khozeimeh, F.; Dehghan, P.; Maheronnaghsh, M.; Zarei, Z. Antifungal Effect of Atorvastatin against Candida Species in Comparison to Fluconazole and Nystatin. *Med. Pharm. Rep.* **2019**, *98*, 368–373. [CrossRef]
20. Didone, L.; Scrimale, T.; Baxter, B.K.; Krysan, D.J. A High-Throughput Assay of Yeast Cell Lysis for Drug Discovery and Genetic Analysis. *Nat. Protoc.* **2010**, *5*, 1107–1114. [CrossRef]
21. Butts, A.; DiDone, L.; Koselny, K.; Baxter, B.K.; Chabrier-Rosello, Y.; Wellington, M.; Krysanb, D.J. A Repurposing Approach Identifies Off-Patent Drugs with Fungicidal Cryptococcal Activity, a Common Structural Chemotype, and Pharmacological Properties Relevant to the Treatment of Cryptococcosis. *Eukaryot. Cell* **2013**, *12*, 278–287. [CrossRef]
22. Kim, K.; Zilbermintz, L.; Martchenko, M. Repurposing FDA Approved Drugs against the Human Fungal Pathogen, Candida Albicans. *Ann. Clin. Microbiol. Antimicrob.* **2015**, *14*, 32. [CrossRef] [PubMed]
23. Wall, G.; Chaturvedi, A.K.; Wormley, F.L.; Wiederhold, N.P.; Patterson, H.P.; Patterson, T.F.; Lopez-Ribot, J.L. Screening a Repurposing Library for Inhibitors of Multidrug-Resistant Candida Auris Identifies Ebselen as a Repositionable Candidate for Antifungal Drug Development. *Antimicrob. Agents Chemother.* **2018**, *62*, e01084-18. [CrossRef] [PubMed]
24. Miró-Canturri, A.; Ayerbe-Algaba, R.; Smani, Y. Drug Repurposing for the Treatment of Bacterial and Fungal Infections. *Front. Microbiol.* **2019**, *10*, 41. [CrossRef] [PubMed]
25. Pereira de Mello, T.; Nunes Silva, L.; de Souza Ramos, L.; Freire Frota, H.; Branquinha, M.H.; Sousa dos Santos, A.L. Drug Repurposing Strategy against Fungal Biofilms. *Curr. Top. Med. Chem.* **2020**, *20*, 509–516. [CrossRef]
26. Wall, G.; Lopez-Ribot, J.L. Screening Repurposing Libraries for Identification of Drugs with Novel Antifungal Activity. *Antimicrob. Agents Chemother.* **2020**, *64*, e00924-20. [CrossRef] [PubMed]
27. Santos, R.; Ursu, O.; Gaulton, A.; Bento, A.P.; Donadi, R.S.; Bologa, C.G.; Karlsson, A.; Al-Lazikani, B.; Hersey, A.; Oprea, T.I.; et al. A Comprehensive Map of Molecular Drug Targets. *Nat. Rev. Drug. Discov.* **2016**, *16*, 19–34. [CrossRef] [PubMed]
28. Berkes, C.; Franco, J.; Lawson, M.; Brann, K.; Mermelstein, J.; Laverty, D.; Connors, A. Kinase Inhibitor Library Screening Identifies the Cancer Therapeutic Sorafenib and Structurally Similar Compounds as Strong Inhibitors of the Fungal Pathogen Histoplasma Capsulatum. *Antibiotics* **2021**, *10*, 1223. [CrossRef]

29. Macreadie, I.G.; Johnson, G.; Schlosser, T.; Macreadie, P.I. Growth Inhibition of Candida Species and Aspergillus Fumigatus by Statins. *FEMS Microbiol. Lett.* **2006**, *262*, 9–13. [CrossRef]

30. Hao, W.; Qiao, D.; Han, Y.; Du, N.; Li, X.; Fan, Y.; Ge, X.; Zhang, H. Identification of Disulfiram as a Potential Antifungal Drug by Screening Small Molecular Libraries. *J. Infect. Chemother.* **2021**, *27*, 696–701. [CrossRef]

31. Antypenko, L.; Sadykova, Z.; Meyer, F.; Garbe, L.A.; Steffens, K. Tacrolimus as Antifungal Agent. *Acta Chim. Slov.* **2019**, *66*, 784–791. [CrossRef]

32. Tu, B.; Yin, G.; Li, H. Synergistic Effects of Vorinostat (SAHA) and Azoles against Aspergillus Species and Their Biofilms. *BMC Microbiol.* **2020**, *20*, 28. [CrossRef] [PubMed]

33. Ohren, J.F.; Chen, H.; Pavlovsky, A.; Whitehead, C.; Zhang, E.; Kuffa, P.; Yan, C.; McConnell, P.; Spessard, C.; Banotai, C.; et al. Structures of Human MAP Kinase Kinase 1 (MEK1) and MEK2 Describe Novel Noncompetitive Kinase Inhibition. *Nat. Struct. Mol. Biol.* **2004**, *11*, 1192–1197. [CrossRef]

34. Han, J.; Liu, Y.; Yang, S.; Wu, X.; Li, H.; Wang, Q. MEK Inhibitors for the Treatment of Non-Small Cell Lung Cancer. *J. Hematol. Oncol.* **2021**, *14*, 1. [CrossRef]

35. Bechman, K.; Galloway, J.B.; Winthrop, K.L. Small-Molecule Protein Kinases Inhibitors and the Risk of Fungal Infections. *Curr. Fungal Infect. Rep.* **2019**, *13*, 229–243. [CrossRef]

36. Dong-mei, M.A.; Ya-juan, J.I.; Fang, Y.; Wei, L.I.U.; Zhe, W.A.N.; Ruo-yu, L.I. Effects of U0126 on Growth and Activation of Mitogen-Activated Protein Kinases in Aspergillus Fumigatus. *Chin. Med. J.* **2013**, *126*, 220–225. [CrossRef]

37. FAN, Y.; GU, S.; DONG, J.; DONG, B. Effects of MEK-Specific Inhibitor U0126 on the Conidial Germination, Appressorium Production, and Pathogenicity of Setosphaeria Turcica. *Agric. Sci. China* **2007**, *6*, 78–85. [CrossRef]

38. Singh, A.; Ruan, Y.; Tippett, T.; Narendran, A. Targeted Inhibition of MEK1 by Cobimetinib Leads to Differentiation and Apoptosis in Neuroblastoma Cells. *J. Exp. Clin. Cancer Res.* **2015**, *34*, 104. [CrossRef]

39. Lee, K.W.; Kang, N.J.; Rogozin, E.A.; Kim, H.G.; Cho, Y.Y.; Bode, A.M.; Lee, H.J.; Surh, Y.J.; Bowden, G.T.; Dong, Z. Myricetin Is a Novel Natural Inhibitor of Neoplastic Cell Transformation and MEK1. *Carcinogenesis* **2007**, *28*, 1918–1927. [CrossRef]

40. Iverson, C.; Larson, G.; Lai, C.; Yeh, L.T.; Dadson, C.; Weingarten, P.; Appleby, T.; Vo, T.; Maderna, A.; Vernier, J.M.; et al. RDEA119/BAY 869766: A Potent, Selective, Allosteric Inhibitor of MEK1/2 for the Treatment of Cancer. *Cancer Res.* **2009**, *69*, 6839–6847. [CrossRef]

41. Salama, A.K.S.; Kim, K.B. Trametinib (GSK1120212) in the Treatment of Melanoma. *Expert Opin. Pharmacother.* **2013**, *14*, 619–627. [CrossRef]

42. Hatzivassiliou, G.; Haling, J.R.; Chen, H.; Song, K.; Price, S.; Heald, R.; Hewitt, J.F.M.; Zak, M.; Peck, A.; Orr, C.; et al. Mechanism of MEK Inhibition Determines Efficacy in Mutant KRAS- versus BRAF-Driven Cancers. *Nature* **2013**, *501*, 232–236. [CrossRef] [PubMed]

43. Yeh, T.C.; Marsh, V.; Bernat, B.A.; Ballard, J.; Colwell, H.; Evans, R.J.; Parry, J.; Smith, D.; Brandhuber, B.J.; Gross, S.; et al. Biological Characterization of ARRY-142886 (AZD6244), a Potent, Highly Selective Mitogen-Activated Protein Kinase Kinase 1/2 Inhibitor. *Clin. Cancer Res.* **2007**, *13*, 1576–1583. [CrossRef] [PubMed]

44. Dong, Q.; Dougan, D.R.; Gong, X.; Halkowycz, P.; Jin, B.; Kanouni, T.; O'Connell, S.M.; Scorah, N.; Shi, L.; Wallace, M.B.; et al. Discovery of TAK-733, a Potent and Selective MEK Allosteric Site Inhibitor for the Treatment of Cancer. *Bioorg. Med. Chem. Lett.* **2011**, *21*, 1315–1319. [CrossRef]

45. Wishart, D.S.; Knox, C.; Guo, A.C.; Shrivastava, S.; Hassanali, M.; Stothard, P.; Chang, Z.; Woolsey, J. DrugBank: A Comprehensive Resource for in Silico Drug Discovery and Exploration. *Nucleic Acids Res.* **2006**, *34*, D668–D672. [CrossRef] [PubMed]

46. Bateman, A.; Martin, M.J.; Orchard, S.; Magrane, M.; Agivetova, R.; Ahmad, S.; Alpi, E.; Bowler-Barnett, E.H.; Britto, R.; Bursteinas, B.; et al. UniProt: The Universal Protein Knowledgebase in 2021. *Nucleic Acids Res.* **2021**, *49*, D480–D489. [CrossRef]

47. Bernstein, F.C.; Koetzle, T.F.; Williams, G.J.B.; Meyer, E.F.; Brice, M.D.; Rodgers, J.R.; Kennard, O.; Shimanouchi, T.; Tasumi, M. The Protein Data Bank: A Computer-Based Archival File for Macromolecular Structures. *J. Mol. Biol.* **1977**, *112*, 535–542. [CrossRef]

48. Moreno, E.; León, K. Geometric and Chemical Patterns of Interaction in Protein-Ligand Complexes and Their Application in Docking. *Proteins Struct. Funct. Genet.* **2002**, *47*, 1–13. [CrossRef]

49. Humphrey, W.; Dalke, A.; Schulten, K. VMD: Visual Molecular Dynamics. *J. Mol. Graph.* **1996**, *14*, 33–38. [CrossRef]

50. Altschul, S.F.; Gish, W.; Miller, W.; Myers, E.W.; Lipman, D.J. Basic Local Alignment Search Tool. *J. Mol. Biol.* **1990**, *215*, 403–410. [CrossRef]

51. Waterhouse, A.; Bertoni, M.; Bienert, S.; Studer, G.; Tauriello, G.; Gumienny, R.; Heer, F.T.; de Beer, T.A.P.; Rempfer, C.; Bordoli, L.; et al. SWISS-MODEL: Homology Modelling of Protein Structures and Complexes. *Nucleic Acids Res.* **2018**, *46*, W296–W303. [CrossRef]

52. Tecle, H.; Shao, J.; Li, Y.; Kothe, M.; Kazmirski, S.; Penzotti, J.; Ding, Y.H.; Ohren, J.; Moshinsky, D.; Coli, R.; et al. Beyond the MEK-Pocket: Can Current MEK Kinase Inhibitors Be Utilized to Synthesize Novel Type III NCKIs? Does the MEK-Pocket Exist in Kinases Other than MEK? *Bioorg. Med. Chem. Lett.* **2009**, *19*, 226–229. [CrossRef] [PubMed]

53. Morris, G.M.; Ruth, H.; Lindstrom, W.; Sanner, M.F.; Belew, R.K.; Goodsell, D.S.; Olson, A.J. Software News and Updates AutoDock4 and AutoDockTools4: Automated Docking with Selective Receptor Flexibility. *J. Comput. Chem.* **2009**, *30*, 2785–2791. [CrossRef] [PubMed]
54. Trott, O.; Olson, A.J. AutoDock Vina: Improving the Speed and Accuracy of Docking with a New Scoring Function, Efficient Optimization, and Multithreading. *J. Comput. Chem.* **2009**, *31*, 455–461. [CrossRef] [PubMed]

Article

Computational Simulation of Colorectal Cancer Biomarker Particle Mobility in a 3D Model

Esteban Vallejo Morales [1,*,†], Gustavo Suárez Guerrero [2,*,†] and Lina M. Hoyos Palacio [3,†]

1   Grupo de Investigación Sobre Nuevos Materiales, Universidad Pontificia Bolivariana, Medellín 050031, Colombia
2   Grupo de Investigación e Innovación en Energía, Institución Universitaria Pascual Bravo, Medellín 050036, Colombia
3   Grupo de Investigación en Biología de Sistemas, Universidad Pontificia Bolivariana, Medellín 050031, Colombia
*   Correspondence: esteban.vallejomo@upb.edu.co (E.V.M.); gustavo.suarez@pascualbravo.edu.co (G.S.G.); Tel.: +57-300-769-8329 (E.V.M.); +57-313-685-2123 (G.S.G.)
†   These authors contributed equally to this work.

Abstract: Even though some methods for the detection of colorectal cancer have been used clinically, most of the techniques used do not consider the in situ detection of colorectal cancer (CRC) biomarkers, which would favor in vivo real-time monitoring of the carcinogenesis process and consequent studies of the disease. In order to give a scientific and computational framework ideal for the evaluation of diagnosis techniques based on the early detection of biomarker molecules modeled as spherical particles from the computational point of view, a computational representation of the rectum, stool and biomarker particles was developed. As consequence of the transport of stool, there was a displacement of CRC biomarker particles that entered the system as a result of the cellular apoptosis processes in polyps with a length lower than 1 cm, reaching a maximum velocity of 3.47×10^{-3} m/s. The biomarkers studied showed trajectories distant to regions of the polyp of origin in 1 min of simulation. The research results show that the biomarker particles for CRC respond to the variations in the movements of the stool with trajectories and speeds that depend on the location of the injury, which will allow locating the regions with the highest possibilities of catching particles through in situ measurement instruments in the future.

Keywords: colorectal cancer; biomarker particles; early detection; in situ

Citation: Vallejo Morales, E.; Suárez Guerrero, G.; Hoyos Palacio, L.M. Computational Simulation of Colorectal Cancer Biomarker Particle Mobility in a 3D Model. *Molecules* **2023**, *28*, 589. https://doi.org/10.3390/molecules28020589

Academic Editor: Jahir Orozco Holguín

Received: 1 October 2022
Revised: 29 December 2022
Accepted: 30 December 2022
Published: 6 January 2023

1. Introduction

According to the World Health Organization [1], and as can be observed in Figure 1, for 2018, colorectal cancer held third place at a worldwide level for the estimated number of new cases in human beings with 10.2 % of the total of new cases, considering both sexes and any age. This number was only surpassed by breast cancer, which was in second place, and by lung cancer, which was in first place on the list. Even though the number of new incidences of colorectal cancer placed it in third place, the same did not happen with the proportion that it represented in the number of deaths in the same year. This can be evidenced by Figure 2, where the number of deaths due to colorectal cancer represented 9.2% of the total deaths caused by cancer in general, positioning colorectal cancer as the second-deadliest cancer in 2018 when considering both sexes and any age [1].

Notwithstanding the reported degree of morbidity and mortality, colorectal cancer generally develops as a result of the neoplastic progression of adenomas to adenocarcinomas, and this can take decades, which enables having an opportune advantage for the early detection of CRC [2] to prevent the complete development of the tumor and improve the prognosis.

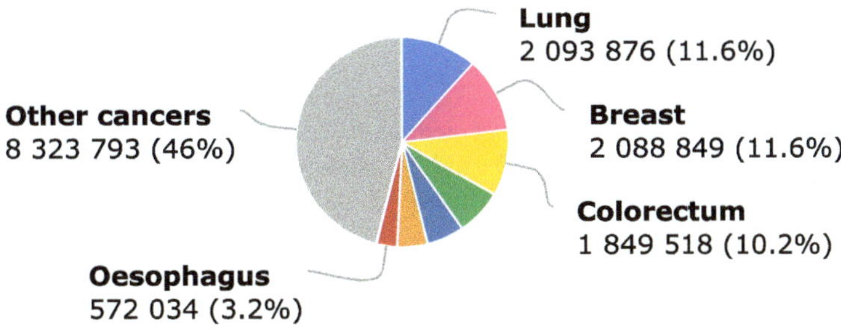

Figure 1. Estimated number of new cancer cases in 2018 worldwide for both sexes and all ages. Here, 33% of the incidence of cancer in the world population during 2018 corresponded to lung, breast and colorectal cancer, which was in third place, which is why the impact of better early diagnosis methods would help a significant portion of the population. Recovered from [1].

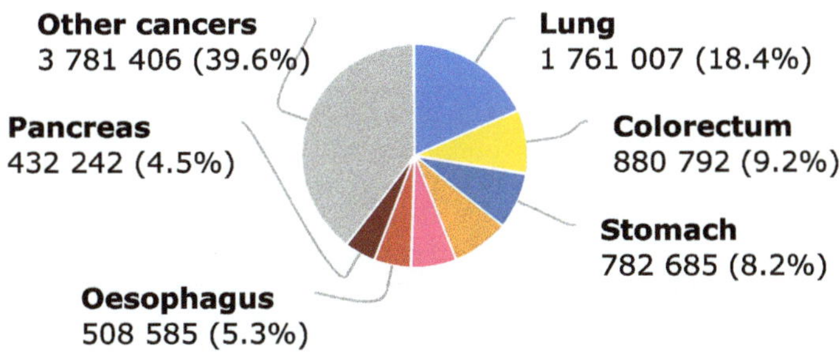

Figure 2. Estimated number of deaths due to cancer in 2018 globally for both sexes and all ages. Colorectal cancer is positioned as the second cancer with the highest mortality during the year 2018. This is due in part to late diagnosis of the disease. Earlier detection could help reduce the 9.2% death rate. Recovered from [1].

With the aim of contributing to the reduction of mortality caused by colorectal cancer, some investigations have focused on improving diagnostic techniques, concentrating their efforts on the identification and sensing of biomarkers associated with the onset of the disease and its corresponding progression.

Even though the study of biomarker molecules in tissues or blood is frequently mentioned by several examples of research [3–6], there is also a special interest in the biomarker molecules resulting from exfoliation processes in precancerous injuries and CRC where the cell-free DNA (cfDNA) molecules are liberated to the stool, especially in treatments with noninvasive diagnoses [3,7–9]. This is because it has been stated that DNA analysis in stool is much more effective than DNA analysis in blood [7].

The first report on DNA hypomethylation about the general contents of 5-metilcitosina in CRC tissues was reported in 1983 by Feinberg and Vogelstein, mentioning that hypo methylation is accompanied by hypermethylation and transcriptional silencing of tumor suppression genes or genes that codify DNA repair proteins [10].

The detection of colorectal neoplasms through a stool DNA test is naturally facilitated by several biological factors, among them being an imbalance in the intestinal microbiota, especially in some bacteria that cause the induction of carcinogenesis through mutations in cells or epigenetic changes that promote cell growth [11], while on the other hand, the unproportioned copious exfoliation and the functional surface area become notably large in neoplasias [12].

Regarding research in the field of computational simulation, most of the studies reviewed focused on the modeling of nanostructured biosensors [13–15], showing mathematical models and the results of simulations of the characteristics of the biosensor without specifically mentioning their application in the early detection of diseases. On the other hand, there were studies found on cancer evolution models based on agents and implementing high-performance computing (HPC) [16,17], as well as the progression of melanoma [18], evolution of metastatic processes, tumor angiogenesis [16], treatment response [19,20] and simulation of the adherence of therapeutic particles [21], evidencing efforts that are much more oriented toward the treatment and description of the evolution of CRC, which is oriented toward the early diagnosis of the description of the behavior of biomarkers in a dynamic domain.

Despite this, Tóth Kinga et al. [22] published a research article with the objective of obtaining quantitative data to explore possible correlations between the presence of the biomarker mSEPT9 in the tissue and in the plasma of the same patients, especially in patients with adenoma. There was not any research found that completely coincided with the focus of this work, which describes the behavior of colorectal cancer biomarker molecules that are released by tumors in the processes of cell apoptosis, and it focuses on demonstrating that these particles respond to the movements of feces with trajectories and speeds that depend on the location of the injury, which will help on-site test tools find the best places to capture particles.

2. Results

A 3D computation model of the rectum was created using ANSYS® software [23], and it enabled us to simulate the interaction between two peristaltic type II waves, the stool and the CRC biomarker molecules, which were modeled as spherical particles from the point of view of computational simulation. The model was tested for different positions of polyps within the domain of this study, showing different behaviors in the function of the location of the particle exfoliation site.

2.1. Behavior of the Stool during the Passing of the Contraction Waves

Figure 3 identifies the field of speeds generated by the contraction of the rectum wall due to the passing of the type II wave for $10 \leq t \leq 60$ s. The encounter between the retrograde and anterograde speeds can be appreciated in the viscous material between 10 and 60 seconds, while for the tiles of 20 s, 30 s and 50 s, there was a field whose components in −z showed a uniform reverse of the stool. At t = 40 s (and other times), a divergent behavior could be appreciated in the lower part of the domain, where the higher portion showed a retrograde direction and the lower portion was directed toward the exit border of the domain. The maximum velocity of the stool reached by the simulations corresponded to 3.47×10^{-3} m/s. The following are the speeds for two of the times that evidenced projection in the trajectory of the stool coming from the region of liberation of the PT 4 biomarker particles.

For t = 21 s in Figure 4a, a retrograde trajectory of the stool was detected coming from the cfDNA PT 4 exfoliation region, and its changes in velocity can be appreciated in Figure 4b, where there is an evident acceleration toward half of the path and a decrease in velocity as it finished its trajectory. According to this figure, the time required for the displacement of the material from the PT 4 point toward the reached region was 102 s.

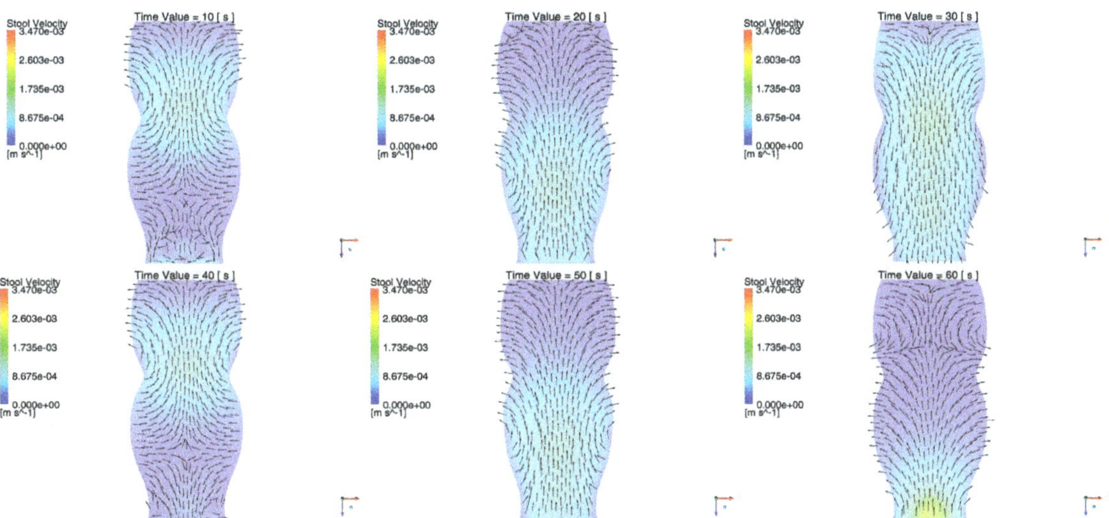

Figure 3. Stool velocity during the type II waves between 10 s and 60 s. The fecal velocity field can be seen in the anterograde and retrograde directions, which implies consistency with the physical behavior described by the fecal mass after the passage of a peristaltic wave.

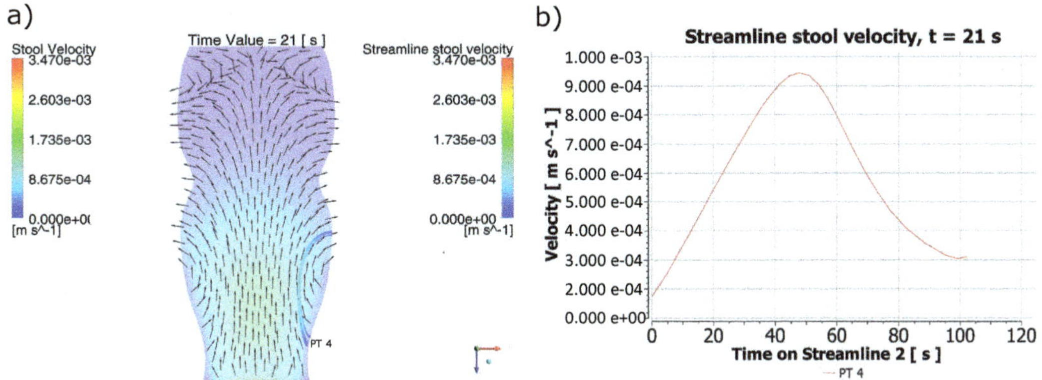

Figure 4. Behavior of the stool for t = 21 s. (**a**) Stool flow lines from the region of the polyp PT 4 for t = 21 s. (**b**) Velocity for the stool flow lines. It can be identified that in the PT4 particle release area, the direction of the fecal mass formed a retrograde trajectory. However, due to changes in the position of the peristaltic contraction, there was a variation in the magnitude of the maximum velocity reached during a new predicted trajectory, indicating a region much further from the point of origin of the biomarker particles.

For t = 48 s in Figure 5a, a retrograde trajectory of the stool was detected coming from the cfDNA PT 4 exfoliation region, and its changes in speed can be appreciated in Figure 5b. Since the highest speeds of the stool were at a higher distance from the PT 4 polyp, the maximum velocity reached by the current line corresponds to a time higher than that in Figure 4b. Notwithstanding this, the possibility of reaching regions farther from the polyp is evident for this distribution of speeds. The time required to reach the region of destiny with the speeds obtained for t = 48 s was 281.88 s.

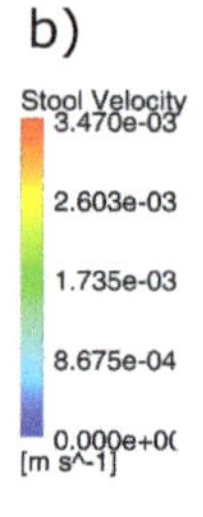

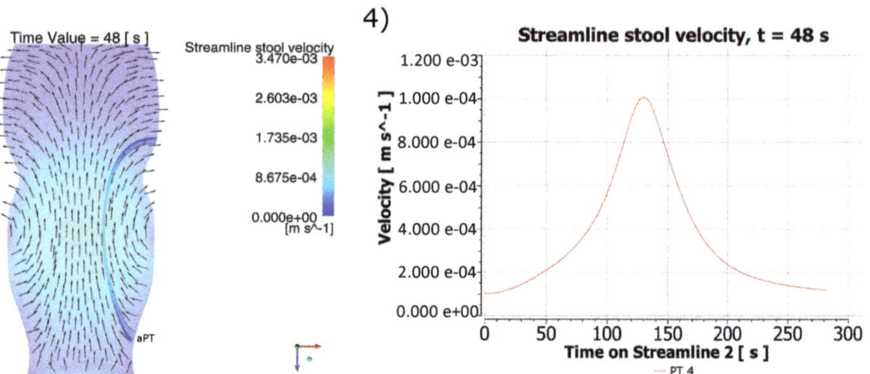

Figure 5. Behavior of the stool for t = 48 s. (**a**) Stool flow lines from the region of the PT 4 polyp for t = 48 s. (**b**) Velocity for the stool flow lines. It can be identified that in the PT4 particle release area, the direction of the fecal mass formed a retrograde trajectory. However, due to changes in the position of the peristaltic contraction, there was a variation in the magnitude of the maximum velocity reached during a new predicted trajectory, indicating a region much further from the point of origin of the biomarker particles, which allows us to state that the direction of the trajectory is retrograde for the type of contraction contemplated, but the site of impact with the predicted colonic wall varies depending on the position of the contraction.

2.2. Behavior of Biomarker Particles during the Passing of the Contraction Waves

The computational simulation of coupled domains enabled us to obtain the behavior of the cfDNA modeled as spherical particles in four different liberation zones. In Figure 6, the trajectories of the particles liberated in the regions established for the studied polyps can be identified.

Due to the anterograde and retrograde directions of the field of speeds of the stool, the biomarker particles liberated exhibited a trajectory similar to an ellipsis. Such behavior is especially visible in Figure 6d, since its displacement direction obeyed to the velocity vectors of the stool corresponding to t = 60.

The maximum velocity obtained by the biomarker particles corresponded to 3.19×10^{-3} m/s, which oscillated in value according to the behavior of the medium that contained them. Figure 7a shows the results of the velocities obtained by the computational simulation.

Figure 7b identifies the traveling of the particles in each region of liberation over 60 s. The maximum travel distance reached by a cfDNA particle in the time of the simulation corresponded to 1.27×10^{-2} m for the PT 4 injection region, while the region whose particle liberation showed the lowest travel distance corresponded to 1.01×10^{-2} m for PT 2, with this being a significant distance in relation to the size of the particle and the time that passed.

After comparing the changes in velocity of the particles in the PT 4 region obtained in Figure 7a and the position of the maximum velocities reached by the stool for t = 10, 20, 30, 40, 50 and 60 s, an increase and decrease could be identified in the velocities of the biomarker particles in proportion to the increase and decrease in the stool velocity, which evidences a successful coupling between the study domains.

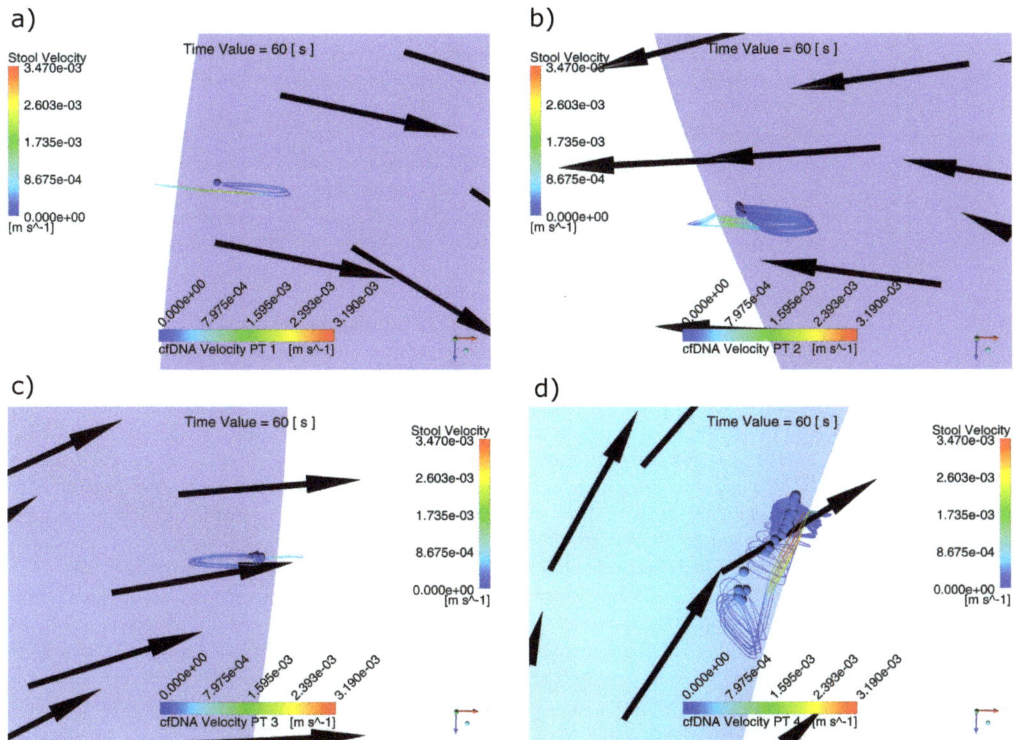

Figure 6. Behavior of the cfDNA particles for t = 60 s: (**a**) region PT1, (**b**) region PT 2, (**c**) region PT 3 and (**d**) region PT 4. The behavior of almost circular trajectories with an advance in the retrograde direction can be explained by the constant change in the direction and magnitude of the velocity field generated by the peristaltic wave.

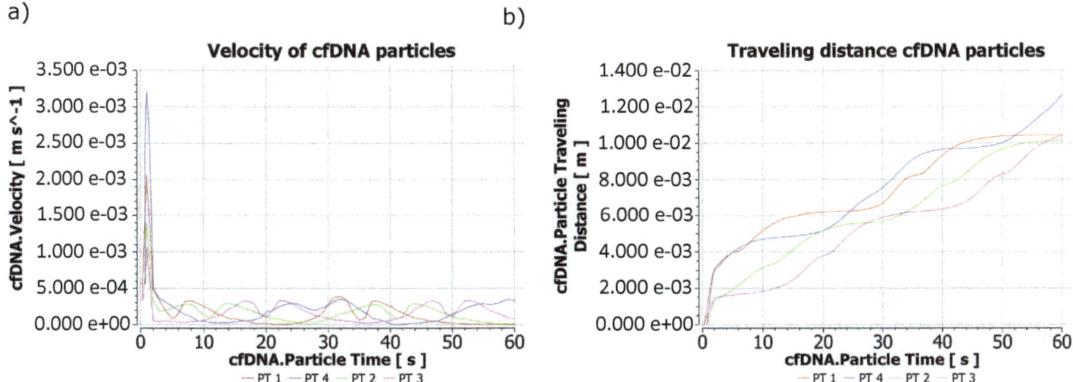

Figure 7. Behavior of the cfDNA particles during 60 s of simulation: (**a**) Velocity of cfDNA particles and (**b**) Traveling distance of cfDNA particles. When observing the line that represents each particle in different places of a colon sample, it is possible to identify the reason why some particles did not seem to have a linear traveled distance, since fluctuations in the velocity magnitude during the simulated time could slow down or accelerate each of the studied particles.

3. Discussion

This research was able to identify the behavior of the biomarker particles of colorectal cancer at an early stage through a dynamic and similar model to the real biological conditions. By using the interaction of coupled domains, it was possible to identify the behavior of the cfDNA liberated into the stool under functionality conditions, because the dynamics of the model was generated by contraction waves whose parameterization came from clinical studies on the magnitude of pressures generated on the colon's motility.

The incorporation of the peristaltic waves as the transportation mechanism for the rectum contents enabled us to obtain the velocities of the stool and biomarkers since, with this mechanism, it was guaranteed that the movement obtained was as a consequence of the colon's motility.

In the dissemination process of the biomarker particles through the stool, there was verification that such the analyte moved to distant regions of the injury in only one minute of simulation. The strategic location of an on-site diagnosis device would enable higher effectiveness in the catchment of the analyte, since the predictive nature of the simulation showed those zones of the intestinal tract where stool would eventually arrive with cfDNA particles, and since the results were in terms of the trajectory, velocity and travel distance for the biomarker particles, future studies can use these findings, as the parameters calculated for the construction of on-site biomarker measurement instruments maximized the capitation of cfDNA particles.

The inclusion of elements that were not considered in other investigations [13–16,21] allowed us to obtain a more adjusted model in terms of the functionality conditions. Some of the results concurred with behaviors that have already been described in other studies, while others enabled us to identify values that did not correspond with the results reported by other authors due to their simplifications.

Determining the trajectories of the biomarkers enables the development of new noninvasive and moderately invasive diagnosis and treatment techniques, increasing the control of the development of cancer in morbid persons and achieving higher effectiveness in treatments for people suffering from colon cancer, which corroborates the usefulness of computational simulations in medical diagnoses as a tool for the solution of complex problems, whose analytic solution is difficult, unknown or impossible to obtain.

Due to the lack of in vivo clinical techniques for the simultaneous observation of biomarker nanoparticles and the stool in a colon in motion, it was necessary to build mathematical models to obtain the parameters necessary for the simulation, which became a contribution for future research. In the same manner, due to the lack of in vivo experimental results, the results obtained in this research were analyzed in light of similar computational simulations, supporting most of the considerations taken since they supported their simplifications.

4. Materials and Methods

Computational implementation was performed using Ansys CFX software [23]. Its meshing tools were used, as well as its ability to perform mechanical and fluid analysis, incorporating the discrete phase to represent the release of colorectal cancer biomarkers.

4.1. Material Properties: Rectum Wall

The hyperelastic approach postulates the existence of a function of W energy, which relates the traveling of the colon's tissue with the corresponding tension values. The deformation energy function represents the energy stored by a system under deformation. When the load is removed, the deformation energy is gradually liberated into the system in such a manner that it returns to its original form [24].

A homogeneous and isotropic material was assumed to represent the rectum walls, with the argument from several works that models of the colon's behavior toward contractions caused by peristaltic movement or toward interaction with a colonoscopy should use these simplifications [24–26].

For a homogeneous material, the deformation energy is the only function of the deformation gradient F, defined as

$$F = \frac{\partial x}{\partial X}.$$

(1)

where x denotes a point in the current configuration and X denotes a point in the reference configuration. For isotropic materials, the deformation energy $W = W(F)$ is also a function of the invariants I_1, I_2 and I_3, which are defined as follows:

$$I_1 = trace(C) = \lambda_1^2 + \lambda_2^2 + \lambda_3^2.$$

(2)

$$I_2 = \frac{1}{2}trace(C) = \lambda_1^2\lambda_2^2 + \lambda_1^2\lambda_3^2 + \lambda_2^2\lambda_3^2.$$

(3)

$$I_3 = \lambda_1^2\lambda_2^2\lambda_3^2.$$

(4)

where λ_1, λ_2 and λ_3 are the stretching coefficients and C is defined as the Cauchy–Green deformations tensor, which represents the deformation measure of a point in a direction i and has the products of the deformation gradients. This tensor is determined through the following equation:

$$C = F^T F$$

(5)

The colon tissue is incompressible; that is, the volume of the material remains constant during the deformation. For incompressible materials, the following is true:

$$det(C) = \lambda_1^2\lambda_2^2\lambda_3^2,$$

(6)

making the deformation energy a function of only two invariants:

$$W = W(I_1, I_2).$$

(7)

There are various mathematical models for representing hyperelastic isotropic materials, including the neo-Hookean, Arruda–Boyce, Gent, Mooney–Rivlin, Ogden, polynomial, Yeoh and Ogden approaches. Polynomial forms are most popular for the constitutive modeling of biological tissues due to their simplicity and, therefore, efficiency in the calculation processes [27,28].

In this work, the Mooney–Rivlin three-parameter model was used as in [29], whose behavior can be appreciated in Figure 8. The mathematical model corresponding to the Mooney–Rivlin three-parameter formulation consists of the following equation:

$$W = C_{10}(I_1 - 3) + C_{01}(I_2 - 3) + C_{11}(I_1 - 3)(I_2 - 3) + \frac{1}{d}(J - 1)^2$$

(8)

where C_{10}, C_{01} and C_{11} are the material rigidity constants, d is the incompressibility parameter and k is the volume module.

Since for an incompressible material $J = \lambda_1\lambda_2\lambda_3 = 1$, W can be simplified as follows:

$$W = C_{10}(I_1 - 3) + C_{01}(I_2 - 3) + C_{11}(I_1 - 3)(I_2 - 3)$$

(9)

The work of Xuehuan [24] enabled the correct parameterization of the model through the values given in Table 1.

Table 1. Rigidity constants of the isotropic hyperelastic material.

Constant	Value (Pa)
C_{10}	652.01
C_{01}	42,835.25
C_{11}	219,120.30

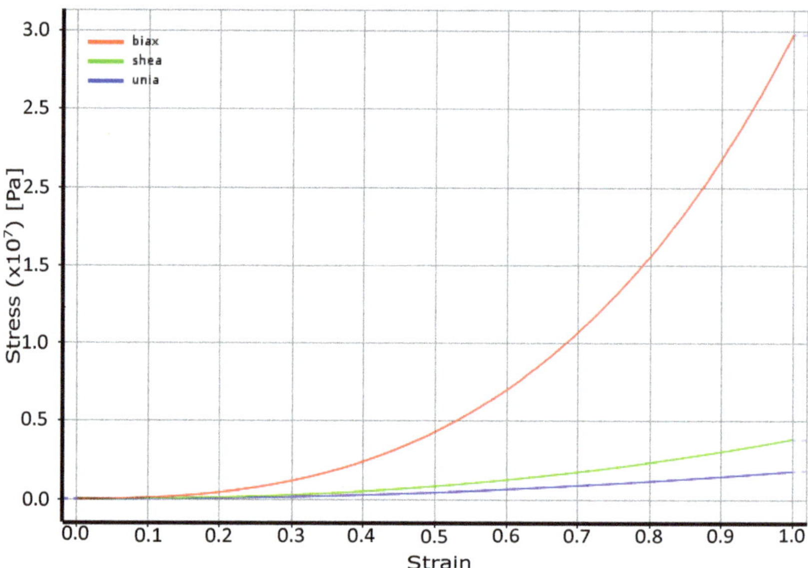

Figure 8. Behavior of the material according to the three-parameter Mooney–Rivlin model.

According to Mcintosh and Anderson [30] in their work "A Comprehensive Tissue Properties Database Provided for the Thermal Assessment of a Human At Rest", the density of the human colon corresponds to the value $\rho = 1132 \ \text{kg/m}^3$.

The incompressibility parameter d could be found based on the following equation:

$$d = \frac{1 - 2\nu}{c_{10} + c_{01}} \tag{10}$$

This value was obtained from the Poisson coefficient reported in [31] and the optimized constants reported in Table 1. Its value corresponded to $d = 4.60 \times 10^{-7}$.

Figure 8 illustrates the tension deformation obtained with the three-parameter Mooney–Rivlin formulation.

4.2. Peristaltic Movement

In adults, the rectum has a resting pressure of 6 mmHg (799.934 Pa) [32,33] and shows three motion patterns:

1. Type I contractions: "simple monophasic waves of low width and short duration. These contractions form holes on the surface creation pressures of 5 to 10 cm of H_2O (490.333–980.665 Pa), their duration varies from 5 to10 s, and their frequency is 8 to 12/min" [34].
2. Type II contractions: "These have a greater width 8, 15 to 30 cm H_2O (784.532, 1471, 2942 Pa), and last longer (25 to 30 s), their frequency is 2/min; both contractions act to mix the stool" [34].
3. Type III contractions: "These represent a change in the base pressure, generally lower than 10 cm H_2O (980.665 Pa), with superposition of type I and II waves" [34].

The domain Ω_s was defined to note the structure formed by the colon walls. Such a structure is deformed by a contraction wave perpendicular to the external surface of the geometry that travels through the longitude of the rectum on the z-axis. The contraction wave that mimics the muscular contraction of the peristaltic movement is represented in Figure 9 and is given by

$$p(z,t) = p_0 + p_1 sech\left[\frac{b_1(z-ct)-b_2}{L}\right] \tag{11}$$

where b_1 and b_2 are constant as well as p_0 and p_1, which are the resting pressure and active pressure, respectively [35], and L is the length of the chosen colorectal section. The constant c represents the velocity of the wave, whose value corresponding to 1 cm/s [36] is adapted to 0.5 cm/s in order to reach the frequency of the type II waves (2 waves/min) mentioned in [34].

Equation (11) can also be expressed as

$$p(z,t) = p_0 + p_1\left\{\frac{2}{e^{\frac{b_1(z-ct)-b_2}{L}}+e^{-\frac{b_1(z-ct)-b_2}{L}}}\right\} \tag{12}$$

The active pressure was determined to be $p_1 = 3999.67$ Pa, which was in the range of pressures reported in the colon and the rectal sigmoid section [37,38], and the rest pressure was $p_0 = (799.934\ Pa)$ [32,33]. The length of the rectum was $L = 1.50 \times 10^{-1}$ m. The values for the remaining parameters were $b_1 = 14$ and $b_2 = 0$ for the first wave and $b_2 = 2.1$ for the second wave.

4.3. Stool

The importance of this domain in the research is because of it being the medium in which the cfDNA biomarker particles are liberated. Such particles respond to the changes in velocity generated by the interaction of the peristaltic movement of the colon walls and the stool under study.

"The stool is made up of proteins, fats, fiber, bacterial biomass, inorganic materials, and carbohydrates. Its chemical and physical characteristics vary widely depending on the person's health and diet" [39]. Nevertheless, for practical effects, parameters such as the density and viscosity of the stool were taken from the research conducted by R. Penn et al. [39] and corresponded to 1060 kg/m^3 and 5.50 Pa s, respectively.

For this research, the stool was modeled as a highly viscous incompressible fluid with Newtonian behavior through the conservation laws of Navier–Stokes through Equation (15), supported by the argument that there was analysis in one section of the colon destined for storage, and therefore, most of the changes in the viscosity were in sections prior to the rectal section. In addition, "The length of the intestine modeled is considerably reduced (15 cm), restricting the possible variation of the viscosity" [40].

The flow was described in terms of the Cauchy tensor [41–43]:

$$\overline{\overline{T}} = 2\mu D(\vec{u}) - pI \tag{13}$$

where: μ is the viscosity, p is the pressure and $\overline{\overline{D(\vec{u})}}$ is the deformation tensor [41–43], a term that can be expressed as

$$\overline{\overline{D(\vec{u})}} = \frac{1}{2}(\nabla\vec{u} + \nabla^T\vec{u}) \tag{14}$$

The fluid dynamics can be described by a transitory domain through the expression

$$\rho\left(\frac{\partial\vec{u}}{\partial t} + (\vec{u}\cdot\nabla)\vec{u}\right) - \mu\Delta\vec{u} + \nabla p = \vec{F} \tag{15}$$

We considered the condition of the incompressible fluid as follows:

$$\nabla\cdot\vec{u} = 0 \tag{16}$$

where: ρ and μ are the density and viscosity of the flow, respectively, $\vec{F}$ represents the exterior forces and $\vec{u}$ is the Euler velocity of the fluid. This model was valid only if the fluid was viscous ($\mu > 0$).

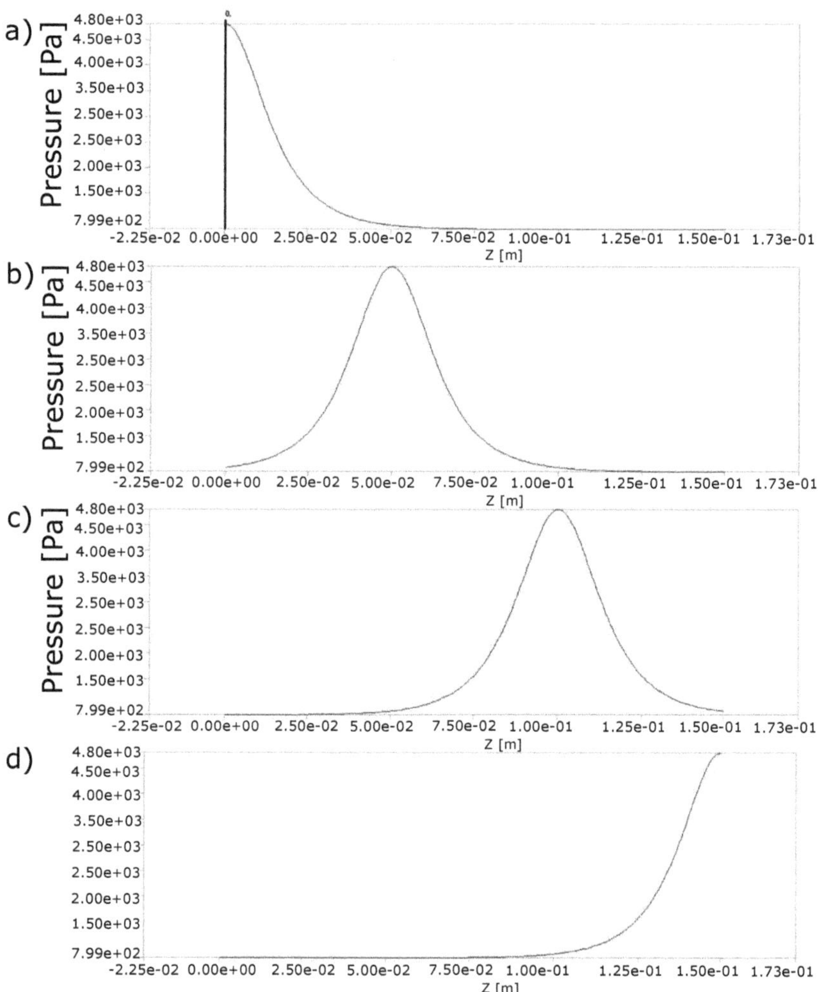

Figure 9. Travel of the peristaltic wave on the z-axis over time: (**a**) 0 s, (**b**) 10 s, (**c**) 20 s and (**d**) 30 s, which were repeated for another 30 s after passing the second wave. Changes in pressure as a consequence of muscle contractions contributed to the transport of the fecal mass and with it the biomarker particles that were released in the process of cell apoptosis.

The first two terms in Equation (15) refer to the inertial mass forces of the fluid. The third term describes the viscous forces of the fluid, while the pressure gradient expressed in the fourth term represents the pressure forces, and the fifth term refers to the external forces. In such equation, the expression $((\vec{u} \cdot \nabla)\vec{u})$ is known as the convective non-linear term, what contributes to the difficulty to reach the convergence of the solution.

For the problem studied, the temperature considered was 37 °C (310.15 K) and invariable in time and space.

The terms Ω_f and Γ_f were defined to represent the domains of the fluid and its borders, respectively. The border conditions imposed were as follows:

- A Dirichlet-type condition for the fluid velocity at the entry border given by

$$\vec{u} = 0 \quad \text{over} \quad \Gamma_f \tag{17}$$

- A Neumann condition given by

$$\overline{\overline{T}} \cdot n = \vec{\sigma} \quad \text{over} \quad \Gamma_f \tag{18}$$

This carried the velocity gradients $\vec{u}$ through the fluid tension tensor and showed a pushing effect over the edge of the fluid.

4.4. Mesh

Figure 10 presents a section that allows visualizing the structured mesh generated for the lumen of the colon portion studied, while Figure 11 reports the quality of the mesh in Figure 10. On the other hand, Figures 12 and 13 refer to to the mesh associated with the colon wall and its corresponding quality report respectively.

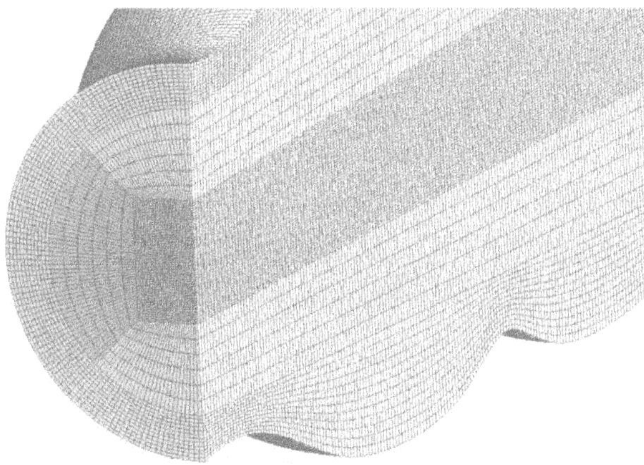

Figure 10. Structured mesh of the rectal lumen model. This type of structured meshing contributes significantly to the stability and convergence of the solution.

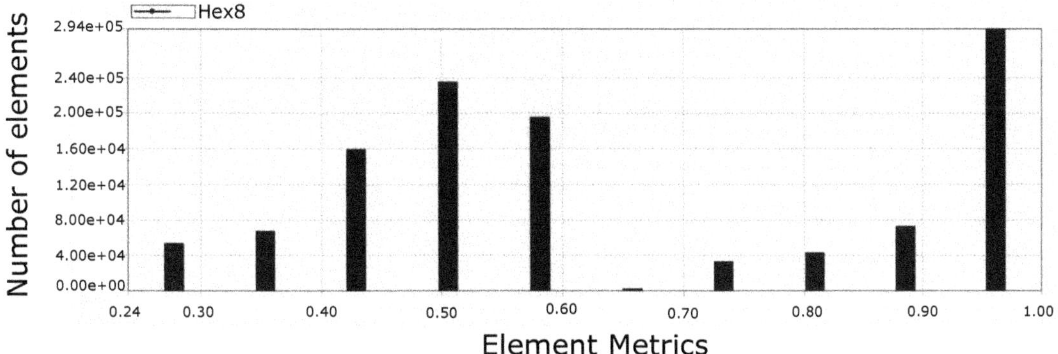

Figure 11. Mesh quality report of rectal lumen model, where 0 is poor element quality and 1 is perfect element quality. The predominance of elements with a metric of 1 contributed to the success of the simulation and generated confidence in the results obtained. Note: Hex8 is the type of element used by the mesh and represents a hexahedron with 8 nodes.

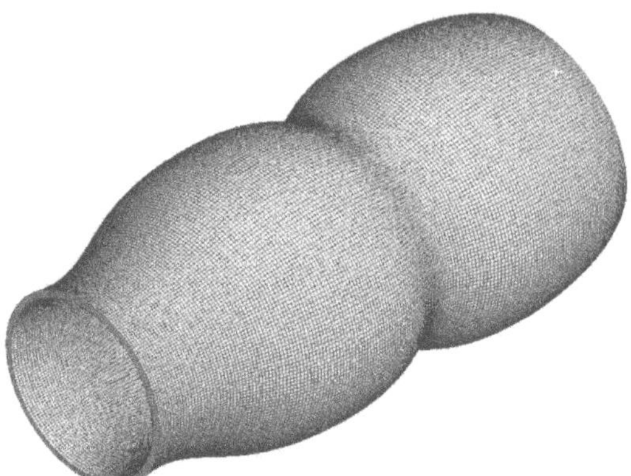

Figure 12. Structured mesh of the rectal wall portion of the colon.

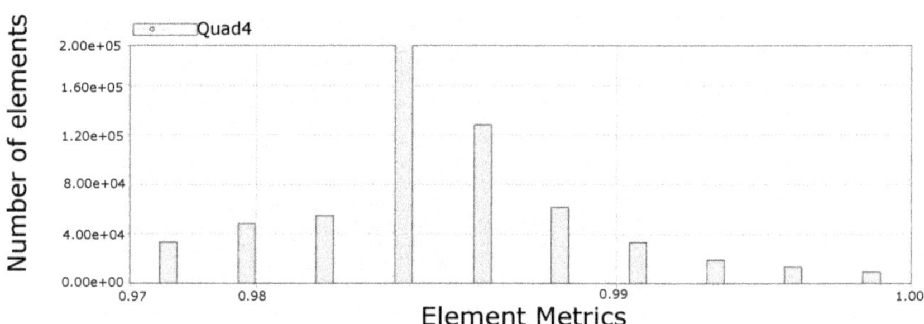

Figure 13. Mesh quality of the rectal wall portion, where 0 is poor element quality and 1 is perfect element quality. The absence of elements with a quality lower than 0.97 accounted for an adequate reticulation of the colon wall, which allowed adequately representing the deformation of the material. Note: Quad4 is the type of element used by the mesh and represents a quadrilateral with 4 nodes.

4.5. Description of the Biomarker Particles

Circulating extracellular DNA (cfDNA) is present in the blood or liberated in the lumen of the large intestine. The amount of cfDNA circulating in serum and plasma seems to be significantly higher in patients with tumors than in healthy controls, especially for those with tumors in advanced states compared with tumors in early stages. Samanta Salvi et al. [44] carried out several studies destined to correlate the reordering in tissue and plasma samples, which coincided to confirm that the analysis of circulating DNA can be used as a diagnosis tool. It has been suggested that cfDNA in healthy people is mainly of a hematopoietic origin. Nevertheless, cfDNA in cancer patients also results from apoptotic and necrotic processes characteristic of tumor cells with high cellular refill. Apoptosis produces DNA fragments of approximately 180 base pairs (bp) or corresponding multiples, while necrosis produces much larger fragments [9,44–46].

According to the studies performed in the field of DNA biomarkers in colorectal cancer, an average molecular weight is commonly accepted for each base pair of DNA of 1.02×10^{-24} kg. According to several studies [9,44,46], the cfDNA fragments have an amount of base pairs that ranges from 115 to 247 base pairs (bp) [9,45,47]. By taking 200 bp as a recurrent value in such texts, it was determined that the mass of dfDNA particles was 2.04×10^{-22} kg.

According to [48], the length between base pairs is used to find the total length of a DNA fragment, since it is mentioned that for a length of 2000 bp, the theoretical length is 680 nm. Supported by the data of such research, it was determined that for 200 bp, the length was 68 nm. Such a calculated value was used to model the size and form of the particle, assuming that cfDNA is a sphere whose diameter corresponds to 68 nm. A spherical shape was assumed for the 3D particle, so the volume of a sphere representing a cfDNA particle was 9.26×10^{-23} m^3.

Once the dimensions and mass of a cfDNA particle were clear, it was determined that, for this study, the density of a fragment of a DNA biomarker for colorectal cancer corresponded to approximately $\rho \approx 2.20$ kg/m^3.

Table 2 summarizes the mass, density and dimensions of the particles.

Table 2. Characterization of the biomarker particles.

Length (bp)	Length (nm)	Mass (kg)	Volume (m^3)	Density (kg/m^3)
200	68	2.04×10^{-22}	9.26×10^{-23}	2.20

4.5.1. Colon Epithelial Cells

The epithelial cells of the intestines have been very difficult to cultivate in vitro as primary cells. Due to this, over the last four decades, the preferred model of the intestine epithelium has been transformed, having CaCo-2 cells as the main reference, whose composition and behavior are particularly adequate [49].

According to Jung et al., it has been estimated that in patients with colon cancer and a tumor of 100 g, close to 3.3 % of the tumor's DNA is liberated daily in the blood flow [45].

The volume of a CaCo-2 cell is $V = 1.40 \times 10^{-15}$ m^3. Assuming that all enterocytes are diploid, their DNA content is 5.70×10^9 base pairs per cell [50,51]. According to the results found in [52], the mass of a cell is 3.00×10^{-12} kg. Such magnitudes have been estimated based on the similarities between enterocytes and CaCo-2 cells validated through [49], and they were used in this research for the calculation of the diffuse flow of cfDNA particles.

4.5.2. Exfoliation Processes

Neoplasias abundantly exfoliate the dysplastic cells and their components in the colon, granting a regular supply of analyte for stool analysis. Since direct histological observation has stated that such exfoliation seems to be continuous [12], such results were used as the starting point to assume a constant injection flow of cfDNA toward the the colon for this model.

The reason why the analysis of the cfDNA liberated toward the stool showed much more significant results for the early detection of CRC than those obtained in the tests of blood hidden in feces is the fact that hidden blood is the result of circulation through a hemorrhage, which is intermittent and often absent in precancerous injuries [12].

4.5.3. Calculation of Parameters for the Injection of Particles through the Rectum Wall

In order to adequately define the injection of cfDNA particles, certain parameters need to be calculated. In this section, a mathematical model is built for the calculation of mass flow, the amount of particles injected per second and the magnitude of the injection velocity of the particles.

For the calculation of the mentioned parameters, it was required to establish a geometric model that represented the structure of the polyp in a simplified manner. Figure 14 represents a neoplastic polyp of a tubular adenoma type, which is made up of a height cylinder H representing the stem and a sphere with a radius R representing the higher part of the polyp. The height of the structure H_{polyp} was used to calculate the total area through which a detachment of DNA fragments from the cells that made up such a structure was produced.

The area of the Figure 14 was calculated based on the following thought:

A sphere is defined and located over the higher face of the height of a cylinder $H = 3.00 \times 10^{-3}$ m and with a base radius of $r = 1.50 \times 10^{-3}$ m. The intersection between both figures generated a spherical cap with a radius $r = 1.50 \times 10^{-3}$ cm and with a height of h. A total height $H_{polyp} = 7.00 \times 10^{-3}$ m was defined for the complete structure, so the diameter of the sphere was $D_s = 7.00 \times 10^{-3} - 3.00 \times 10^{-3} + h$ m; that is, $D_s = 4.00 \times 10^{-3} + h$ m, and therefore, its radius was $R = \frac{D_s}{2}$.

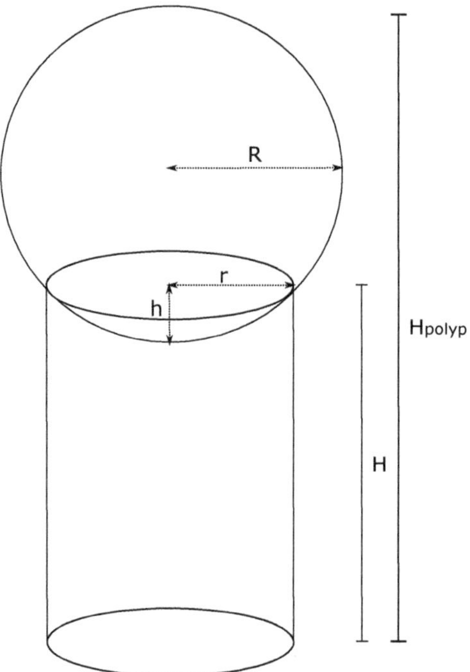

Figure 14. Simplified representation of a neoplastic polyp of the tubular adenoma type.

The total area of the polyp corresponded to the addition of the area of the spherical portion and the stem, obtaining $A_{polyp} = 8.67 \times 10^{-5}$ m^2 as a result.

The volume of the proposed polyp can be calculated as follows:

$$V_{polyp} = (V_{sphere} - V_{cap}) + V_{stem} \tag{19}$$

where

$$V_{sphere} = \pi r^2 H \tag{20}$$

$$V_{cas} = \frac{\pi h (3r^2 + h^2)}{6} \tag{21}$$

Therefore, after performing such calculations, it was found that $V_{polyp} = 1.42 \times 10^{-5}$ m^3.

If the volume of the polyp was divided into the previously reported volume of a CaCo-2 cell, the result was the amount of cells that made up such a polyp. With such a calculation, the total mass of the proposed polyp could be estimated under the supposition that such neoplasia was totally made up of only CaCo-2 cells:

$$PolypCellAmount = \frac{V_{polyp}}{V_{CaCo-2}} = 1.02 \times 10^{10} \tag{22}$$

Therefore, the mass of the polyp was 0.03 kg.

Once the area of the polyp was calculated, it was necessary to determine the velocity for the injection of particles into the system. Due to the specific nature of the data, and after an extensive review of the pertinent literature, it was found that such a value was not explicitly reported in the research dedicated to analysis of the exfoliation of biomarker particles in a neoplastic polyp. As an answer to such a question, a previous simulation without the injection of particles was proposed in order to determine the minimum velocity reached (different from 0 m/s) by part of the stool after a peristaltic movement with a duration of 15 s in the rectum. The results in terms of the velocity were exactly the same in the presence or absence of particles since, for the interest of this study, and also considering the size of the particles, the effect that these had on the simulated viscous mass was not taken into account.

After an evaluation of the average velocity reached for each time step in the simulation, the research proceeded with the representation of such data in time, the exportation of the results and selection of the minimum value. Once this was performed, it was determined that the velocity assigned at the entrance of the particles to the system was 5.57×10^{-4} m/s.

With all the parameters identified, the mass flow was defined as

$$\dot{M} = \frac{\Delta M}{\Delta t} = \rho \frac{\Delta V}{\Delta t} = \rho A_{polyp} |\vec{v}| \tag{23}$$

where ρ is the density expressed in Table 2 and v is the entrance velocity stipulated for the injection of particles.

For a magnitude with a particle entry velocity $v = 5.57 \times 10^{-4}$ m/s, the mass flow would then be $\dot{M} = 0.06 \times 10^{-6}$ kg/s.

The mass flow and particle size were considered to calculate the real (physical) number of particles. It was assumed that each numerical particle was a group of real particles that behaved in the same manner. In order to know how many particles hid behind the numerical particles, the mass flow and sizes must be supplied.

The particle flow rate was defined as

$$T_{fp} = \frac{\dot{M}}{M_{cfDNA}} = 2.99 \times 10^{14} \; particles/s \tag{24}$$

where M_{cfDNA} is the mass of a cfDNA particle, expressed in Table 2.

Nevertheless, such a value is very elevated and requires an excessive amount of resources from the system. Therefore, such a value was divided by the amount of particles intended to be represented within only one particle that represented them graphically.

To obtain the parameters associated with the diffusion of biomarkers in the stool, the information supplied by Jung et al. was used, who estimated that in patients with colon cancer with a tumor size of 100 g, approximately 3.3% of the tumor's DNA was liberated daily into the blood flow [45]. Given that, in the literature reviewed, there was no explicit explanation of an amount for a polyp of 0.03 kg, the percentage reported by Jung et al. was used to establish a linear relation and estimate the percentage of cfDNA liberated daily for the studied neoplasia as follows:

$$cfDNA_{percentage} = \frac{0.03 \; \text{kg} \times 0.03}{0.10 \; \text{kg}} = 0.99\% \tag{25}$$

This percentage of cfDNA was analyzed as follows.

An average molecular weight of 1.02×10^{-24} kg is commonly accepted for each base pair (bp) of DNA [53]. Given that a cell contains 5.70×10^9 bp, the mass of the cfDNA present in a cell within a polyp of 0.03 kg could be calculated as follows:

$$M_{DNA} = 5.70 \times 10^9 \; \frac{\text{bp}}{\text{cell}} \left(1.02 * 10^{-24} \; \frac{\text{kg}}{\text{bp}}\right) = 5.81 \times 10^{-15} \; \frac{\text{kg}}{\text{cell}} \tag{26}$$

If the liberation of cfDNA was assumed to be continuous during one day (86,400 s), and 0.99% of the tumor's DNA was liberated daily in the blood flow, then the flow of liberated cfDNA could be calculated as follows:

$$\varphi = 5.81 \times 10^{-15} \; \frac{\text{kg}}{\text{cell}} (1.01 \times 10^{10} \; \text{cell}) \left(\frac{0.99 \; \%}{\text{day}}\right) \left(\frac{1 \; \text{day}}{86400 \; \text{s}}\right) = 6.72 \times 10^{-12} \; \frac{\text{kg}}{\text{s}} \tag{27}$$

Finally, according to Fick's first law, the density of diffusive flow is expressed as

$$J = \frac{\varphi}{A_{polyp}} \tag{28}$$

Therefore, we have

$$J = \frac{6.72 \times 10^{-12}\,\frac{\text{kg}}{\text{s}}}{8.67 \times 10^{-5}\,\text{m}^2} = 7.79 \times 10^{-8}\,\frac{\text{kg}}{\text{m}^2\text{s}} \tag{29}$$

5. Conclusions

This research contributes new knowledge to advance the search for solutions to the early detection of colorectal cancer with treatments that can decrease morbidity. The development of mathematical and computational models implemented to solve biological situations, especially in cases where there is interaction between macroscopic and nanoscopic mediums, such as in this research, offers opportunities for the generation of new developments that can be used to analyze other cases of the same disease or other diseases, which contributes to the implementation of solutions for the monitoring of cfDNA concentrations in real time, since this can be one of the most effective techniques for the early diagnosis of CRC, given that diverse studies highlight the diagnostic and predictive importance of the analysis of biomarkers in the stool.

The lack of scientific information on some parameters that were characterized and used computationally in this research became an opportunity to provide mathematical models to calculate their values. This contributed to appropriate mathematical modeling and computational implementation, and there were findings that showed similarities with a real biological situation, what increased the reliability of the results.

The future possibility of implanting a biosensor coupled to the rectum for patients with risk factors would represent a great advantage in terms of a reduction in the time required for the collection and analysis of the samples, as well as the preparations necessary for the following colonoscopies.

It is concluded that this research will be useful for future studies associated with the early diagnosis of colon cancer, due to the multidisciplinary approach to the phenomenon, the computational simulation tools and the mathematical modeling based on scientific evidence that supported its development.

Finally, this research is a computational study in which there is no experimental evidence. The validity of the findings obtained must be verified based on references from other future research oriented toward the analysis of biomarkers in diseases.

Author Contributions: Conceptualization, E.V.M. and G.S.G.; formal analysis, E.V.M. and G.S.G.; funding acquisition, L.M.H.P.; investigation, E.V.M., G.S.G. and L.M.H.P.; methodology, E.V.M.; resources, L.M.H.P.; software, E.V.M.; supervision, L.M.H.P.; visualization, E.V.M.; writing—original draft, E.V.M.; writing—review and editing, E.V.M. and G.S.G. All authors have read and agreed to the published version of the manuscript.

Funding: This research and APC was funded by MINCIENCIAS, MINEDUCACIÓN, MINCIT and ICETEX, through the Program Ecosistema Científico, grant number FP44842-211-2018 Project number 58962.

Institutional Review Board Statement: Not applicable.

Informed Consent Statement: Not applicable.

Data Availability Statement: Not applicable.

Acknowledgments: The authors would like to thank the Centro de Investigación para el Desarrollo y la Innovación (CIDI) of Universidad Pontificia Bolivariana and the program Colombia Científica in the project Nanobiocáncer for the valuable funding and technical and administrative support.

Conflicts of Interest: The authors declare no conflict of interest.

References

1. World Health Organization. *Estimated Number of Incident Cases, Both Sexes, Worldwide (Top 10 Cancer Sites) in 2018*; World Health Organization: Geneva, Switzerland, 2018.
2. Metkar, S.K.; Girigoswami, K. Diagnostic biosensors in medicine—A review. *Biocatal. Agric. Biotechnol.* **2019**, *17*, 271–283. [CrossRef]
3. Liu, Y.; Wu, H.; Zhou, Q.; Song, Q.; Rui, J.; Zou, B.; Zhou, G. Digital quantification of gene methylation in stool DNA by emulsion-PCR coupled with hydrogel immobilized bead-array. *Biosens. Bioelectron.* **2017**, *92*, 596–601. [CrossRef] [PubMed]
4. Oussalah, A.; Rischer, S.; Bensenane, M.; Conroy, G.; Filhine-Tresarrieu, P.; Debard, R.; Forest-Tramoy, D.; Josse, T.; Reinicke, D.; Garcia, M.; et al. Plasma mSEPT9: A Novel Circulating Cell-free DNA-Based Epigenetic Biomarker to Diagnose Hepatocellular Carcinoma. *EBioMedicine* **2018**, *30*, 138–147. [CrossRef] [PubMed]
5. Song, L.; Jia, J.; Yu, H.; Peng, X.; Xiao, W.; Gong, Y.; Zhou, G.; Han, X.; Li, Y. The performance of the mSEPT9 assay is influenced by algorithm, cancer stage and age, but not sex and cancer location. *J. Cancer Res. Clin. Oncol.* **2017**, *143*, 1093–1101. [CrossRef] [PubMed]
6. Lee, H.S.; Hwang, S.M.; Kim, T.S.; Kim, D.W.; Park, D.J.; Kang, S.B.; Kim, H.H.; Park, K.U. Circulating Methylated Septin 9 Nucleic Acid in the Plasma of Patients with Gastrointestinal Cancer in the Stomach and Colon. *Transl. Oncol.* **2013**, *6*, 290-IN4. [CrossRef]
7. Deeb, D.; Gao, X.; Jiang, H.; Arbab, A.S.; Dulchavsky, S.A.; Gautam, S.C. The Stool DNA Test is More Accurate than the Plasma Septin 9 Test in Detecting Colorectal Neoplasia. *Anticancer Res.* **2010**, *30*, 3333–3339.
8. Ahlquist, D.A.; Taylor, W.R.; Mahoney, D.W.; Zou, H.; Domanico, M.; Thibodeau, S.N.; Boardman, L.A.; Berger, B.M.; Lidgard, G.P. The Stool DNA Test Is More Accurate Than the Plasma Septin 9 Test in Detecting Colorectal Neoplasia. *Clin. Gastroenterol. Hepatol.* **2012**, *10*, 272–277.e1,
9. De Maio, G.; Rengucci, C.; Zoli, W.; Calistri, D. Circulating and stool nucleic acid analysis for colorectal cancer diagnosis. *World J. Gastroenterol.* **2014**, *20*, 957–967. [CrossRef]
10. Vatandoost, N.; Ghanbari, J.; Mojaver, M.; Avan, A.; Ghayour-Mobarhan, M.; Nedaeinia, R.; Salehi, R. Early detection of colorectal cancer: From conventional methods to novel biomarkers. *J. Cancer Res. Clin. Oncol.* **2016**, *142*, 341–351. [CrossRef]
11. Saeed, M.; Shoaib, A.; Kandimalla, R.; Javed, S.; Almatroudi, A.; Gupta, R.; Aqil, F. Microbe-based therapies for colorectal cancer: Advantages and limitations. *Semin. Cancer Biol.* **2022**, *86*, 652–665. [CrossRef]
12. Hlquist, D.A.A.A. Stool DNA screening for colorectal neoplasia: Biological and technical basis for high detection rates colorectal cancer screening. *Pathology* **2012**. *44*, 80–88. [CrossRef]
13. Yousefi, A.T.; Ikeda, S.; Rusop Mahmood, M.; Yousefi, H.T. Simulation of Nano Sensor Based on Carbon Nanostructures in Order to Form Multifunctional Delivery Platforms. *Adv. Mater. Res.* **2013**, *832*, 778–782. [CrossRef]
14. Ravi, A.; Krishna, R.M.A.; Christen, J.B. Modeling and Simulation of Dual Application Capacitive MEMS Sensor. In Proceedings of the 2014 COMSOL Conference, Arizona State University, Tempe, AZ, USA, 2014; pp. 1–4.
15. Lacatus, E.; Alecu, G.C.; Tudor, A. Models for Simulation Based Selection of 3D Multilayered Graphene Biosensors. In Proceedings of the 2015 COMSOL Conference, Grenoble, France, 14–16 October 2015; pp. 3–8. [CrossRef]
16. Kang, G.; Márquez, C.; Barat, A.; Byrne, A.T.; Prehn, J.H.M.; Sorribes, J.; César, E. Colorectal tumour simulation using agent based modelling and high performance computing. *Future Gener. Comput. Syst.* **2017**, *67*, 397–408. [CrossRef]
17. Mustapha, K.; Gilli, Q.; Frayret, J.M.; Lahrichi, N.; Karimi, E. Agent-based Simulation Patient Model for Colon and Colorectal Cancer Care Trajectory. *Procedia Comput. Sci.* **2016**, *100*, 188–197. [CrossRef]
18. Dzwinel, W.; Kłusek, A.; Vasilyev, O.V. Supermodeling in simulation of melanoma progression. *Procedia Comput. Sci.* **2016**, *80*, 999–1010. [CrossRef]
19. Bethge, A.; Schumacher, U.; Wedemann, G. Simulation of metastatic progression using a computer model including chemotherapy and radiation therapy. *J. Biomed. Inform.* **2015**, *57*, 74–87. [CrossRef]
20. Wedemann, G.; Bethge, A.; Haustein, V.; Schumacher, U. Computer simulation of the metastatic progression. In *Methods in Molecular Biology*, 2nd ed.; Dwek, M., Schumacher, U., Brooks, S.A., Eds.; Springer: New York, NY, USA, 2014; Chapter 8; Volume 1070, pp. 107–116. [CrossRef]
21. Liu, Y.; Shah, S.; Tan, J. Computational Modeling of Nanoparticle Targeted Drug Delivery. *Rev. Nanosci. Nanotechnol.* **2012**, *1*, 66–83. [CrossRef]
22. Tóth, K.; Wasserkort, R.; Sipos, F.; Kalmár, A.; Wichmann, B.; Leiszter, K.; Valcz, G.; Juhász, M.; Miheller, P.; Patai, Á.V.; et al. Detection of methylated Septin 9 in tissue and plasma of colorectal patients with neoplasia and the relationship to the amount of circulating cell-free DNA. *PLoS ONE* **2014**, *9*, e115415. [CrossRef]
23. ANSYS Inc. *ANSYS Help*; ANSYS Inc.: Canonsburg, PA, USA, 2019.
24. He, X. Modeling of the Interaction between Colon and Colonoscope during a Colonoscopy. Ph.D. Thesis, University of Minnesota, Minneapolis, MN, USA, 2018.
25. Sokolis, D.P.; Sassani, S.G. Microstructure-based constitutive modeling for the large intestine validated by histological observations. *J. Mech. Behav. Biomed. Mater.* **2013**, *21*, 149–166. [CrossRef]
26. Yazdanpanh-Ardakani, K.; Niroomand-Oscuii, H. New Approach in Modeling Peristaltic Transport of Non-Newtonian Fluid. *J. Mech. Med. Biol.* **2013**, *13*, 1350052. [CrossRef]

27. Yu, Y. Numerical Methods for Fluid-Structure Interaction: Analysis and Simulations. Ph.D. Thesis, Brown University, Providence, RI, USA, 2014.
28. TermehYousefi, A.; Bagheri, S.; Shahnazar, S.; Rahman, M.H.; Kadri, N.A. Computational local stiffness analysis of biological cell: High aspect ratio single wall carbon nanotube tip. *Mater. Sci. Eng. C* **2016**, *59*, 636–642. [CrossRef] [PubMed]
29. Guerrero, G.S.; Maday, Y.; Osorno, J.B. Solución del Problema de Dominios Acoplados con Interaccion Fluido-Estructura (F-E) en un Dispositivo de Asistencia Ventricular Cardiaca; Ph.D. Thesis, Universidad Pontificia Bolivariana, Medellin, Colombia, 2011.
30. Mcintosh, R.L.; Anderson, V. Erratum: "A Comprehensive Tissue Properties Database Provided for the Thermal Assessment of a Human At Rest". *Biophys. Rev. Lett.* **2013**, *8*, 99–100. [CrossRef]
31. Christensen, M.B.; Oberg, K.; Wolchok, J.C. *Tensile Properties of the Rectal and Sigmoid Colon: A Comparative Analysis of Human and Porcine Tissue*; Springer: Berlin/Heidelberg, Germany, 2015; Volume 4. [CrossRef]
32. Omari, T.I.; Rudolph, C.D. Chapter 112—Gastrointestinal Motility. In *Fetal and Neonatal Physiology*, 3rd ed.; Polin, R.A., Fox, W.W., Abman, S.H., Eds.; W.B. Saunders: Philadelphia, PA, USA, 2004; pp. 1125–1138. . [CrossRef]
33. Baker, M.L.; Tune, J.M.; Hightower, D. Intraluminal Pressure Measurements During Barium Enema. *Am. J. Roentgenol.* **1981**, *137*, 217–221.
34. Arróniz, M.A.C.; Palacios, A.C.; Zebadúa, Ó.A. Anatomía y fisiología de colon. In *Gastroenterología*; Torres, E.P., Francis, J.M.A., Sahagún, F.B., Stalnikowitz, D.K., Eds.; McGraw-Hill Education: New York, NY, USA, 2015.
35. Toklu, E. A new mathematical model of peristaltic flow on esophageal bolus transport. *Sci. Res. Essays* **2011**, *6*, 6606–6614. [CrossRef]
36. Sarna, S.K. Giant migrating contractions and their myoelectric correlates in the small intestine. *Am. J. Physiol.-Gastrointest. Liver Physiol.* **1987**, *253*, G697–G705. [CrossRef]
37. Painter, N.S.; Truelove, S.C. The intraluminal pressure patterns in diverticulosis of the colon Part I Resting patterns of pressure. *Gut* **1964**, *5*, 201–207. [CrossRef]
38. Chen, J.H.; Yu, Y.; Yang, Z.; Yu, W.Z.; Chen, W.L.; Yu, H.; Kim, M.J.M.; Huang, M.; Tan, S.; Luo, H.; et al. Intraluminal pressure patterns in the human colon assessed by high-resolution manometry. *Sci. Rep.* **2017**, *7*, 1–16. [CrossRef]
39. Penn, R.; Ward, B.J.; Strande, L.; Maurer, M. Review of synthetic human faeces and faecal sludge for sanitation and wastewater research. *Water Res.* **2018**, *132*, 222–240. [CrossRef]
40. Sinnott, M.D.; Cleary, P.W.; Arkwright, J.W.; Dinning, P.G. Investigating the relationships between peristaltic contraction and fluid transport in the human colon using Smoothed Particle Hydrodynamics. *Comput. Biol. Med.* **2012**, *42*, 492–503. [CrossRef]
41. Bourgault, Y.; Ethier, M.; Leblanc, V.G. ESAIM: Mathematical Modelling and Numerical Analysis Simulation of Electrophysiological Waves with an Unstructured Finite Element Method *ESAIM Math. Model. Numer. Anal.* **2003**. *37*, 649–661. [CrossRef]
42. Heil, M. An efficient solver for the fully coupled solution of large-displacement fluid-structure interaction problems. *Comput. Methods Appl. Mech. Eng.* **2004**, *193*, 1–23. [CrossRef]
43. Olivella, X.; Agelet de Saracíbar, C. *Mecánica de Medios Continuos Para Ingenieros*; Universitat Politecnica de Catalunya: Barcelona, Spain, 2002; p. 329.
44. Salvi, S.; Gurioli, G.; De Giorgi, U.; Conteduca, V.; Tedaldi, G.; Calistri, D.; Casadio, V. Cell-free DNA as a diagnostic marker for cancer: Current insights. *OncoTargets Ther.* **2016**, *9*, 6549–6559. [CrossRef]
45. Jung, K.; Fleischhacker, M.; Rabien, A. Cell-free DNA in the blood as a solid tumor biomarker-A critical appraisal of the literature. *Clin. Chim. Acta* **2010**, *411*, 1611–1624. [CrossRef]
46. Hao, T.B.; Shi, W.; Shen, X.J.; Qi, J.; Wu, X.H.; Wu, Y.; Tang, Y.Y.; Ju, S.Q. Circulating cell-free DNA in serum as a biomarker for diagnosis and prognostic prediction of colorectal cancer. *Br. J. Cancer* **2014**, *111*, 1482–1489. [CrossRef]
47. Tanić, M.; Beck, S. Epigenome-wide association studies for cancer biomarker discovery in circulating cell-free DNA: Technical advances and challenges. *Curr. Opin. Genet. Dev.* **2017**, *42*, 48–55. [CrossRef]
48. Salomo, M.; Kegler, K.; Gutsche, C.; Reinmuth, J.; Skokow, W.; Kremer, F.; Hahn, U.; Struhalla, M. The elastic properties of single double-stranded DNA chains of different lengths as measured with optical tweezers. *Colloid Polym. Sci.* **2006**, *284*, 1325–1331. [CrossRef]
49. Venema, K. The TNO In Vitro Model of the Colon (TIM-2). In *The Impact of Food Bioactives on Health*; Springer International Publishing: Cham, Switzerland, 2015; pp. 293–304.
50. Wiśniewski, J.R.; Ostasiewicz, P.; Duś, K.; Zielińska, D.F.; Gnad, F.; Mann, M. Extensive quantitative remodeling of the proteome between normal colon tissue and adenocarcinoma. *Mol. Syst. Biol.* **2012**, *8*, 611. [CrossRef]
51. Collins, F.S.; Lander, E.S.; Rogers, J. International human genome sequencing consortium, finishing the euchromatic sequence of the human genome. *Nature* **2004**, *431*, 931–945. [CrossRef]
52. Martínez-Martín, D.; Fläschner, G.; Gaub, B.; Martin, S.; Newton, R.; Beerli, C.; Mercer, J.; Gerber, C.; Müller, D.J. Inertial picobalance reveals fast mass fluctuations in mammalian cells. *Nature* **2017**, *550*, 500–505. [CrossRef]
53. Fonseca, J.C.; Marques, J.C.; Paiva, A.A.; Freitas, A.M.; Madeira, V.M.; Jørgensen, S.E. Nuclear DNA in the determination of weighing factors to estimate exergy from organisms biomass. *Ecol. Model.* **2000**, *126*, 179–189. [CrossRef]

Review

Glycan-Based Electrochemical Biosensors: Promising Tools for the Detection of Infectious Diseases and Cancer Biomarkers

Danilo Echeverri and Jahir Orozco *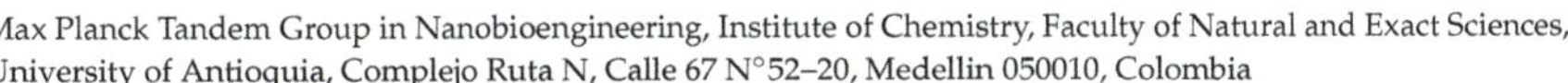

Max Planck Tandem Group in Nanobioengineering, Institute of Chemistry, Faculty of Natural and Exact Sciences, University of Antioquia, Complejo Ruta N, Calle 67 N°52–20, Medellin 050010, Colombia
* Correspondence: grupotandem.nanobioe@udea.edu.co

Abstract: Glycan-based electrochemical biosensors are emerging as analytical tools for determining multiple molecular targets relevant to diagnosing infectious diseases and detecting cancer biomarkers. These biosensors allow for the detection of target analytes at ultra-low concentrations, which is mandatory for early disease diagnosis. Nanostructure-decorated platforms have been demonstrated to enhance the analytical performance of electrochemical biosensors. In addition, glycans anchored to electrode platforms as bioreceptors exhibit high specificity toward biomarker detection. Both attributes offer a synergy that allows ultrasensitive detection of molecular targets of clinical interest. In this context, we review recent advances in electrochemical glycobiosensors for detecting infectious diseases and cancer biomarkers focused on colorectal cancer. We also describe general aspects of structural glycobiology, definitions, and classification of electrochemical biosensors and discuss relevant works on electrochemical glycobiosensors in the last ten years. Finally, we summarize the advances in electrochemical glycobiosensors and comment on some challenges and limitations needed to advance toward real clinical applications of these devices.

Keywords: glycan; electrochemical glycobiosensor; infectious disease; cancer biomarker

Citation: Echeverri, D.; Orozco, J. Glycan-Based Electrochemical Biosensors: Promising Tools for the Detection of Infectious Diseases and Cancer Biomarkers. *Molecules* **2022**, *27*, 8533. https://doi.org/10.3390/molecules27238533

Academic Editor: Lothar Elling

Received: 17 October 2022
Accepted: 24 November 2022
Published: 3 December 2022

Publisher's Note: MDPI stays neutral with regard to jurisdictional claims in published maps and institutional affiliations.

1. Introduction

Glycan-based electrochemical biosensors are analytical devices that use glycans as biorecognition elements immobilized on electrode surfaces to detect a target molecule electrochemically. Furthermore, glycan-based biosensors include biosensing platforms for detecting glycan-based biomarkers [1]. Electrochemical biosensors offer a valuable alternative for biomarker detection with high specificity, sensitivity, low cost, and the possibility of implementation in decentralized settings. In addition, they can be user-friendly, portable, amenable to miniaturization, and cost-affordable [2,3].

Glycans are carbohydrate molecules in free form or attached to other molecules such as proteins and lipids [4]. Glycans play a fundamental biological role because they participate in cell signaling and cell–cell adhesion, provide specific receptors for microorganisms, toxins, or antibodies, and regulate protein functions glycosylation dependently [5,6]. In addition, glycans are involved in cell growth and development, host–pathogen interactions and the progress of infections, immune recognition/response, tumor growth, and metastasis [5–7]. All these characteristics make glycans useful for the diagnosis/prognosis of diseases [8]. For example, glycans can be used as bioreceptors in sensing platforms for the detection of infectious diseases and as cancer biomarkers, including glycolipids such as glycosylphosphatidylinositol (GPI) for the diagnosis of parasites [9,10] and glycoproteins such as the carcinoembryonic antigen (CEA) and the carbohydrate antigen (CA 19-9) as biomarkers for the diagnosis of colon cancer [11,12]. In addition, other glycoproteins, such as the β-1,4-galactosyltransferase-V, have been reported as novel biomarkers of colon cancer [13–15].

Currently, the clinical diagnosis of some infectious diseases and cancer uses serological tests for biomarkers detection, including immunoassays [16,17] and chromatographic-based methods [18,19]. However, these methods have limitations, including laborious experimental procedures, limited multiplexing options, the need for sophisticated and centralized laboratory equipment, and skilled personnel [20,21]. In this context, glycan-based biosensors offer an alternative for biomarker detection in the clinical diagnosis of diseases, with high specificity, sensitivity, rapid response, low cost, and opportunity for miniaturization and portability [1,22]. These attributes make glycan-based biosensors ideal candidates for device-based disease diagnosis close to the patient, for example, in low- and middle-income settings with limited resources. Furthermore, they offer the advantage of short assay time, high portability, and multiplexing ability, enabling the possibility of implementation at the point-of-care (POC) [23].

This work aimed to review glycan-based electrochemical biosensors for detecting infectious diseases and cancer biomarkers. First, relevant aspects of structural glycobiology and the concepts of electrochemical glycobiosensors are briefly described. In addition, representative electrochemical glycobiosensors for detecting infectious diseases and cancer biomarkers reported in the last ten years are discussed to provide a more comprehensive background. This work contributes to the state of the art of glycan-based biosensors, especially toward their application in the clinical diagnosis of infectious diseases and detection of cancer biomarkers, with a particular focus on colorectal cancer and emphasis on electrochemical impedance spectroscopy (EIS) and electrochemical capacitive spectroscopy (ECS) transduction.

2. Structural Glycobiology

Glycans are carbohydrate compounds (polyhydroxy -aldehydes or -ketones) in their free form or monosaccharide units covalently linked by glycosidic bonds in the form of oligosaccharides and polysaccharides. In addition, glycoconjugate refers generically to any carbohydrate or assembly of carbohydrates covalently attached to another molecule (mainly proteins and lipids) [24].

Both eukaryotic and prokaryotic cells synthesize glycoconjugates, including glycoproteins, GPI-anchored glycoproteins, proteoglycans, and glycosphingolipids [25]. Glycoproteins are the main class of glycoconjugates [26] and consist of polypeptides that have glycans covalently linked to asparagine and serine/threonine residues (N-glycans and O-glycans, respectively) and through C-mannosylation, where a covalent bond between carbon one of the mannose and carbon two of the indole ring of tryptophan is formed [5].

GPIs are glycolipids with a conserved core structure of phosphatidylinositol-lipid linked to no-acetylated N-acetyl glucosamine and three mannose residues followed by an ethanolamine [27]. Generally, GPIs have the function of anchoring proteins on the cell membrane in eukaryotes and archaea; additionally, non-protein (linked free GPIs) are abundant on the surface of several protozoan parasites, such as *Trypanosoma brucei*, *Plasmodium falciparum*, and *Toxoplasma gondii* [5]. In addition, GPI-anchored proteins participate in cell signaling and adhesion and are related to health and disease processes [28].

Proteoglycans are macromolecules that consist of a protein backbone to which glycosaminoglycan chains (GAGs) and N- and or O-linked oligosaccharides are covalently attached. GAGs are linear and negatively charged polysaccharides composed of repeating disaccharides of acetylated hexosamines (N-acetyl-galactosamine or N-acetyl-glucosamine) and mainly by uronic acids (D-glucuronic acid or L-iduronic acid) sulfonated at various positions [29]. Ceramides (Cer) are the core structure of sphingolipids composed of a fatty acid linked by an amide bond to the unsaturated amino alcohol sphingosine. In addition, ceramides may have glycans attached and other polar phosphate-containing head groups [30]. Proteoglycans and sphingolipids are involved in cell signaling and, therefore, are related to the development of different human diseases [29,31–33].

2.1. Biological Functions of Glycans

The chemical diversity and structural complexity of glycans imply that they have diverse biological functions [5]. For example, glycans participate in cell–cell adhesion and cell signaling, provide specific receptors for microorganisms, toxins, or antibodies, and modulate protein functions in a glycosylation-dependent manner [7]. Furthermore, they participate in the folding of many proteins and protein trafficking, differentiate blood groups, participate in different signaling pathways, and play a fundamental role in the infectivity process of many pathogenic bacteria and viruses [6].

2.2. Principles of Glycan Recognition

Glycans can interact with different types of receptors, including biomolecules such as proteins (antibodies and lectins) [13,14]; aptamers, which are short, single-stranded oligonucleotides (DNA or RNA) [15]; and other smaller synthetic receptors such as boronic acids [16]. Overall, lectin–glycan interactions and antibody– and aptamer–carbohydrate complexes are held together by hydrogen bonds, CH-π, Van der Waals, and electrostatic interactions [17]. For example, it has been reported that antibodies can interact with glycans via tryptophan residues of antibodies and OH groups of glycans mediated by OH-π and CH-π interactions. In addition, hydrogen bonds among mainly polarized residues of aspartate, histidine, lysine, and threonine of antibodies with the mannose of the glycans also contribute to their recognition specificities [34]. On the other hand, in the case of nucleic acids, highly polar glycans can interact with oligonucleotides by stacking on cytosine–guanine base pairs through CH-π interactions. In addition, when hydrophobic contacts are available, apolar glycans can interact with loop DNA bases through hydrogen bonding [35]. In summary, the predominant type of interaction will depend on the structure of the glycan and the receptor to which it binds. These interactions between glycans and other biomolecules establish the principle of biomolecular recognition events on sensing surfaces that are converted into a readout signal by the transducer. For this reason, it is mandatory to know the type of biomolecular interactions between the bioreceptor and analyte to design the sensing surface.

2.3. Glycan-Based Biomarkers for Diagnosis of Diseases

According to the International Program on Chemical Safety, a biomarker is any substance, structure, or process that can be measured in the body or its products and influence or predict the incidence of outcome or disease [36]. As mentioned above, glycans are biomolecules regulating human physiology and pathology, including cell signal transduction and microbial infections. Therefore, diagnosing infectious diseases and cancer often uses glycan-based biomarkers [8].

2.3.1. Infectious Diseases

Identifying pathogens that cause infections, such as viruses, bacteria, fungi, and protozoa, is possible by detecting glycan-based biomarkers. Glycans are present on the outermost surface of viruses, mainly in glycoproteins [37]. Therefore, detecting structural glycoproteins on viruses, or quantifying titers of antibodies in the host against viral antigenic glycoproteins, allows the diagnosis of infections by viruses [37]. Some examples of virus determination based on glycoprotein detection are coronaviruses by detecting Spike glycoprotein [38,39], E1 and E2 glycoproteins in hepatitis C and Chikungunya viruses, and NS1 glycoprotein in dengue virus [37]. Like viruses, an array of glycans covers the bacteria cells, comprising their cell wall [24]. In addition, there are sugars with a limited expression on pathogenic bacteria; for example, *Neisseria meningitides* utilizes 2,4-diacetamido-2,4,6-trideoxyhexose, *Pseudomonas aeruginosa* installs N-acetylfucosamine residues, and *Bacteroides fragilis* appends 2-acetamido-4-amino-2,4,6-trideoxy-galactose into its cell surface polysaccharides. These are specific sugars of pathogenic bacteria and are promising biomarkers for their detection in clinical diagnosis [40]. Moreover, pathogenic fungi determination is also possible by detecting glycan on the cell surface. Some examples of pathogen fungi diagnosis based on glycan-type antigens include the detection of the antigen mannan in fungi of the genus *Candida* and *Cryptococcus* and galactomannan in *Aspergillus* [41].

On the other hand, an alternative for infectious disease diagnosis is glycan microarrays. Glycan microarrays are arrangements of multiple glycans, or glycoconjugates, immobilized onto a solid phase platform for screening with glycan-binding proteins (GBPs, also known as lectins), antibodies, bacteria, viruses, and other microorganisms. The GBPs or microorganisms are either directly fluorescently labeled or labeled with a tag such as biotin that can be indirectly detected [42]. Glycan arrays are a tool to elucidate carbohydrate interactions with different GBPs, including soluble proteins such as immune toxins, lectins, and microbial and mammalian surface receptors [43]. Furthermore, serum antibodies and human GBPs directed against cell surface glycans have been applied for the detection of diverse pathogens, including protozoa such as *Toxoplasma gondii* and *Plasmodium falciparum*, and for a wide range of pathogenic viruses and bacteria such as influenza viruses, human immunodeficiency virus (HIV), *Salmonella* and *Burkholderia pseudomallei*, respectively [43,44].

2.3.2. Cancer Biomarkers

Glycans can be used as biomarkers in cancer since they have different biological functions, especially in cell signaling pathways and post-translational protein modifications, which are related to the development and progression of this disease [45]. In this context, glycoproteins such as CEA and CA 19-9 have been used conventionally as biomarkers for cancer diagnosis, especially colon cancer [46–48]. CEA is a GPI-cell surface-anchored glycoprotein whose specialized sialofucosylated glycoforms serve as functional colon carcinoma L-selectin and E-selectin ligands, which may be critical to the metastatic dissemination of colon carcinoma cells [49]. CA 19-9 antigen is a tetrasaccharide carbohydrate termed sialyl Lewis-a, synthesized by gastrointestinal epithelium and overexpressed in colorectal cancer [50]. In addition, other glycoproteins, such as sialic-acid-containing glycoproteins are involved in cancer initiation, progression, and metastasis and are used as biomarkers for the disease diagnosis/prognosis [51–53].

Other biomarkers for cancer diagnosis related to glycan-type molecules are the enzymes involved in glycosylation and their reaction products. Glycosylation is the enzymatic process that produces glycosidic linkages of saccharides to other saccharides, proteins, or lipids [45]. A large family of enzymes called glycosyltransferases synthesizes the carbohydrate motifs of glycoconjugates. These enzymes catalyze glycosidic bond formation using sugar donors containing a nucleoside phosphate or a lipid phosphate leaving group [54,55]. Alterations in protein glycosylation are among the main molecular events accompanying oncogenic transformations in the gastric and colorectal tracts [56]. In addition, there is a report of increased glycosphingolipid levels due to aberrant glycosylation and metabolism in colorectal cancer [57]. It is relevant to emphasize that β-1,4-galactosyltransferase-V (β-1,4-Galt-V) catalyzes the glycosylation of glucosylceramide and the N-acetylglucosamine β-1-6 mannose group of the highly branched N-glycans, which are overexpressed in colorectal tumor cells [58]. This evidence indicates that β-1,4-GalT-V may serve as a diagnostic biomarker for the progression of human colorectal cancer [13].

3. Electrochemical Biosensors for Biomarker Detection

Electrochemical biosensing enables the detection of different analytes with high sensitivity; the equipment is simple, affordable, and amenable to miniaturization; and the electrode surface chemistry can adapt to specific applications [22,59]. Electrochemical biosensors use several electroanalytical techniques, including voltammetric techniques—such as cyclic voltammetry (CV), differential pulse voltammetry (DPV), square wave voltammetry (SWV), and amperometry—as well as potentiometric, conductometric, and spectroscopic techniques such as EIS and ECS [22,60,61]. Regarding analytical performance, EIS- and ECS-based glycan biosensors are highly sensitive compared with voltammetric and amperometric biosensors. Therefore, they are promising techniques for developing devices for detecting analytes at ultralow concentrations [62].

3.1. Definitions and Classification of Electrochemical Biosensors

An electrochemical biosensor is a self-contained integrated device that can provide specific quantitative or semi-quantitative analytical information of molecular recognition events into an analytically valuable signal using a biological recognition element (bioreceptor) that is retained in direct spatial contact with an electrochemical transduction element (Figure 1A) [47,48]. The main function of the transducer is to convert a molecular biorecognition event into a measurable signal proportional to the analyte concentration [49]. According to the type of bioreceptor, there are biosensors based mainly on whole cells, enzymes, antigens, antibodies, nucleic acids, aptamers, lectins, and glycans [49–51]. The bioreceptor is the element that confers specificity and selectivity to the biosensor, which are relevant characteristics of these devices, and refers to the ability to detect a specific analyte in a mixture that contains interferences [50] and differentiate the target from homologous counterparts.

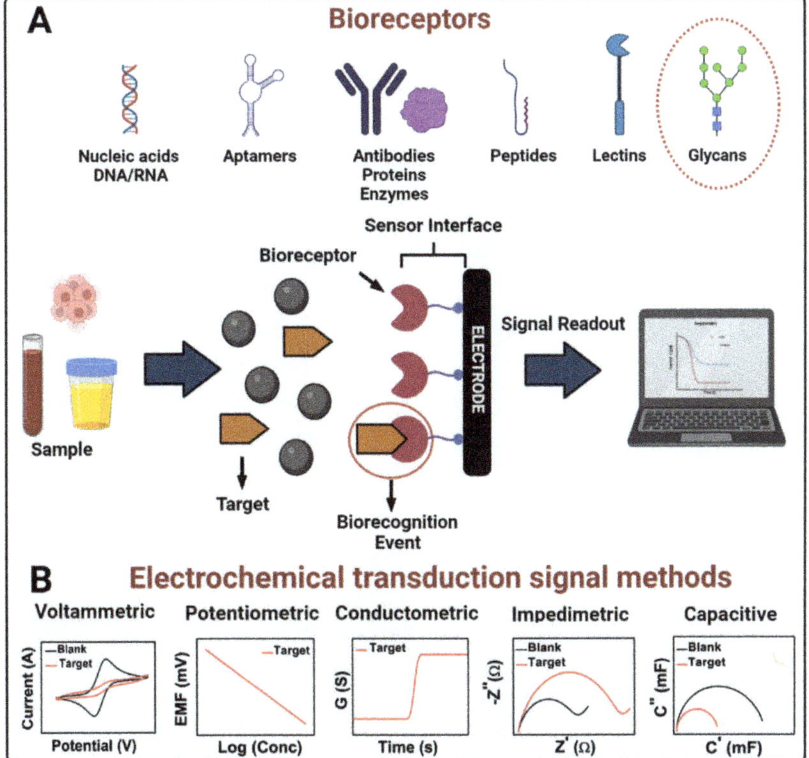

Figure 1. (**A**) Scheme of a generic electrochemical biosensor. A bioreceptor (nucleic acids, aptamers, antibodies, proteins, enzymes, peptides, lectins, glycans, etc.) attached to the electrode surface recruits the molecular target (analyte present in a sample) onto the sensor interface by an affinity reaction. After the bioreceptor binding with the target (biorecognition event), the transducer converts the binding event into a measurable signal proportional to the concentration of the target (signal readout). (**B**) Electrochemical transduction signal methods in biosensors: voltammetric, potentiometric, conductometric, impedimetric, and capacitive. Adapted from [22] with permission. Copyright Elsevier 2022.

Other characteristics of biosensors are sensitivity, linear range, reproducibility, and stability. Sensitivity is the slope of the calibration curve and is related to the limit of detection (LOD). The LOD is the lowest concentration of an analyte in a sample that can be detected, with reasonable certainty, for a given analytical procedure [63]. The linear range is the concentration range over which the signal output is directly proportional to the concentration of the analyte and is often correlated with a straight line [64]. Reproducibility is the closeness and agreement between independent results obtained with the same method on identical test material but under different conditions (different operators, apparatus, laboratories, or time intervals) [63]. Finally, stability is the degree of biosensor susceptibility to ambient disturbances. One way to assess the stability is by continuously or sequentially performing biosensor exposure to analyte solution or by measuring the change in the baseline or sensitivity over a fixed period [65,66]. These characteristics make electrochemical biosensors affordable, accurate, rapid, and sensitive analytical platforms for detecting multiple disease biomarkers [67].

A convenient classification of electrochemical biosensors is according to the mode of signal transduction, as mentioned in Section 3.1 and Figure 1B [22,68]. Moreover, another classification of electrochemical biosensors is according to the bioreceptor type that recognizes the analyte [68]. Therefore, biosensors based on antibodies, or fragments of these, are affinity biosensors (immunosensors); enzyme-based biosensors use enzymes as bioreceptors; genosensors use nucleic acids; aptasensors use aptamers; glycobiosensors use lectins or glycans; and cytosensors whole cells [1,22]. Since this work focused on developing electrochemical biosensors, and we will tackle the glycan-based biosensors in Section 2.3, the following classification zooms in according to the mode of signal transduction.

In summary, electroanalytical methods allow the precise determination of multiple analytes with high sensitivity and fast response. Table 1 helps sort out electrochemical glycobiosensors by comparing electroanalytical methods based on analytical performance and the detection principle.

Table 1. Table of electrochemical glycobiosensors and analytical performance.

Electrochemical Technique	Sensing Principle	The Typical Range of the Limit of Detection (LOD)	Reference
CV	Application of a time-dependent potential to an electrochemical cell and measuring the resultant current as a function of the applied potential	10^{-6}–10^{-15} M	[69]
DPV			
SWV			
Amperometry			
Potentiometry	Perturbation of the potential of the electrochemical cell	10^{-3}–10^{-6} M	[59]
Conductometry	Quantification of the conductance change in the electrochemical cell		
EIS	Application of a small sinusoidal voltage perturbation in a range of frequencies while monitoring the resulting current	10^{-9}–10^{-18} M	[22]
ECS			

Abbreviations: CV: cyclic voltammetry. DPV: differential pulse voltammetry. ECS: electrochemical capacitance spectroscopy. EIS: electrochemical capacitance spectroscopy. SWV: square wave voltammetry.

3.2. Characterization of Electrochemical Glycobiosensors

The electrode surface of electrochemical glycobiosensors is usually characterized in terms of surface chemistry, morphology, and electrochemical performance. Atomic force microscopy (AFM) is a technique used to characterize the electrode surface morphology in glycobiosensors; it can operate in multiple modes, such as electrochemical AFM, which enables the analysis of electrochemical reactions occurring at the electrode [70]. Morphology and surface chemical composition are characterized via scanning electron microscopy (SEM) coupled with energy-dispersive X-ray spectroscopy (EDX) [71,72]. Furthermore, infrared (FT-IR) and X-ray photoelectron spectroscopic (XPS) techniques are used to analyze surface chemical composition [73]. In addition, thermogravimetric analysis (TGA) is applied to analyze the materials used to modify the electrode surface and implies studying sample mass change under programmed conditions. Therefore, TGA is mainly used to analyze certain thermal events, such as absorption, adsorption, desorption, vaporization, sublimation, decomposition, oxidation, and reduction [74].

Similarly, differential scanning calorimetry (DSC) measures the amount of energy absorbed or released by the sample when it is heated or cooled. TGA is a versatile technique used to study the self-assembly of supramolecular nanostructures such as glycopolymers, latent heat of melting, denaturalization temperatures, and compositional analysis [75]. On the other hand, the optical properties are mainly characterized via ultraviolet-visible (UV-vis) spectroscopy and photoluminescence (PL) techniques [76]. Finally, electrochemical performance is characterized via voltammetric techniques such as CV and DPV and spectroscopic techniques such as EIS and ECS [77,78], as commented on in Section 3.2. The affinity constants of glycan-based biorecognition elements can be determined using the Biacore system [79]. Biacore uses surface plasmon resonance (SPR) as a label-free detection technique to monitor the interaction between biomolecules in real-time. The biorecognition molecule is immobilized on a sensor chip's surface. At the same time, the sample containing its ligand is injected over the surface at a constant flow rate through a microfluidic channel system. The changes in mass concentrations at the surface of the sensor chip due to molecule association/dissociation are measured as an SPR response and displayed as a time function [80]. Table 2 shows the most common characterization techniques of electrochemical glycobiosensors.

Table 2. Characterization techniques of electrochemical glycobiosensors.

Characterization Technique	Properties	Technique Principle	References
AFM	Morphology	Measurement of intermolecular forces and "seeing" atoms by using probe surfaces.	[70]
SEM/EDX	Morphology Composition	Application of kinetic energy to produce signals from the interaction of the electrons (secondary, backscattered, and diffracted backscattered). Secondary and backscattered electrons are used to visualize the morphology, and backscattered are related to composition.	[71,72]
FT-IR	Surface chemical composition	Measurement of the vibrations of atoms, and from this, functional groups are determined.	[73]
XPS	Surface chemical composition	The sample is irradiated with an X-ray, and some electrons become excited enough to escape from the atoms. The photo-ejected electrons are collected by an electron analyzer that measures their kinetic energy, allowing the element to be identified.	[73]
TGA, DSC	Sorption Composition	The sample is heated or cooled under controlled conditions and changes in some physical properties are measured.	[74,75]
UV-vis, PL	Optical	Light absorption and scattering by a sample.	[76]
CV, DPV, EIS, ECS	Electron transfer kinetics	Perturbation of the electrode by applying an electric potential and recording the resulting current.	[77,78]
Biacore	Bioreceptors affinity	The change in SPR response is measured after association/dissociation of a bioreceptor and ligand, respectively, with the sample flow in a microfluidic channel.	[80]

Abbreviations: AFM: atomic force microscopy. CV: cyclic voltammetry. DPV: differential pulse voltammetry. DSC: differential scanning calorimetry. ECS: electrochemical capacitance spectroscopy. EIS: electrochemical impedance spectroscopy. EDX: energy-dispersive X-ray spectroscopy. FT-IR: Fourier transform infrared spectroscopy. PL: photoluminescence. SEM: scanning electron microscopy. TGA: thermogravimetric analysis. UV-vis: ultraviolet-visible spectroscopy. XPS: X-ray photoelectron spectroscopy.

3.3. Nanostructured Electrochemical Glycobiosensors

Nanostructured biosensors are analytical devices that integrate nano- and bio-materials platforms for trace detection of biomolecules or chemical analytes [81]. Due to their size-dependent properties, such as a large surface area with improved conductivity and reactivity, nanomaterials are used for developing highly sensitive biosensors [82]. As a result, a wide range of nanomaterials has been incorporated onto the electrode surface to improve the biosensor analytical performance, such as carbon-based nanomaterials, noble metals, metal oxides, metal chalcogenides, magnetic nanoparticles, and conductive polymers, among others [83].

Likewise, nanomaterials are similar in size to most biological entities such as proteins, nucleic acids, lipids, cells, viruses, glycans, etc., making them ideal interfaces between these entities and signal transduction surfaces as those used in biosensors [84–87]. Furthermore, stability, biocompatibility, and the advantage of modulating the nanomaterial's surface chemistry make them suitable for conjugating multiple chemical species and biomolecules [83]. One general advantage of all nanomaterials is the high specific surface area that enables a high surface loading of biorecognition elements on the electrode surface and their resultant improved electron transfer and electrocatalytic activity ability [83,88]. Their combination with suitable bioreceptors such as glycans could originate synergistic effects eliciting unforeseen benefits [89]. Therefore, an essential issue for nanobiosensor development is the size, structure, chemical composition, shape, and nanomaterial's surface modification [87].

3.3.1. Synthesis of Nanomaterials and Surface Biofunctionalization

There are different methods to synthesize nanomaterials depending on their type and nature. In summary, the two main methods to synthesize nanomaterials are "top-down" and "bottom-up" approaches [88]. In the top-down approach, the nanomaterial synthesis uses the size reduction from bulk materials down to the nanoscale. Unlike the top-down method, the bottom-up synthesis of nanomaterials consists of obtaining nanostructures from elementary-level building blocks of atomic or molecular size [88]. The most common methods used to synthesize nanomaterials are the chemical vapor deposition method, thermal decomposition, hydrothermal synthesis, solvothermal method, pulsed laser ablation, templating method, combustion method, microwave synthesis, gas phase method, and conventional sol–gel method [88].

As mentioned above, glycan-based biorecognition elements confer specificity and selectivity to glycobiosensor devices, recognizing the target analyte and binding it to the sensor surface for transduction [66]. Nanomaterials are supporting platforms for glycan-based bioreceptors attachment by physical and chemical methods [90] and enhancing analytical performance [91]. Such bioreceptors are immobilized to the nanomaterials by physical methods without chemical bond formation through physical entrapment, microencapsulation, adsorption, and sol–gel techniques [92]. Unlike physical methods, chemical approaches form covalent bonds in the presence of two mutually reactive chemical groups from the bioreceptors and the substrate surface [90,93]. One of the more common approaches involves amide bond formation in the presence of carboxylic acids (-COOH) and primary amines ($-NH_2$). This approach requires activation of -COOH with 1-ethyl-3-(3-dimethylaminopropyl) carbodiimide (EDC) and N-hydroxysuccinimide (NHS) or sulfo-NHS, producing esters, which react with primary amines to form amides. Another approach couples the bioreceptor via the free thiols (-SH), which react stoichiometrically with maleimides through a Michael addition reaction [93]. The glycans also can be functionalized with thiols and disulfide-bearing linkers for the direct formation of self-assembled monolayers (SAMs) on metallic surfaces, i.e., gold. Alternatively, some linkers can be pre-assembled on the transducer surface for subsequent attachment of glycans via reactive terminal groups [94,95].

Bioreceptors, like glycans, are also immobilized onto conducting polymers with redox properties, such as polystyrene sulfonate, polyvinyl ferrocene, polythiophene, polyaniline, and quinone polymers, which results in an enhancement of the biosensor electrochemical performance [96]. Furthermore, there are immobilization strategies based on affinity interactions, such as the biotin–avidin interaction or immobilizing binding proteins such as Protein A or G onto the electrode surface, followed by the subsequent capture of antibodies and blocking of the nonspecific adsorption sites steps with bovine serum albumin (BSA), casein, or other blocking agents [90,93,97–100].

3.3.2. Operation Modes of Electrochemical Nanobiosensors

Electrochemical biosensors that integrate nanostructures and suitable biorecognition elements on the electrode surface are highly sensitive and specific. These features enable the detection and quantification of disease biomarkers at ultra-low concentrations, which is a requisite for the early diagnosis of diseases [88,101]. On the other hand, regarding the detection of molecular biorecognition events on the electrode surfaces, various labeling strategies are used to amplify the detection signal on electrochemical biosensors. Label-based approaches can involve avidin–biotin conjugation with redox enzymes, covalent attachment, intercalation, or electrostatic interaction of small molecules, particles, or ions with the biorecognition elements responsible for generating the electrochemical signal. In contrast, label-free biosensors directly transduce a molecular binding event into a physically measurable quantity, i.e., without needing an additional antibody, enzymatic, fluorescent, or electroactive label, or any other amplification strategy, to provide a response that is proportional to the concentration of bound molecules [102]. Label-free electrochemical biosensors measure interfacial electrical property changes, such as charge transfer resistance or electrochemical capacitance, through the EIS or the ECS techniques and by measuring current changes via CV, DPV, or SWV [22,77]. Figure 2 shows a label-free and label-based nanobiosensors setup.

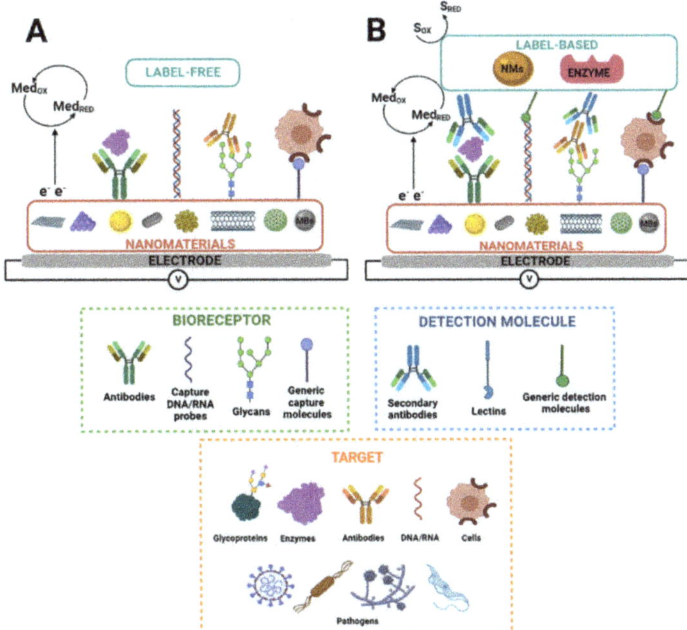

Figure 2. Scheme of nanostructured electrochemical biosensors. S$_{RED}$ and S$_{OX}$ and Med$_{RED}$ and Med$_{OX}$ indicate substrate reduction and oxidation and mediator reduction and oxidation, respectively. MBs indicate magnetic beads. (**A**) Label-free and (**B**) label-based nanobiosensors set up. Adapted from [83].

In summary, electrochemical nanobiosensors based on glycans offer exceptional attributes for disease diagnosis, such as being affordable, sensitive, specific, user-friendly, portable, rapid, robust, simple to construct, equipment-free, and deliverable to all people in need [2,103]. In addition, nanostructured electrodes provide a large surface area to immobilize a high load of bioreceptors on the electrode surface and enhance the electron transfer and electrocatalytic activity ability, resulting in biosensors of enhanced analytical performance [82].

3.3.3. Glycans as Biorecognition Elements in Electrochemical Biosensors

A myriad of bioreceptors, such as antibodies, nucleic acids, aptamers, peptides, enzymes, etc., have been used in biosensors as biorecognition elements for detecting multiple biomarkers [68,104–106]. However, the application of glycans as bioreceptors is less explored in biosensor platforms to develop new diagnosis/prognosis tests [107].

The most common types of electrochemical glycobiosensors use lectins as selective biorecognition elements. Lectins are natural proteins that recognize and reversibly bind to specific free carbohydrates and terminal groups on glycans of glycoconjugates [59,67,108,109]. Lectin-based biosensors have been used for detecting CEA, P-glycoprotein, prostate-specific antigen (PSA), and viruses [67,110,111].

On the other hand, there are some applications where glycans are attached to the electrode surface as biorecognition elements [62]. In two of these applications, a glycan sialyllactose was immobilized on SAM-modified gold electrodes via amine coupling to detect proteins present on the envelope of influenza viruses [112,113]. Similarly, another application uses a mannose glycan immobilized on SAM-modified gold electrodes for bacteria detection [114]. Furthermore, there is a report for the immobilization of Tn antigen (N-galactosamine attached to serine) on SAM-modified gold electrodes for binding a tumor-associated antibody [115]. In addition, there are reports describing glycans immobilization with a built-in redox center. In these works, glycans were attached to quinone moieties and applied to detect intact bacterial and cancerous cells using graphene-modified electrodes [116–118].

In summary, glycan immobilization on the electrode surface depends on the glycan structure and the electrode surface chemistry. Glycans specifically recognize molecular targets and confine them on the electrode surface. The biorecognition molecular event changes the interfacial electrical properties, and the analyte concentration correlates with the changes in electrical properties measured by electrochemical techniques.

The bioreceptor affinity is inversely related to the dissociation constant K_D. It describes the binding strength between a bioreceptor, such as a lectin or an antibody, and its ligand [119]. K_D is in the nM–mM range for lectin–glycan interactions and pM-nM for oligonucleotide hybridization and antibody–antigen interactions [120–123]. Lectins have multivalence to recognize glycans, allowing significant affinity amplification and even reaching the subnanomolar range [121]. In addition, engineered glycomaterials allow for overcoming affinity and selectivity challenges in glycan-based molecular binding events achieving affinity in the picomolar range [124]. For this reason, glycans are promising bioreceptors for electrochemical biosensors, as mentioned.

4. Electrochemical Glycobiosensors for Infectious Disease Diagnosis and Cancer Biomarker Detection

Regarding different applications of glycobiosensors, there are some reports in the literature, mainly for the clinical diagnosis of infectious diseases caused by pathogens (such as viruses and bacteria) and the detection of cancer biomarkers. Therefore, below, this section describes representative works on electrochemical glycan-based biosensors for pathogens and colorectal cancer biomarkers in the last ten years.

4.1. Glycobiosensors for Infectious Disease Diagnosis

Glycan derivatives have been applied to constructing glycan biosensors for virus detection. For example, Hushegyi et al. developed an impedimetric glycan biosensor for detecting lectins and influenza hemagglutinins (HAs) down to attomolar concentrations (aM). The biosensor was assembled onto modified gold electrodes with a mixed self-assembled monolayer of 11-mercaptoundecanoic acid (MUA) and 6-mercaptohexanol (MCH). Next, an amine-terminated glycan was coupled through EDC/NHS chemistry to form an amide covalent bond. As a result, a wide linear concentration range was obtained from attomolar to nanomolar for lectins and influenza HAs [112]. Similarly, Hushegyi et al. increased the selectivity of H3N2 influenza virus detection by the glycobiosensor using surface chemistry based on a mixed SAM composed of thiols bearing oligoethylene glycol (OEG) moieties resisting nonspecific interactions. The glycobiosensor was applied to detect H3N2 viruses, achieving a LOD of 13 viral particles in 1 µL, and was highly selective, enabling differentiation of H3N2 from the H7N7 influenza viruses [113]. Recently, *Soto* et al. developed an impedimetric peptide-based biosensor for detecting Spike protein, a SARS-CoV-2 envelope glycoprotein. The electrochemical biosensor was based on gold electrodes modified with a synthetic thiolated peptide bonded to Spike glycoprotein. The biosensor demonstrated a linear response of 0.05–1 µg mL^{-1} and a LOD of 18 ng mL^{-1}. The biosensor could differentiate between positive and negative nasopharyngeal swab samples concerning the controls and differentiate the viral loading in clinical samples [39]. Santos et al. developed a capacitive biosensor to detect nonstructural glycoprotein 1 (NS1) related to dengue virus infection. The capacitive biosensor consisted of gold electrodes modified with a mixed SAM composed of 11-(ferrocenyl)-undecanethiol (11Fc) and polyethylene glycol-thiol (PEG-SH). An antibody was used as a biorecognition element to capture the NS1 glycoprotein on the electrode surface. The capacitive biosensor detected the molecular recognition event by a decrease in redox capacitance (C_r) in a linear range from 1 to 5000 ng mL^{-1} and with a LOD of 340 pg mL^{-1} [125].

In the case of bacteria detection, Ma et al. reported polythiophene (PTPh) interface containing fused quinone moieties, which were then mannosylated to form a carbohydrate platform for *E. coli* detection. The bacteria detection was developed using Pili-Man and lipopolysaccharide (LPS)-ConA-Man binding approaches, being more sensitive than just the LPS-ConA-Man approach. Pili are multi-protein structures with adhesive properties related to the infectious ability of *E. coli* and LPS is the major component of the outer layer of the outer membrane of Gram-negative bacteria, such as *E. coli* and *Salmonella typhimurium* [126,127]. The electrochemical technique used to transduce the biorecognition event was SWV and demonstrated a LOD of 25 cells mL^{-1} and better selectivity and stability compared to the presently available technologies [128].

Cui et al. developed a label-free impedimetric glycobiosensor for quantitatively assessing interactions between pathogenic bacteria and mannose (Man). The sensing platform was based on gold electrodes modified with a SAM of Man/MUA/MCH and was applied to capture *E. coli* and *Salmonella typhimurium* bacteria. The sensing surface had a better binding affinity for *S. typhimurium* in a linear range of 50–1000 CFU mL^{-1} and LOD of 50 CFU mL^{-1} [129]. Similarly, Dechtrirat et al. detected GBPs and *E. coli* using an electrochemical displacement sensor based on ferrocene boronic acid as an electroactive reporter molecule and immobilized glycan. The sensor was based on gold electrodes modified with a SAM of thiolated-Man/OE conjugate and a ferrocene boronic acid (FcBA) pre-assembled as a reporter molecule onto the mannose surface. Upon the binding of GBP to the Man, the reporter molecule was displaced, and the decrease in SWV signal could be correlated to the GBP concentration. The sensor detected *E. coli* in a linear range from 6×10^2 to 6×10^5 cells mL^{-1} with a LOD of 6×10^2 cells mL^{-1}. In addition, it was highlighted that the sensor could complete a rapid analysis within 15 min [115]. Figure 3 shows a scheme of biosensing platforms for detecting viruses and bacteria mentioned above.

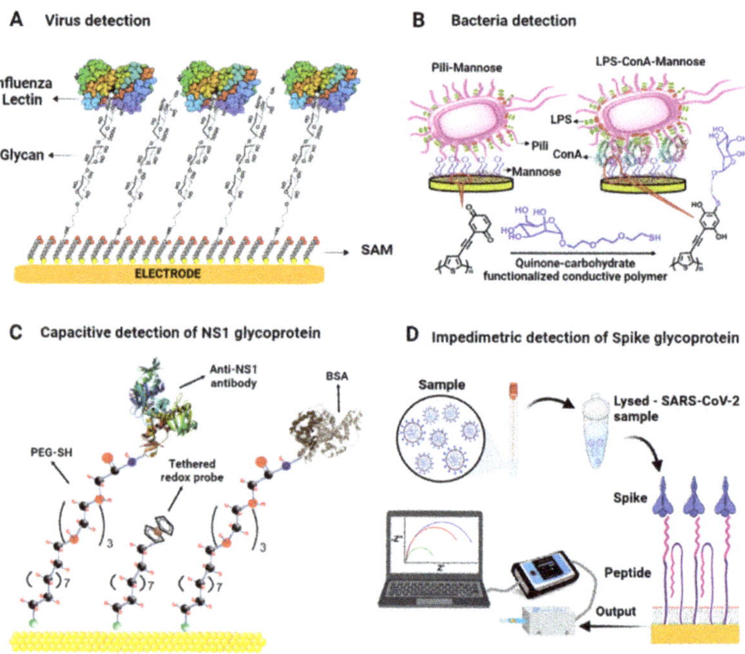

Figure 3. Scheme of glycobiosensors for pathogen detection. (**A**) Glycan-based biosensor for the detection of Influenza lectin [112]. (**B**) Biosensor based on redox-active conductive glycopolymer for *E. coli* detection. Adapted with permission from [128]. Copyright 2015 American Chemical Society. (**C**) Scheme of capacitive biosensing platform for detecting NS1 glycoprotein. Adapted from [125] with permission. Copyright Elsevier 2018. (**D**) Scheme of impedimetric biosensing platform for detecting Spike glycoprotein of SARS-CoV-2. Reproduced from [39] with permission. Copyright Elsevier 2022.

4.2. Glycobiosensors for Cancer Biomarker Detection

4.2.1. CEA and CA 19-9 Glycoproteins

CEA and CA-19-9 glycoproteins are cancer biomarkers detected using electrochemical glycobiosensors via label-free and label-based approaches. For label-free detection of CEA glycoprotein, Liu et al. synthesized a conducting polymer (poly (2-amino thiophenol), PATP) with incorporated Au nanoparticles (AuNPs), in which they adsorbed anti-CEA antibody for the sensitive and label-free electrochemical detection of CEA. The recognition of CEA was characterized using DPV with a linear range from 1 fg mL^{-1} to 10 ng mL^{-1} and LOD of 0.015 fg mL^{-1}. Moreover, the results of CEA detection in real serum samples were consistent with those determined by a conventional immunoassay, showing the practical utility of the biosensor [98]. Likewise, Zhao et al. developed an electrochemical biosensing platform to detect CEA that employed lectin as a bioreceptor anchored at AuNPs and enzymatic catalysis for signal amplification. The biosensor was made up of electrodeposition of AuNPs on screen-printed carbon electrodes and self-assembly of cysteamine (Cys) on top of their surface for subsequent covalent coupling of the lectins via the amidation reaction. The target protein was then captured by the lectin-modified platform and made to react with a horseradish peroxidase (HRP) labeled anti-CEA antibody bioconjugate in a sandwich-type assay. The transduction of this molecular recognition event was made via chronoamperometry in the presence of hydroquinone (HQ) and hydrogen peroxide, achieving a linear concentration range of CEA from 0.5 ng mL^{-1} to 7 ng mL^{-1}, with LOD of 0.01 ng mL^{-1} [130]. Figure 4A shows a schematic of the biosensing platform for detecting CEA glycoprotein.

Other researchers developed a polyaniline (PA) electrochemical derivative, a poly-(N,N′-diphenyl-p-phenylenediamine)-Au/Pt highly electroactive nanocomposite, as a platform to link the anti-CA 19-9 antibody for the label-free and sensitive detection of CA 19-9. The analytical performance of this immunosensor was tested via the SWV technique thanks to its intrinsic electrocatalytic activity, with H_2O_2 for signal recording. A wide linear range was obtained from 0.001 to 40 U mL^{-1} and ultralow LOD of 2.3×10^{-4} U mL^{-1}. In addition, the biosensor was used to analyze clinical serum samples. The results agreed with those from the ELISA standard analysis, suggesting the potential application of this immunosensor for the clinical diagnosis of this biomarker [131]. Furthermore, Thapa et al. developed a highly sensitive capacitive biosensor to detect CA 19-9 glycoprotein based on gold electrodes modified with polyethyleneimine (PEI) and CNTs with an anti-CA 19-9 antibody immobilized covalently on the surface. The biosensor could detect CA 19-9 glycoprotein in a linear range of 0.05–60 U mL^{-1} with a LOD of 0.35 U mL^{-1} [132]. Figure 4A shows a schematic of the biosensing platform for detecting CEA glycoprotein.

4.2.2. Protein Glycosylation
STn Antigen and Anti-STn Antibodies

The protein glycosylation can be altered in many diseases, including cancer. An example is the mucin-type O-glycoproteins that express the truncated glycans Thomsen-nouvelle (Tn) and sialyl-Tn (STn) [133]. Silva et al. developed an impedimetric biosensor based on the SNA I lectin immobilized on 16-mercaptohexadecanoic acid (MHDA) SAM-modified-gold electrodes to detect cancer-associated STn antigen. Lectin biosensor could detect transferrin, an STn-containing glycoprotein, in a linear range from 20 to 70 ng and LOD of 20 ng. The biosensor could detect the glycan from serum glycoproteins of cancer patients, and the complete assay took about 10 min [134]. Furthermore, Kveton et al. developed a glycobiosensor for detecting a Tn-associated antibody. The biosensor was built on an electrochemically activated/oxidized graphene screen-printed electrode (GSPE) for covalent attachment of human serum albumin (HSA). HSA acted as a natural nanoscaffold for covalent immobilization of Tn antigen to be fully available for affinity interaction with a tumor-associated antibody. The molecular binding of antibody and Tn antigen was monitored via DPV, achieving a linear range of 10 aM–10 pM and a LOD of 10 aM [135]. Figure 4B shows a schematic of the biosensing platform for detecting a tumor-associated antibody.

α2,3-Sialylated Glycans

Alteration of α2,3-sialylation is related to the development of certain cancers, in which sialylated glycoproteins can be released into the blood during apoptosis and detected circulating in serum [136]. There have been reports of some biosensors for detecting sialylated glycans useful for cancer diagnosis and clinical research. For example, Niu et al. used polyamidoamine (PAMAM) dendrimer conjugated with carboxyl-functionalized multiwalled carbon nanotubes (c-MWCNTs) for sensitive detecting of α2,3-, and α2,6-sialylated glycans in serum via DPV. 1,4-phenylene diisothiocyanate (PDITC) was used as a green homobifunctional cross-linker for SNA I and MAL lectin immobilization. Under optimal detection conditions, the linear range for α2,3-sialylated glycans was 10 fg mL^{-1}-50 ng mL^{-1} and for α2,6-sialylated glycans 10 fg mL^{-1}–50 ng mL^{-1}, respectively. The LOD was 3 fg mL^{-1} for both biosensors [137]. Similarly, for the determination of α2,3-sialylated glycans in serum, Niu et al. developed a biosensor using a GCE modified with c-MWCNTs, a PAMAM dendrimer, and the MAL lectin immobilized by the PDITC linker. The biosensor had a linear range of 10 fg mL^{-1}–50 ng mL^{-1} with a LOD of 3 fg mL^{-1}, and serum recovery experiments demonstrated high precision (around 100%) [138].

Yuan et al. assembled a sandwich-type biosensor for detecting α2,3-sialylated glycans based on fullerene–palladium–platinum alloy (n-C$_{60}$-PdPt) and 4-mercaptophenylboronic acid (4-MPBA) nanoparticle hybrids coupled with Au-polymethylene blue-MAL lectin (Au-PMB-MAL) signal amplification. The biosensor was fabricated via surface modification of GCE with amino-functionalized fullerene coupled with n-C$_{60}$-PdPt nanocrystals. The n-C_{60} nanomaterial had a large surface area for the on-site reduction of bimetallic alloy nanoparticles and high electron transfer. 4-MPBA were immobilized on the n-C_{60}-PdPt by chemisorption of the thiol group and captured the $\alpha2,3$-sialylated glycans by coordinating the boron atom of 4-MPBA to the amide group of Neu5Ac in the glycan structure. The MAL-Au-PMB nanocomposites recognized the $\alpha2,3$-sialylated glycans attached to the electrode surface, enabling signal amplification via DPV. The sandwich-type biosensor demonstrated a wide linear range of 10 fg mL^{-1}–100 ng mL^{-1} and a LOD of 3 fg mL^{-1}. Furthermore, the biosensor could detect $\alpha2,3$-sialylated glycans in serum samples, indicating its potential use in clinical applications [139].

α2,6-Sialylated Glycans

In addition, the alteration of $\alpha2,6$-sialylation is also related to cancer development. Apoptotic cells overexpress the $\alpha2,6$-sialylated glycans, which are released into the blood and can be detected in human serum [140]. Gao et al. fabricated an ultrasensitive electrochemical biosensor based on graphite oxide (GO), Prussian blue (PrB), and PTC-NH$_2$ (an ammonolysis product of 3,4,9,10-perylenetetracarboxylic dianhydride) nanocomposite for the selective detection of $\alpha2,6$-sialylated glycans. Glassy carbon electrodes (GCE) were modified with the nanocomposite and adsorbed AuNPs. An SNA I lectin was covalently immobilized onto the AuNPs, which detected $\alpha2,6$-sialylated glycans. The glycobiosensor was applied to detect $\alpha2,6$-sialylated glycans via DPV in human serum, and it worked well over a broad linear range (0.1 pg mL^{-1}–500 ng mL^{-1}) with a LOD of 0.03 pg mL^{-1} [141]. Afterward, Li et al. enhanced the SNA I lectin biosensor sensitivity toward detection in serum of $\alpha2,6$-sialylated glycans. The improved glycobiosensor was based on GCE modified with a nanocomposite of reduced graphene oxide-tetraethylene pentamine (rGO-TEPA) and 1-butyl-3-methylimidazolium hexafluorophosphate (BMIMPF$_6$). Bimetallic gold platinum alloy nanoparticles (AuPtNPs) were adsorbed on the nanocomposite surface, providing a large surface area for the immobilization of SNA I lectin. The lectin–glycan molecular biorecognition event was detected and quantified via amperometry in a linear range of 10 fg mL^{-1}–1 μg mL^{-1} and a LOD of 3 fg mL^{-1} [142].

Niu et al. developed an ultrasensitive dual-type responsive electrochemical biosensor for detecting $\alpha2,6$-sialylated glycans based on gold nanorods functionalized with streptavidin (AuNRs-SA) and SNA-biotinylated lectin that recognized the glycan and enabled detection via DPV. Carboxylated single-walled carbon nanohorns/sulfur-doped platinum nanocluster (c-SWCNHs/S-PtNC) was used as a signal label, showing excellent catalytic performance and allowing for the expansion of the ultrasensitive detection of $\alpha2,6$-sialylated glycans. The c-SWCNHs/S-PtNC was functionalized with phenylboronic acid bound to the glycan–lectin complex in a sandwich-type format and significantly amplified the electrochemical signal recorded by amperometry. The sandwich-type biosensor possessed a wide linear range from 1 fg mL^{-1} to 100 ng mL^{-1} with a LOD of 0.69 fg mL^{-1}. Furthermore, the biosensor exhibited excellent stability and recovery in serum samples, indicating its potential for use in clinical applications [143]. Zhao et al. developed a signal amplification system for the sensitive determination of $\alpha2,6$-sialylated glycans in serum based on gold electrodes modified with SNA I lectin, which captured the $\alpha2,6$-sialylated glycans and Fe-based metal–organic frameworks (Fe-MOFs) decorated with silver nanoparticles (AgNPs). Ag/Fe-MOFs nanocomposite was functionalized with SH-PEG-COOH to bind with 3-aminophenylboronic acid (M-APBA) via an amide bond, which recognized the $\alpha2,6$-sialylated glycans. Ag/Fe-MOFs nanocomposite exhibited excellent redox properties enabling amplification signal via DPV; the linear range was 1 fg mL^{-1}–1 ng mL^{-1} with a LOD of 0.09 fg mL^{-1} [144]. Figure 4C shows a schematic of the biosensing platform for detecting $\alpha2,6$-sialylated glycans.

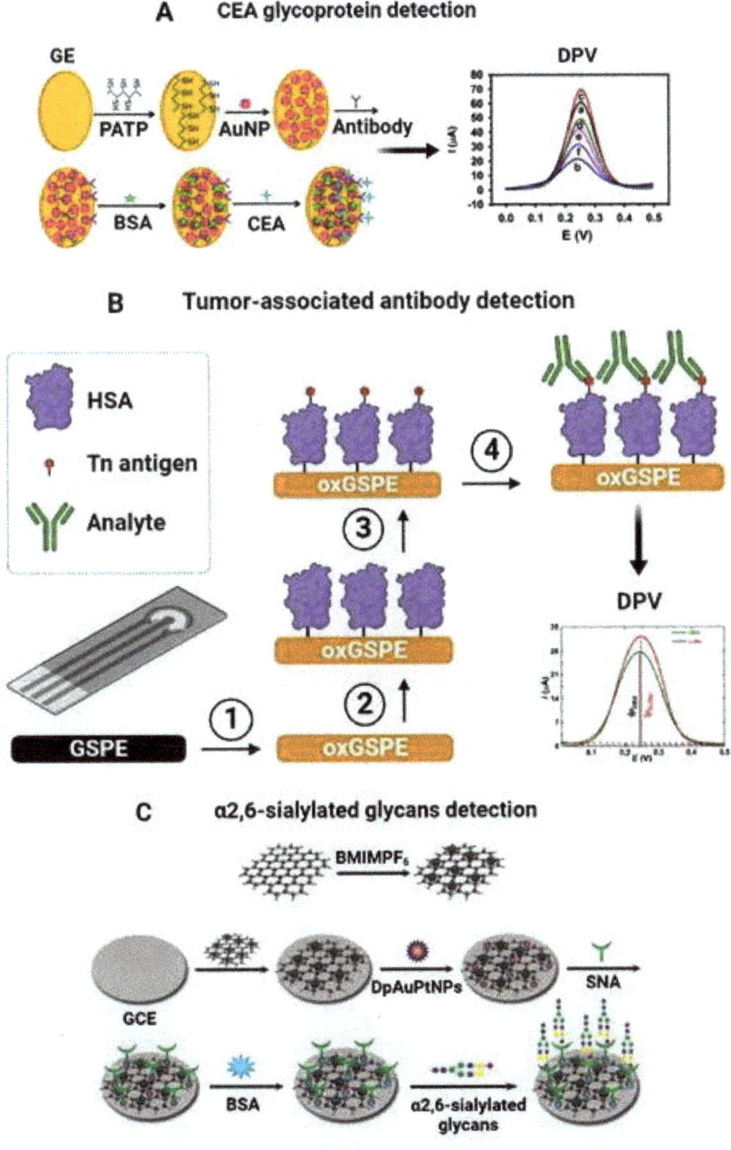

Figure 4. (**A**) Scheme of the different steps involved in developing a label-free nanoimmunosensor for CEA detection. (a) Bare gold electrodes and modified with (b) PATP, (c) PATP–AuNPs, (d) PATP–AuNPs–anti-CEA, (e) PATP–AuNPs–anti-CEA blocked with BSA, and (f) PATP–AuNPs–anti-CEA incubated with CEA after being blocked with BSA. Adapted from [98] with permission. Copyright Elsevier 2013. (**B**) Modification of graphene screen-printed electrode (GSPE) by electrochemical oxidation (step 1), covalent immobilization of human serum albumin (HSA) as a natural nanoscaffold (step 2), and covalent immobilization of a Tn antigen to HSA (step 3). The final step is incubation with the analyte protein (step 4) [135]. (**C**) Schematic representation of the electrochemical biosensor based on GCE modified with rGO-EPA/BMIMPF₆/AuPtNPs. Reproduced from [142] with permission. Copyright Elsevier 2015.

Glycosyltransferases

Glycosyltransferases (GTs) are the enzymes that glycosylate proteins and other molecules. Specific GTs are overexpressed during the tumorigenesis process, making them suitable biomarkers for diagnosis/prognosis [145]. Biosensors based on electrochemiluminescence (ECL) for detecting glycosyltransferases have been published. Xie et al. developed a biosensor to analyze β-1,4-galactosyltransferase (Gal-T) activity based on a graphitic carbon nitride (g-C_3N_4) and polystyrene microsphere nanoprobe functionalized with a lectin. The strategy in the biosensor development involved conjugation of N-acetylglucosamine-BSA (GlcNAc-BSA) to g-C_3N_4 modified glassy carbon electrode, which exhibited a strong ECL signal. In the presence of Gal T and UDP-Gal as a co-substrate, galactose was transferred to the GlcNAc-BSA, and the ECL signal decreased slightly. Next, the signal decreased significantly by the lectin–galactose interaction because the nanoprobe's poor conductivity inhibited the electron transfer at the electrode interface. The biosensor displayed high sensitivity for Gal-T activity detection with a low LOD of 7×10^{-5} U mL^{-1} [146]. Another similar strategy was developed by Chen et al., who used the same format of GlcNAc-BSA immobilized on gold electrodes and galactose conjugation to GlcNAc-BSA by Gal-T. The galactose was then specifically recognized by *Artocarpus integrifolia* lectin immobilized on gold nanorods conjugated to xanthine oxidase. The LOD obtained was 9×10^{-4} U mL^{-1} [147]. Figure 5A shows a schematic of the biosensing platform for detecting β-1,4-galactosyltransferase.

4.3. Ultrasensitive Impedimetric and Capacitive Biosensors for the Detection of Glycan-Based Biomarkers

Impedimetric and capacitive biosensors are among the most sensitive label-free analytical devices available [148,149]. In general, impedimetric and capacitive biosensors are sensitive to femtomolar to picomolar concentrations of the molecular target, and some publications have reported limits significantly below this [149–151]. Bertok et al. developed an ultrasensitive impedimetric glycobiosensor capable of detecting glycoproteins down to the aM level. The biosensing platform was assembled on aminoalkanethiol (SH-$(CH_2)_{11}$-NH_2)/AuNPs-modified gold electrodes, and a lectin from *Sambucus nigra* (SNA I) was covalently immobilized onto the nanostructured surface. After that, the nonspecific binding sites were blocked with polyvinyl alcohol (PVA). Finally, the glycobiosensor was applied to detect glycoproteins fetuin and asialofetuin containing sialic acid. The linear range was from 1 aM to 10 pM and a LOD of 1 aM (e.g., 24 glycoprotein molecules in 40 µL of a sample or 40 yoctomoles) [152]. In addition, Bertok et al. covalently immobilized an SNA I lectin on a mixed-SAM of MUA and betaine-terminated thiol to resist nonspecific interactions. The glycobiosensor detected the fetuin glycoprotein within a linear range from 100 fM to 100 nM and a LOD of 100 fM [153]. Figure 5B shows a schematic of the ultrasensitive biosensing platform for detecting fetuin glycoprotein.

Pihíková et al. developed an ultrasensitive impedimetric biosensor for glycoprofiling PSA. The biosensor was based on gold electrodes modified with a SAM of MUA/MCH and an anti-PSA antibody covalently immobilized on the surface. After the antibody–antigen binding event, the biosensor was incubated with the lectin (SNA I) that recognized the glycan part of PSA (α-2,6 linked sialic acid). The impedimetric biosensor responded linearly from 3.4 aM to 380 pM with a LOD of 4 aM [154]. In addition, Pihíková et al. assembled the glycobiosensor on a SAM of MUA/MCH, optimized a blocking agent (carbo-free commercial solution), and glycoprofiled PSA with *Maackia amurensis* lectin (MAL, recognizing α-2,3-sialic acid) in serum samples. As a result, the glycobiosensor could detect PSA glycoprotein in serum in a linear range of 100 ag mL^{-1} to 1 µg mL^{-1} with a LOD of 100 ag mL^{-1} and discriminated serum samples from healthy individuals of prostate cancer patients [155]. Figure 5C shows a schematic of the biosensing platform for PSA protein and its glycoprofiling.

Chocholova et al. developed another glycoprofiling platform using an antifouling zwitterionic layer-based impedimetric biosensor for determining human epidermal growth factor receptor (HER2) in human serum. The biosensor was assembled on screen-printed carbon electrodes modified with zwitterionic hydrogels that resisted nonspecific protein adsorption and allowed covalent attachment of an anti-HER2 antibody. The biosensor could detect the analyte via EIS in a linear range from 0.1 to 10 ng mL^{-1} with a LOD of 5 pg mL^{-1} (77 fM). In addition, the results demonstrated a significant difference between a high-risk woman without breast cancer (BCa) vs. a woman with the second stage of BCa via glycan profiling using erythroagglutinin lectin from *Phaseolus vulgaris*, specific for complex and branched structures of GlcNAc linked to mannose [156].

Capacitive biosensors offer novelty applications to detect glycoproteins related to viruses and cancer biomarkers [149]. Wang et al. developed a capacitive biosensor using microwires coated with Zika or Chikungunya virus envelope antigen. The biosensor could detect ten antibody molecules in a 30 µL volume and could be used to determine the isotypes present in a serum sample [157]. Oliveira et al. developed a capacitive nanobiosensing interface to detect interleukin-6 (IL-6) glycoprotein. The capacitive biosensor consisted of carbon electrodes nanostructured with graphene oxide (GO) and the Prussian blue (PrB) redox-active compound. The capacitive nanobiosensor detected the molecular recognition event between an antibody and IL-6 by a decrease in C_r in a linear range from 0.2 ng mL^{-1} to 20 µg mL^{-1} and a LOD of 5.6 ng mL^{-1} [158]. Table 3 shows the summary of electrochemical glycobiosensors mentioned above.

In summary, electrochemical glycobiosensing is a versatile tool for detecting multiple disease biomarkers, as summarized in Table 3. Electrochemical glycobiosensors meet the REASSURED criteria (real-time connectivity, ease of specimen collection, affordable, sensitive, specific, user friendly, rapid, equipment free, and delivered to those who need it); these characteristics make them suitable devices for the point-of-care molecular diagnosis/prognosis of diseases [159]. Overall, voltammetry- and amperometry-based biosensors have advantages such as fast response, ease of use, capability for analyzing two or more analytes simultaneously in the same sample, and more straightforward data analysis. The main shortcoming of voltammetry- and amperometry-based biosensors is relatively low sensitivity and limited precision [160]. Instead, impedance- and capacitance-based biosensors have the main advantage of high sensitivity, and they avoid the need for modification of analyte recognition elements with redox mediators. The main shortcoming of impedance- and capacitance-based biosensors is that data analysis requires extensive knowledge of electrochemistry, and measurements are difficult to perform in portable systems [150]. The reproducibility of these electrochemical glycobiosensors is always a significant limitation. One way to handle this shortcoming is to properly design the biosensing surface to ensure a similar initial signal. This could be achieved by adjusting surface roughness, immobilizing a maximal quantity of bioreceptors on nanostructures, controlling the thickness of the detection film, and using stable nanomaterials and bioreceptors [14,15,161,162].

These shortcomings could be addressed by combining effective point-of-care electrochemical biosensing methods with intelligent software and data processing methods that enable global monitoring of diseases and real-time decision-making [22].

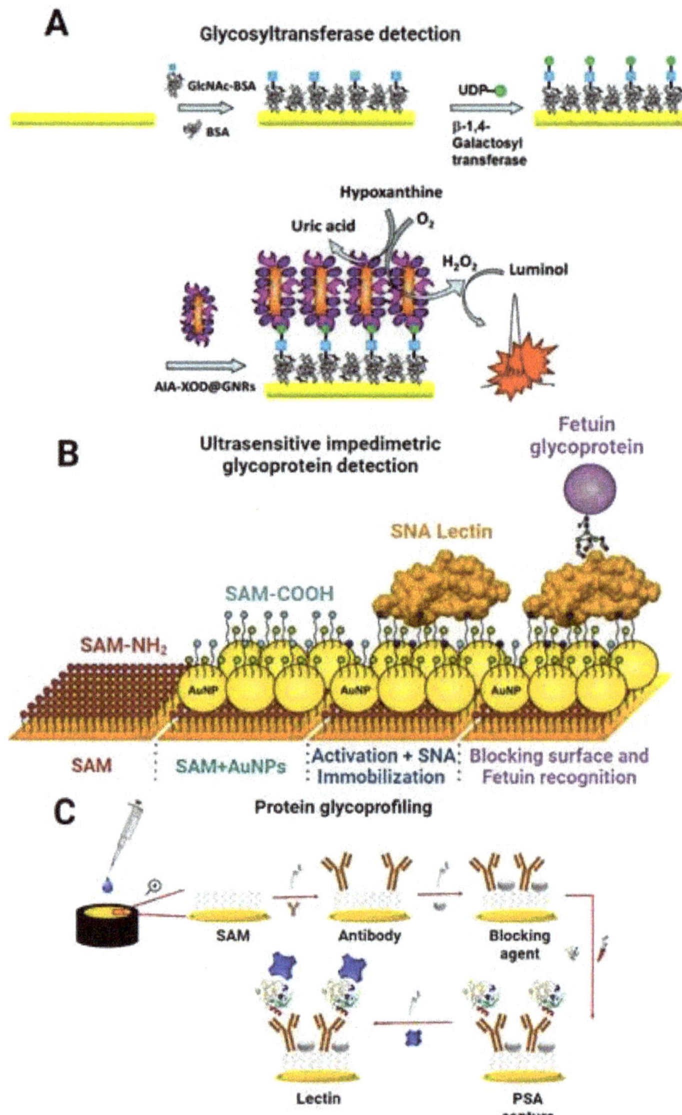

Figure 5. (**A**) Scheme of fabrication steps of biosensor based on a gold electrode modified with a bioconjugate of GlcNAc-BSA. Galactose transferred from UDP-Gal was specifically recognized by *Artocarpus integrifolia* lectin (AIA) immobilized on gold nanorods (GNRs) conjugated to xanthine oxidase (XOD). Reproduced from [147] with permission. Copyright Elsevier 2016. (**B**) Scheme of biosensor construction steps (from left to right): formation of a linker layer (NH₂-terminated alkanethiol-AT) on a gold surface (1st SAM on AuE); deposition of gold nanoparticles (AuNPs) and formation of a 2nd mixed SAM layer consisting of 11-mercaptoundecanoic acid and 6-mercaptohexanol on AuNPs; activation of the carboxyl group, subsequent covalent attachment of SNA I lectin, and finally an application of the lectin biosensor in the biorecognition of a glycoprotein, fetuin (FET). Adapted from [152] with permission. Copyright Elsevier 2013. (**C**) Scheme of construction steps of the biosensor for detecting PSA and glycoprofiling of PSA by application of lectin. Reproduced from [155] with permission. Copyright Elsevier 2013.

Table 3. Glycobiosensors for the detection of pathogens and cancer biomarkers reported in the last ten years.

Glycobiosensor Purpose	Sensing Platform Bioreceptor/Detection Technique	Target	Limit of Detection (LOD)	Linear Range	Ref.
Pathogens	SAM (MUA/MCH) Glycan-NH_2/EIS	HAs	8 aM	8 aM–80 nM	[112]
	SAM (OEG/OEG-COOH) Glycan-NH_2/EIS	H3N2	13 viral particles μL^{-1}	10–10^3 viral particles μL^{-1}	[113]
	SAM (Peptide) Peptide/EIS	SARS-CoV-2	18 ng mL^{-1}	0.05–1 $\mu g\ mL^{-1}$	[39]
	SAM (PEG-SH/11Fc) Antibody/ECS	NS1	340 pg mL^{-1}	1–5000 ng mL^{-1}	[125]
	Microwires/SAM Envelope Protein E/capacitance	Antibodies anti-ZIKV E	10 antibodies in 30 μL	10–10^3 antibodies in 30 μL	[157]
	SAM (MUA/MCH) Mannose/EIS	*S. typhimurium*	50 CFU mL^{-1}	50–10^3 CFU mL^{-1}	[129]
	PTPh-quinone Mannose/SWV	*E. coli*	25 cells mL^{-1}	2×10^3–5×10^4 cells mL^{-1}	[128]
	SAM (OEG/FcBA) Mannose/SWV		6×10^2 cells mL^{-1}	6×10^2–6×10^5 cells mL^{-1}	[115]
Cancer	PATP/AuNPs Antibody/DPV	CEA	0.015 fg mL^{-1}	1 fg mL^{-1}–10 ng mL^{-1}	[98]
	AuNPs/Cys Lectin/chronoamperometry		0.01 ng mL^{-1}	0.5–7 ng mL^{-1}	[130]
	PA/Au-Pt Antibody/SWV	CA-19-9	2×10^{-4} U mL^{-1}	0.001–40 U mL^{-1}	[131]
	PEI/CNTs Antibody/EIS		0.35 U mL^{-1}	0.05–60 U mL^{-1}	[132]
	SAM (MHDA) Lectin/EIS	STn	20 ng	20–70 ng	[134]
	GSPE/HSA Tn antigen/DPV	Antibody anti-Tn	10 aM	10 aM–10 pM	[135]
	PAMAM/c-MWCNTs/PDITC Lectin/DPV	α2,3-sialylated glycans	3 fg mL^{-1}	10 fg mL^{-1}–50 ng mL^{-1}	[137]
	n-C_{60}-PdPt/4-MBPA/Au-PMB Lectin/DPV		3 fg mL^{-1}	10 fg mL^{-1}–100 ng mL^{-1}	[139]
	GO-PrB-PTC-NH_2/AuNPs Lectin/DPV	α2,6-sialylated glycans	0.03 pg mL^{-1}	0.1 pg mL^{-1}–500 ng mL^{-1}	[141]
	rGO-TEPA-BMIMPF_6/AuPtNPs Lectin/amperometry		3 fg mL^{-1}	10 fg mL^{-1}–1 $\mu g\ mL^{-1}$	[142]
	AuNRs-SA/c-SWCNHs/S-PtNC Lectin/amperometry		0.69 fg mL^{-1}	1 fg mL^{-1}–100 ng mL^{-1}	[143]
	Ag/Fe-MOFs/M-APBA Lectin/DPV		0.09 fg mL^{-1}	1 fg mL^{-1}–1 ng mL^{-1}	[144]
	g-C_3N_4/GlcNAc-BSA Lectin/ECL	β-1,4-GalT	7×10^{-5} U mL^{-1}	5×10^{-4}–0.05 U mL^{-1}	[146]
	GlcNAc-BSA Lectin/ECL		9×10^{-4} U mL^{-1}	0.001–0.1 U mL^{-1}	[147]
	SAM (SH-$(CH_2)_{11}$-NH_2)/AuNPs Lectin/EIS	Fetuin	1 aM	1 aM–10 pM	[152]
	SAM (MUA/betaine) Lectin/EIS		100 fM	100 fM–100 nM	[153]
	SAM (MUA/MCH) Antibody/EIS	PSA	4 aM	3.4 aM–380 pM	[154]
	SAM (MUA/MCH/carbo-free) Antibody/EIS		100 ag mL^{-1}	100 ag mL^{-1}–1 $\mu g\ mL^{-1}$	[155]
	Zwitterionic hydrogel Antibody/EIS	HER2	5 pg mL^{-1}	100 pg mL^{-1}–10 ng mL^{-1}	[156]
	GO/PrB Antibody/ECS	IL-6	5.6 ng mL^{-1}	0.2 ng mL^{-1}–20 $\mu g\ mL^{-1}$	[158]

5. Concluding Remarks, Current Challenges and Opportunities

Glycans are biomolecules with relevant biological functions. They participate during infections provoked by pathogens and cellular expression changes related to more complex processes, particularly during cancer development [5,7]. Glycans are structurally diverse and complex. However, glycosylation is typically site specific, and specific types of glycans are present on restricted subsets of glycoproteins [145]. All of these indicate that glycans are helpful as disease biomarkers because cells express specific types of glycoproteins and release them into body fluids during disease progression. Furthermore, glycans are functional as biorecognition elements and can be easily incorporated into electrochemical biosensors to detect multiple analytes. The synergy of glycans as highly specific bioreceptors and proper transduction techniques in electrochemical biosensors enables the sensitive and specific detection of multiple molecular targets. Furthermore, incorporating nanomaterials into the electrode surface may improve the biosensor's analytical performance because these nanostructures have a large surface area for bioreceptor immobilization and can promote a fast electron transfer.

There are different approaches to immobilizing the biomolecules on the electrode, depending on the chemical composition of the biomolecules and electrode surface. The biofunctionalization process is characterized by different physicochemical techniques to confirm successful biosensor assembly. In particular, the biofunctionalization process and the bioreceptor–analyte molecular biorecognition event can be monitored using highly sensitive electrochemical techniques. Electrochemical glycobiosensors can perform similarly or better than conventional clinical methods and offer a practical approach to detecting different molecular targets. Electrochemical glycobiosensors have high sensitivity, specificity, selectivity, rapid response, user-friendly, and cost-effective features, as well as the possibility of miniaturization to deliver devices at the point of care.

Glycan-based electrochemical biosensors still have some weaknesses and limitations. A challenge to overcome in the electrochemical glycobiosensor field is that glycans are biomolecules very sensitive to environmental conditions and can rapidly lose their biological activity over time. For this reason, it is mandatory to incorporate glycan-based biorecognition elements with improved long-term stability into electrochemical biosensors that allow the development of efficient, robust, and low-cost analytical devices delivered to the end user ready to use.

Yet, glycan-based electrochemical biosensors offer tremendous opportunities for ultrasensitive biomarker monitoring at the POC. For example, developing new nanomaterials with improved electrochemical performance and new surface chemical moieties allows convenient and efficient biomolecule immobilization. Furthermore, electrochemical glycobiosensors can be incorporated into microfluidic platforms to develop fast detection assays with minimal sample manipulation by the user. This approach could also detect different molecular targets in a multiplexed format. Multiple working electrodes are individually modified in multiplexing formats with different bioreceptors to detect various analytes simultaneously. This approach enables the determination of multiple levels of molecular markers (e.g., nucleic acids, proteins, and metabolites), paving the way for precision medicine and providing a detailed disease characterization to customize healthcare [3]. Furthermore, the signal can be acquired using portable systems, such as hand potentiostats with a smartphone's signal readout. These attributes pave the way to personalized medicine enabling diagnosis/prognosis of diseases in decentralized settings, at the POC, closer to the patient.

The state of the art reviewed here demonstrates that ultrasensitive glycan-based nanobiosensing interfaces could be promising approaches to detect and quantify molecular targets in body fluids, thus holding considerable potential for determining cancer biomarkers and other infectious diseases in decentralized settings with a minimal reagent consumption and user-friendly operation mode.

Author Contributions: D.E. and J.O.: conceptualization; methodology; formal analysis; investigation; data curation; writing—original draft preparation; writing—review and editing. J.O.: supervision; project administration; funding acquisition. Both authors have read and agreed to the published version of the manuscript. All authors have read and agreed to the published version of the manuscript.

Funding: This research was funded by MINCIENCIAS, MINEDUCACION, MINCIT and ICETEX through the Program Ecosistema Científico Cod. FP44842-211-2018, project number 58536. J.O. thanks support from The University of Antioquia and the Max Planck Society through the cooperation agreement 566–1, 2014.

Institutional Review Board Statement: Not applicable.

Informed Consent Statement: Not applicable.

Data Availability Statement: Not applicable.

Acknowledgments: The authors thank EPM and Ruta N. for hosting the Max Planck Tandem Groups.

Conflicts of Interest: The authors declare no conflict of interest.

Sample Availability: Not applicable.

References

1. Hushegyi, A.; Tkac, J. Are Glycan Biosensors an Alternative to Glycan Microarrays? *Anal. Methods* **2016**, *6*, 6610–6620. [CrossRef]
2. Campuzano, S.; Pedrero, M.; Yáñez-Sedeño, P.; Pingarrón, J.M. New Challenges in Point of Care Electrochemical Detection of Clinical Biomarkers. *Sens. Actuators B Chem.* **2021**, *345*, 130349. [CrossRef]
3. Campuzano, S.; Barderas, R.; Yáñez-Sedeño, P.; Pingarrón, J.M. Electrochemical Biosensing to Assist Multiomics Analysis in Precision Medicine. *Curr. Opin. Electrochem.* **2021**, *28*, 100703. [CrossRef]
4. Varki, A.; Cummings, R.D.; Esko, J.D.; Stanley, P.; Hart, G.W.; Aebi, M.; Mohnen, D.; Kinoshita, T.; Packer, N.H.; Prestegard, J.H.; et al. *Essentials of Glycobiology*; Cold Spring Harbor Laboratory Press: New York, NY, USA, 2022. [CrossRef]
5. Defaus, S.; Gupta, P.; Andreu, D.; Gutiérrez-Gallego, R. Mammalian Protein Glycosylation—Structure versus Function. *Analyst* **2014**, *139*, 2944–2967. [CrossRef]
6. Taron, C.; Rudd, P. Glycomics: A Rapidly Evolving Field with a Sweet Future. *NEB Expr.* **2016**, 2–4. Available online: https://international.neb.com/tools-and-resources/feature-articles/glycomics-a-rapidly-evolving-field-with-a-sweet-future (accessed on 10 October 2022).
7. Dwek, R.A. Glycobiology: "Towards Understanding the Function of Sugars. " *Biochem Soc Trans* **1995**, *23*, 1–25. [CrossRef]
8. Hu, M.; Lan, Y.; Lu, A.; Ma, X.; Zhang, L. *Glycan-Based Biomarkers for Diagnosis of Cancers and Other Diseases: Past, Present, and Future*, 1st ed.; Elsevier Inc.: Amsterdam, The Netherlands, 2019; Volume 162, ISBN 9780128177389.
9. Götze, S.; Azzouz, N.; Tsai, Y.H.; Groß, U.; Reinhardt, A.; Anish, C.; Seeberger, P.H.; Silva, D.V. Diagnosis of Toxoplasmosis Using a Synthetic Glycosylphosphatidyl-Inositol Glycan. *Angew. Chem. Int. Ed.* **2014**, *53*, 13701–13705. [CrossRef]
10. Echeverri, D.; Garg, M.; Varón Silva, D.; Orozco, J. Phosphoglycan-Sensitized Platform for Specific Detection of Anti-Glycan IgG and IgM Antibodies in Serum. *Talanta* **2020**, *217*, 121117. [CrossRef]
11. Thomsen, M.; Skovlund, E.; Sorbye, H.; Bolstad, N.; Nustad, K.J.; Glimelius, B.; Pfeiffer, P.; Kure, E.H.; Johansen, J.S.; Tveit, K.M.; et al. Prognostic Role of Carcinoembryonic Antigen and Carbohydrate Antigen 19-9 in Metastatic Colorectal Cancer: A BRAF-Mutant Subset with High CA 19-9 Level and Poor Outcome. *Br. J. Cancer* **2018**, *118*, 1609–1616. [CrossRef]
12. Vukobrat-Bijedic, Z.; Husic-Selimovic, A.; Sofic, A.; Bijedic, N.; Bjelogrlic, I.; Gogov, B.; Mehmedovic, A. Cancer Antigens (CEA and CA 19-9) as Markers of Advanced Stage of Colorectal Carcinoma. *Med. Arch.* **2013**, *67*, 397. [CrossRef]
13. Chatterjee, S.B.; Hou, J.; Ratnam Bandaru, V.V.; Pezhouh, M.K.; Syed Rifat Mannan, A.A.; Sharma, R. Lactosylceramide Synthase β-1,4-GalT-V: A Novel Target for the Diagnosis and Therapy of Human Colorectal Cancer. *Biochem. Biophys Res. Commun.* **2019**, *508*, 380–386. [CrossRef]
14. Echeverri, D.; Cruz-Pacheco, A.F.; Orozco, J. Capacitive Nanobiosensing of β-1,4-Galactosyltransferase-V Colorectal Cancer Biomarker. *Sens. Actuators B Chem.* **2023**, *374*, 132784. [CrossRef]
15. Echeverri, D.; Orozco, J. β-1,4-Galactosyltransferase-V Colorectal Cancer Biomarker Immunosensor with Label-Free Electrochemical Detection. *Talanta* **2022**, *243*, 123337. [CrossRef]
16. Wellinghausen, N.; Abele-Horn, M.; Mantke, O.D.; Enders, M.; Fingerle, V.; Gärtner, B.; Hagedorn, J.; Rabenau, H.F.; Reiter-Owona, I.; Tintelnot, K.; et al. Immunological Methods for the Detection of Infectious Diseases. Available online: https://www.instand-ev.de/fileadmin/uploads/Wissenschaftliche_Aktivitaeten/INSTAND_Schriftenreihe_Band_I_MiQ_Serologie_2018_en.pdf (accessed on 15 September 2019).
17. Fitzgerald, S.; O'Reilly, J.A.; Wilson, E.; Joyce, A.; Farrell, R.; Kenny, D.; Kay, E.W.; Fitzgerald, J.; Byrne, B.; Kijanka, G.S.; et al. Measurement of the IgM and IgG Autoantibody Immune Responses in Human Serum Has High Predictive Value for the Presence of Colorectal Cancer. *Clin. Colorectal. Cancer* **2019**, *18*, e53–e60. [CrossRef]
18. Moss, C.W. Chromatographic Analysis for Identification of Microorganisms and Diagnosis of Infection: An Introduction. In *Rapid Methods and Automation in Microbiology and Immunology*; Springer: Berlin/Heidelberg, Germany, 1985; p. 231. [CrossRef]
19. Pan, S.; Chen, R.; Aebersold, R.; Brentnall, T.A. Mass Spectrometry Based Glycoproteomics—From a Proteomics Perspective. *Mol. Cell. Proteom.* **2011**, *10*, R110.003251. [CrossRef]
20. Hosseini, S.; Vázquez-Villegas, P.; Rito-Palomares, M.; Martinez-Chapa, S.O. Advantages, Disadvantages and Modifications of Conventional ELISA. In *Enzyme-Linked Immunosorbent Assay (ELISA)*; SpringerBriefs in Applied Sciences and Technology: Singapore, 2018; pp. 67–115. ISBN 9789811067662.
21. Kinoshita, Y.; Uo, T.; Jayadev, S.; Garden, G.A.; Conrads, T.P.; Veenstra, T.D.; Morrison, R.S. Potential Applications and Limitations of Proteomics in the Study of Neurological Disease. *Arch. Neurol* **2006**, *63*, 1692–1696. [CrossRef]

22. Lopes, L.C.; Santos, A.; Bueno, P.R. An Outlook on Electrochemical Approaches for Molecular Diagnostics Assays and Discussions on the Limitations of Miniaturized Technologies for Point-of-Care Devices. *Sens. Actuators Rep.* **2022**, *4*, 100087. [CrossRef]
23. Da Silva, E.T.S.G.; Souto, D.E.P.; Barragan, J.T.C.; de, F. Giarola, J.; de Moraes, A.C.M.; Kubota, L.T. Electrochemical Biosensors in Point-of-Care Devices: Recent Advances and Future Trends. *ChemElectroChem* **2017**, *4*, 778–794. [CrossRef]
24. Varki, A.; Cummings, R.; Esko, J.; Stanley, P.; Hart, G.; Aebi, M.; Darvill, A.; Kinoshita, T.; Packer, N.; Prestegard, J.; et al. *Essentials in Glycobiology*, 3rd ed.; Varki, A., Ed.; Cold Spring Harbor Laboratory Press: New York, NY, USA, 2017; ISBN 9781621821328.
25. Koffas, M.A.G.; Linhardt, R.J. Metabolic Bioengineering: Glycans and Glycoconjugates. *Emerg. Top Life Sci.* **2018**, *2*, 333–335. [CrossRef]
26. Barb, A.W.; Borgert, A.J.; Liu, M.; Barany, G.; Live, D. *Intramolecular Glycan-Protein Interactions in Glycoproteins*, 1st ed.; Elsevier Inc: Amsterdam, The Netherlands, 2010; Volume 478.
27. Saha, S.; Anilkumar, A.A.; Mayor, S. GPI-Anchored Protein Organization and Dynamics at the Cell Surface. *J. Lipid. Res.* **2016**, *57*, 159–175. [CrossRef]
28. Tsai, Y.H.; Liu, X.; Seeberger, P.H. Chemical Biology of Glycosylphosphatidylinositol Anchors. *Angew. Chem. Int. Ed.* **2012**, *51*, 11438–11456. [CrossRef]
29. Theocharis, A.D.; Skandalis, S.S.; Tzanakakis, G.N.; Karamanos, N.K. Proteoglycans in Health and Disease: Novel Roles for Proteoglycans in Malignancy and Their Pharmacological Targeting. *FEBS J.* **2010**, *277*, 3904–3923. [CrossRef]
30. Michaelson, L.V.; Napier, J.A.; Molino, D.; Faure, J.D. Plant Sphingolipids: Their Importance in Cellular Organization and Adaption. *Biochim. Biophys Acta Mol. Cell Biol. Lipids* **2016**, *1861*, 1329–1335. [CrossRef]
31. Poole, A.R. Proteoglycans in Health and Disease: Structures and Functions. *Biochem. J.* **1986**, *236*, 1–14. [CrossRef]
32. Hannun, Y.A.; Obeid, L.M. Sphingolipids and Their Metabolism in Physiology and Disease. *Nat. Rev. Mol. Cell Biol.* **2018**, *19*, 175–191. [CrossRef]
33. Posse de Chaves, E.; Sipione, S. Sphingolipids and Gangliosides of the Nervous System in Membrane Function and Dysfunction. *FEBS Lett.* **2010**, *584*, 1748–1759. [CrossRef]
34. Kusumoto, M.; Ueno-Noto, K.; Takano, K. Systematic Interaction Analysis of Anti-Human Immunodeficiency Virus Type-1 Neutralizing Antibodies with High Mannose Glycans Using Fragment Molecular Orbital and Molecular Dynamics Methods. *J. Comput. Chem.* **2020**, *41*, 31–42. [CrossRef]
35. Gómez-Pinto, I.; Vengut-Climent, E.; Lucas, R.; Aviñó, A.; Eritja, R.; González, C.; Morales, J.C. Carbohydrate–DNA Interactions at G-Quadruplexes: Folding and Stability Changes by Attaching Sugars at the 5′-End. *Chem. A Eur. J.* **2013**, *19*, 1920–1927. [CrossRef]
36. WHO International Programme on Chemical Safety. Biomarkers in Risk Assessment: Validity and Validation. Available online: http://www.inchem.org/documents/ehc/ehc/ehc222.htm (accessed on 10 October 2022).
37. Banerjee, N.; Mukhopadhyay, S. Viral Glycoproteins: Biological Role and Application in Diagnosis. *Virusdisease* **2016**, *27*, 1. [CrossRef]
38. Vásquez, V.; Navas, M.C.; Jaimes, J.A.; Orozco, J. SARS-CoV-2 Electrochemical Immunosensor Based on the Spike-ACE2 Complex. *Anal. Chim. Acta* **2022**, *1205*, 339718. [CrossRef]
39. Soto, D.; Orozco, J. Peptide-Based Simple Detection of SARS-CoV-2 with Electrochemical Readout. *Anal. Chim. Acta* **2022**, *1205*, 339739. [CrossRef]
40. Tra, V.N.; Dube, D.H. Glycans in Pathogenic Bacteria—Potential for Targeted Covalent Therapeutics and Imaging Agents. *Chem. Commun. (Camb)* **2014**, *50*, 4659. [CrossRef]
41. Masuoka, J. Surface Glycans of Candida Albicans and Other Pathogenic Fungi: Physiological Roles, Clinical Uses, and Experimental Challenges. *Clin. Microbiol. Rev.* **2004**, *17*, 281–310. [CrossRef]
42. Song, X.; Heimburg-Molinaro, J.; Smith, D.F.; Cummings, R.D. Glycan Microarrays. In *Chemical Genomics and Proteomics. Methods in Molecular Biology (Methods and Protocols)*; Zanders, E., Ed.; Humana Press: Totowa, NJ, USA, 2012; Volume 800, pp. 173–186. ISBN 978-1-61779-348-6.
43. Geissner, A.; Reinhardt, A.; Rademacher, C.; Johannssen, T.; Monteiro, J.; Lepenies, B.; Thépaut, M.; Fieschi, F.; Mrázková, J.; Wimmerova, M.; et al. Microbe-Focused Glycan Array Screening Platform. *Proc. Natl. Acad. Sci. USA* **2019**, *116*, 1958–1967. [CrossRef]
44. Geissner, A.; Anish, C.; Seeberger, P.H. Glycan Arrays as Tools for Infectious Disease Research. *Curr. Opin. Chem. Biol.* **2014**, *18*, 38–45. [CrossRef]
45. Pinho, S.S.; Reis, C.A. Glycosylation in Cancer: Mechanisms and Clinical Implications. *Nat. Rev. Cancer* **2015**, *15*, 540–555. [CrossRef]
46. Tang, Y.; Cui, Y.; Zhang, S.; Zhang, L. *The Sensitivity and Specificity of Serum Glycan-Based Biomarkers for Cancer Detection*, 1st ed.; Elsevier Inc.: Amsterdam, The Netherlands, 2019; Volume 162, ISBN 9780128177389.
47. Zhai, H.; Huang, J.; Yang, C.; Fu, Y.; Yang, B. Serum CEA and CA19-9 Levels Are Associated with the Presence and Severity of Colorectal Neoplasia. *Clin. Lab.* **2018**, *64*, 351–356. [CrossRef]
48. Svarovsky, S.A.; Joshi, L. Cancer Glycan Biomarkers and Their Detection-Past, Present and Future. *Anal. Methods* **2014**, *6*, 3918–3936. [CrossRef]
49. Dasgupta, A.; Wahed, A. Tumor Markers. In *Clinical Chemistry, Immunology and Laboratory Quality Control*; Elsevier Inc.: Amsterdam, The Netherlands, 2014; pp. 229–246. ISBN 9780124078215.

50. Scarà, S.; Bottoni, P.; Scatena, R. CA 19-9: Biochemical and Clinical Aspects. In *Advances in Cancer Biomarkers—From Biochemistry to Clinic for a Critical Revision*; Scatena, R., Ed.; Springer Netherlands: Rome, Italy, 2015; pp. 247–260. ISBN 9789401772150.
51. Pearce, O.M.T.; Läubli, H. Sialic Acids in Cancer Biology and Immunity. *Glycobiology* **2016**, *26*, 111–128. [CrossRef]
52. Julien, S.; Delannoy, P. Sialic Acid and Cancer. In *Glycoscience: Biology and Medicine*; Spinger: Berlin/Heidelberg, Germany, 2015; pp. 1419–1424. [CrossRef]
53. Dobie, C.; Skropeta, D. Insights into the Role of Sialylation in Cancer Progression and Metastasis. *Br. J. Cancer* **2020**, *124*, 76–90. [CrossRef]
54. Qasba, P.; Ramakrishnan, B.; Boeggeman, E. Structure and Function of β-1,4-Galactosyltransferase. *Curr. Drug Targets* **2008**, *9*, 292–309. [CrossRef]
55. Lairson, L.L.; Henrissat, B.; Davies, G.J.; Withers, S.G. Glycosyltransferases: Structures, Functions, and Mechanisms. *Annu. Rev. Biochem.* **2008**, *77*, 521–555. [CrossRef]
56. Ferreira, J.A.; Magalhães, A.; Gomes, J.; Peixoto, A.; Gaiteiro, C.; Fernandes, E.; Santos, L.L.; Reis, C.A. Protein Glycosylation in Gastric and Colorectal Cancers: Toward Cancer Detection and Targeted Therapeutics. *Cancer Lett.* **2017**, *387*, 32–45. [CrossRef]
57. García-Barros, M.; Coant, N.; Truman, J.P.; Snider, A.J.; Hannun, Y.A. Sphingolipids in Colon Cancer. *Biochim. Biophys. Acta Mol. Cell Biol. Lipids* **2014**, *1841*, 773–782. [CrossRef]
58. Shirane, K.; Sato, T.; Segawa, K.; Furukawa, K. Involvement of β-1,4-Galactosyltransferase V in Malignant Transformation-Associated Changes in Glycosylation. *Biochem. Biophys Res. Commun.* **1999**, *265*, 434–438. [CrossRef]
59. Sánchez-Pomales, G.; Zangmeister, R.A. Recent Advances in Electrochemical Glycobiosensing. *Int. J. Electrochem.* **2011**, *2011*, 825790. [CrossRef]
60. Dziąbowska, K.; Czaczyk, E.; Nidzworski, D. Application of Electrochemical Methods in Biosensing Technologies. A Prospective Way for Rapid Analysis. In *Biosensing Technologies for the Detection of Pathogens*; Rinken, T., Ed.; InTechOpen: London, UK, 2018; ISBN 978-953-51-4087-0.
61. Wongkaew, N.; Simsek, M.; Griesche, C.; Baeumner, A.J. Functional Nanomaterials and Nanostructures Enhancing Electrochemical Biosensors and Lab-on-a-Chip Performances: Recent Progress, Applications, and Future Perspective. *Chem. Rev.* **2019**, *119*, 120–194. [CrossRef]
62. Blsakova, A.; Kveton, F.; Tkac, J. Glycan-Modified Interfaces in Biosensing: An Electrochemical Approach. *Curr. Opin. Electrochem.* **2019**, *14*, 60–65. [CrossRef]
63. IUPAC. *Compendium of Chemical Terminology*, 2nd ed.; (the "Gold, Book"); McNaught, A.D., Wilkinson, A., Eds.; Blackwell Scientific Publications: Oxford, UK, 1997; ISBN 0-9678550-9-8. [CrossRef]
64. Bhalla, N.; Jolly, P.; Formisano, N.; Estrela, P. Introduction to Biosensors. *Essays Biochem.* **2016**, *60*, 1–8. [CrossRef]
65. Lee, Y.H.; Mutharasan, R. Biosensors. In *Sensor Technology Handbook*; Wilson, J.S., Ed.; Newnes, Elsevier: Amsterdam, The Netherlands, 2005; pp. 161–180. ISBN 978-0-7506-7729-5.
66. Karunakaran, C.; Bhargava, K.; Benjamin, R. Introduction to Biosensors. In *Biosensors and Bioelectronics*; Elsevier: Amsterdam, The Netherlands, 2015; pp. 2–66. ISBN 9780128031018.
67. Belicky, T.; Katrlik, J.; Tkac, J. Glycan and Lectin Biosensors. *Essays Biochem.* **2016**, *60*, 37–47. [CrossRef]
68. Thevenot, D.; Toth, K.; Durst, R.; Wilson, G.; Thevenot, D.; Toth, K.; Durst, R.; Wilson, G. Electrochemical Biosensors: Recommended Definitions and Classification. *Biosens. Bioelectron.* **2001**, *16*, 121–131. [CrossRef] [PubMed]
69. Mahato, K.; Kumar, A.; Maurya, P.K.; Chandra, P. Shifting Paradigm of Cancer Diagnoses in Clinically Relevant Samples Based on Miniaturized Electrochemical Nanobiosensors and Microfluidic Devices. *Biosens. Bioelectron.* **2018**, *100*, 411–428. [CrossRef] [PubMed]
70. Sarkar, A. Biosensing, Characterization of Biosensors, and Improved Drug Delivery Approaches Using Atomic Force Microscopy: A Review. *Front. Nanotechnol.* **2022**, *3*, 102. [CrossRef]
71. Akhtar, K.; Khan, S.A.; Khan, S.B.; Asiri, A.M. Scanning Electron Microscopy: Principle and Applications in Nanomaterials Characterization. In *Handbook of Materials Characterization*; Springer: Cham, Switzerland, 2018; pp. 113–145. [CrossRef]
72. Prasek, J.; Trnkova, L.; Gablech, I.; Businova, P.; Drbohlavova, J.; Chomoucka, J.; Adam, V.; Kizek, R.; Hubalek, J. Optimization of Planar Three-Electrode Systems for Redox System Detection. *Int. J. Electrochem. Sci* **2012**, *7*, 1785–1801.
73. Abdulbari, H.A.; Basheer, E.A.M. Electrochemical Biosensors: Electrode Development, Materials, Design, and Fabrication. *ChemBioEng Rev.* **2017**, *4*, 92–105. [CrossRef]
74. Loganathan, S.; Valapa, R.B.; Mishra, R.K.; Pugazhenthi, G.; Thomas, S. Thermogravimetric Analysis for Characterization of Nanomaterials. *Therm. Rheol. Meas. Tech. Nanomater. Charact.* **2017**, *3*, 67–108. [CrossRef]
75. Abraham, J.; Mohammed, A.P.; Kumar, M.P.A.; George, S.C.; Thomas, S. Thermoanalytical Techniques of Nanomaterials. In *Characterization of Nanomaterials: Advances and Key Technologies*; Elsevier: Amsterdam, The Netherlands, 2018; pp. 213–236. [CrossRef]
76. Nair, A.K.; Mayeen, A.; Shaji, L.K.; Kala, M.S.; Thomas, S.; Kalarikkal, N. Optical Characterization of Nanomaterials. In *Characterization of Nanomaterials: Advances and Key Technologies*; Elsevier: Amsterdam, The Netherlands, 2018; pp. 269–299. [CrossRef]
77. Wang, J. *Analytical Electrochemistry*, 3rd ed.; John Wiley and Sons: Hoboken, NJ, USA, 2006; ISBN 9780471678793.
78. Bueno, P.R.; Mizzon, G.; Davis, J.J. Capacitance Spectroscopy: A Versatile Approach to Resolving the Redox Density of States and Kinetics in Redox-Active Self-Assembled Monolayers. *J. Phys. Chem. B* **2012**, *116*, 8822–8829. [CrossRef]

79. Suenaga, E.; Mizuno, H.; Penmetcha, K.K.R. Monitoring Influenza Hemagglutinin and Glycan Interactions Using Surface Plasmon Resonance. *Biosens. Bioelectron.* **2012**, *32*, 195–201. [CrossRef]
80. Hashimoto, S. Principles of BIACORE. In *Real-Time Analysis of Biomolecular Interactions*; Springer: Tokyo, Japan, 2000; pp. 23–30. [CrossRef]
81. Shandilya, R.; Bhargava, A.; Bunkar, N.; Tiwari, R.; Goryacheva, I.Y.; Mishra, P.K. Nanobiosensors: Point-of-Care Approaches for Cancer Diagnostics. *Biosens. Bioelectron.* **2019**, *130*, 147–165. [CrossRef]
82. Thakur, P.S.; Sankar, M. Nanobiosensors for Biomedical, Environmental, and Food Monitoring Applications. *Mater. Lett.* **2022**, *311*, 131540. [CrossRef]
83. Quinchia, J.; Echeverri, D.; Cruz-Pacheco, A.F.; Maldonado, M.E.; Orozco, J.A. Electrochemical Biosensors for Determination of Colorectal Tumor Biomarkers. *Micromachines* **2020**, *11*, 411. [CrossRef] [PubMed]
84. Curreli, M.; Zhang, R.; Ishikawa, F.N.; Chang, H.K.; Cote, R.J.; Zhou, C.; Thompson, M.E. Real-Time, Label-Free Detection of Biological Entities Using Nanowire-Based FETs. *IEEE Trans. Nanotechnol.* **2008**, *7*, 651–667. [CrossRef]
85. Delbianco, M.; Davis, B.G.; Seeberger, P.H. Glycans in Nanotechnology. In *Essentials of Glycobiology*; Cold Spring Harbor Laboratory Press: New York, NY, USA, 2022. [CrossRef]
86. Lee, Y.K.; Lee, H.; Nam, J.M. Lipid-Nanostructure Hybrids and Their Applications in Nanobiotechnology. *NPG Asia Mater.* **2013**, *5*, e48. [CrossRef]
87. Walcarius, A.; Minteer, S.D.; Wang, J.; Lin, Y.; Merkoçi, A. Nanomaterials for Bio-Functionalized Electrodes: Recent Trends. *J. Mater. Chem. B* **2013**, *1*, 4878–4908. [CrossRef]
88. Khan, F.A. Synthesis of Nanomaterials: Methods & Technology. In *Applications of Nanomaterials in Human Health*; Springer: Singapore, 2020; pp. 15–21. [CrossRef]
89. Arduini, F.; Micheli, L.; Moscone, D.; Palleschi, G.; Piermarini, S.; Ricci, F.; Volpe, G. Electrochemical Biosensors Based on Nanomodified Screen-Printed Electrodes: Recent Applications in Clinical Analysis. *TrAC Trends Anal. Chem.* **2016**, *79*, 114–126. [CrossRef]
90. Putzbach, W.; Ronkainen, N.J. Immobilization Techniques in the Fabrication of Nanomaterial-Based Electrochemical Biosensors: A Review. *Sensors* **2013**, *13*, 4811–4840. [CrossRef]
91. Holzinger, M.; Goff, A.; Cosnier, S. Nanomaterials for Biosensing Applications: A Review. *Front. Chem.* **2014**, *2*, 63. [CrossRef]
92. Martinkova, P.; Kostelnik, A.; Valek, T.; Pohanka, M. Main Streams in the Construction of Biosensors and Their Applications. *Int. J. Electrochem. Sci.* **2017**, *12*, 7386–7403. [CrossRef]
93. Steen Redeker, E.; Ta, D.T.; Cortens, D.; Billen, B.; Guedens, W.; Adriaensens, P. Protein Engineering for Directed Immobilization. *Bioconjug. Chem.* **2013**, *24*, 1761–1777. [CrossRef]
94. O'Neil, C.L.; Stine, K.J.; Demchenko, A.V. Immobilization of Glycans on Solid Surfaces for Application in Glycomics. *J. Carbohydr. Chem.* **2018**, *37*, 225–249. [CrossRef]
95. Bhattarai, J.K.; Neupane, D.; Mikhaylov, V.; Demchenko, A.V.; Stine, K.J. Self-Assembled Monolayers of Carbohydrate Derivatives on Gold Surfaces. In *Carbohydrate*; InTechOpen: London, UK, 2017; pp. 63–97.
96. Zeng, X.; Qu, K.; Rehman, A. Glycosylated Conductive Polymer: A Multimodal Biointerface for Studying Carbohydrate-Protein Interactions. *Acc. Chem. Res.* **2016**, *49*, 1624–1633. [CrossRef]
97. Sharma, S.; Byrne, H.; O'Kennedy, R.J. Antibodies and Antibody-Derived Analytical Biosensors. *Essays Biochem.* **2016**, *60*, 9–18. [CrossRef]
98. Liu, Z.; Ma, Z. Fabrication of an Ultrasensitive Electrochemical Immunosensor for CEA Based on Conducting Long-Chain Polythiols. *Biosens. Bioelectron.* **2013**, *46*, 1–7. [CrossRef]
99. Hermanson, G.T. (Strept)avidin–Biotin Systems. In *Bioconjugate Techniques*, 3rd ed.; Audet, J., Preap, M., Eds.; Elsevier: Whaltam, MA, USA, 2013; p. 56.
100. Mahato, K.; Maurya, P.K.; Chandra, P. Fundamentals and Commercial Aspects of Nanobiosensors in Point-of-Care Clinical Diagnostics. *3 Biotech.* **2018**, *8*, 149. [CrossRef]
101. Pihíková, D.; Kasák, P.; Tkac, J. Glycoprofiling of Cancer Biomarkers: Label-Free Electrochemical Lectin-Based Biosensors. *Open Chem.* **2015**, *13*, 636–655. [CrossRef]
102. Johnson, S.; Krauss, T.F. Label-Free Affinity Biosensor Arrays: Novel Technology for Molecular Diagnostics. *Expert Rev. Med. Devices* **2017**, *14*, 177–179. [CrossRef]
103. Zhu, C.; Yang, G.; Li, H.; Du, D.; Lin, Y. Electrochemical Sensors and Biosensors Based on Nanomaterials and Nanostructures. *Anal. Chem.* **2014**, *87*, 230–249. [CrossRef]
104. Morales, M.A.; Halpern, J.M. Guide to Selecting a Biorecognition Element for Biosensors. *Bioconjug. Chem.* **2018**, *29*, 3231–3239. [CrossRef]
105. Xiao, X.; Kuang, Z.; Slocik, J.M.; Tadepalli, S.; Brothers, M.; Kim, S.; Mirau, P.A.; Butkus, C.; Farmer, B.L.; Singamaneni, S.; et al. Advancing Peptide-Based Biorecognition Elements for Biosensors Using in-Silico Evolution. *ACS Sens.* **2018**, *3*, 1024–1031. [CrossRef]
106. Karunakaran, R.; Keskin, M. Biosensors: Components, Mechanisms, and Applications. In *Analytical Techniques in Biosciences*; Academic Press: Cambridge, MA, USA, 2022; pp. 179–190. [CrossRef]

107. Klukova, L.; Filip, J.; Tkac, J. Graphene-Based Lectin Biosensor for Ultrasensitive Detection of Glycan Structures Applicable in Early Diagnostics. In Proceedings of the 2015 9th International Conference on Sensing Technology (ICST), Auckland, New Zealand, 8–10 December 2015; pp. 40–45.
108. Cunningham, S.; Gerlach, J.Q.; Kane, M.; Joshi, L. Glyco-Biosensors: Recent Advances and Applications for the Detection of Free and Bound Carbohydrates. *Analyst* **2010**, *135*, 2471–2480. [CrossRef]
109. Taylor, M.E.; Drickamer, K.; Schnaar, R.L.; Etzler, M.E.; Varki, A. Discovery and Classification of Glycan-Binding Proteins. In *Essentials of Glycobiology*; Cold Spring Harbor Laboratory Press: New York, NY, USA, 2017. [CrossRef]
110. Paleček, E.; Tkáč, J.; Bartošík, M.; Bertók, T.; Ostatná, V.; Paleček, J. Electrochemistry of Nonconjugated Proteins and Glycoproteins. Toward Sensors for Biomedicine and Glycomics. *Chem. Rev.* **2015**, *115*, 2045–2108. [CrossRef] [PubMed]
111. Pihikova, D.; Pakanova, Z.; Nemcovic, M.; Barath, P.; Belicky, S.; Bertok, T.; Kasak, P.; Mucha, J.; Tkac, J. Sweet Characterisation of Prostate Specific Antigen Using Electrochemical Lectin-Based Immunosensor Assay and MALDI TOF/TOF Analysis: Focus on Sialic Acid. *Proteomics* **2016**, *16*, 3085–3095. [CrossRef] [PubMed]
112. Hushegyi, A.; Bertok, T.; Damborsky, P.; Katrlik, J.; Tkac, J. An Ultrasensitive Impedimetric Glycan Biosensor with Controlled Glycan Density for Detection of Lectins and Influenza Hemagglutinins. *Chem. Commun.* **2015**, *51*, 7474–7477. [CrossRef]
113. Hushegyi, A.; Pihíková, D.; Bertok, T.; Adam, V.; Kizek, R.; Tkac, J. Ultrasensitive Detection of Influenza Viruses with a Glycan-Based Impedimetric Biosensor. *Biosens. Bioelectron.* **2016**, *79*, 644–649. [CrossRef]
114. Cui, F.; Xu, Y.; Wang, R.; Liu, H.; Chen, L.; Zhang, Q.; Mu, X. Label-Free Impedimetric Glycan Biosensor for Quantitative Evaluation Interactions between Pathogenic Bacteria and Mannose. *Biosens. Bioelectron.* **2018**, *103*, 94–98. [CrossRef]
115. Kveton, F.; Blšáková, A.; Hushegyi, A.; Damborsky, P.; Blixt, O.; Jansson, B.; Tkac, J. Optimization of the Small Glycan Presentation for Binding a Tumor-Associated Antibody: Application to the Construction of an Ultrasensitive Glycan Biosensor. *Langmuir* **2017**, *33*, 2709–2716. [CrossRef]
116. Zhu, B.W.; Cai, L.; He, X.P.; Chen, G.R.; Long, Y.T. Anthraquinonyl Glycoside Facilitates the Standardization of Graphene Electrodes for the Impedance Detection of Lectins. *Chem. Cent. J.* **2014**, *8*, 67. [CrossRef]
117. Xie, D.; Feng, X.Q.; Hu, X.L.; Liu, L.; Ye, Z.; Cao, J.; Chen, G.R.; He, X.P.; Long, Y.T. Probing Mannose-Binding Proteins That Express on Live Cells and Pathogens with a Diffusion-to-Surface Ratiometric Graphene Electrosensor. *ACS Appl. Mater. Interfaces* **2016**, *8*, 25137–25141. [CrossRef]
118. He, X.P.; Zhu, B.W.; Zang, Y.; Li, J.; Chen, G.R.; Tian, H.; Long, Y.T. Dynamic Tracking of Pathogenic Receptor Expression of Live Cells Using Pyrenyl Glycoanthraquinone-Decorated Graphene Electrodes. *Chem. Sci.* **2015**, *6*, 1996–2001. [CrossRef]
119. Cleaves, H.J. Affinity Constant. In *Encyclopedia of Astrobiology*; Springer: Berlin/Heidelberg, Germany, 2011; pp. 23–24. [CrossRef]
120. Nagae, M.; Yamaguchi, Y. Biophysical Analyses for Probing Glycan-Protein Interactions. *Adv. Exp. Med. Biol.* **2018**, *1104*, 119–147. [CrossRef]
121. Ardá, A.; Jiménez-Barbero, J. The Recognition of Glycans by Protein Receptors. Insights from NMR Spectroscopy. *Chem. Commun.* **2018**, *54*, 4761–4769. [CrossRef] [PubMed]
122. Patrick Walton, S.; Stephanopoulos, G.N.; Yarmush, M.L.; Roth, C.M. Thermodynamic and Kinetic Characterization of Antisense Oligodeoxynucleotide Binding to a Structured MRNA. *Biophys. J.* **2002**, *82*, 366–377. [CrossRef] [PubMed]
123. Landry, J.P.; Ke, Y.; Yu, G.L.; Zhu, X.D. Measuring Affinity Constants of 1450 Monoclonal Antibodies to Peptide Targets with a Microarray-Based Label-Free Assay Platform. *J. Immunol. Methods* **2015**, *417*, 86–96. [CrossRef] [PubMed]
124. Richards, S.-J.; Gibson, M.I. Toward Glycomaterials with Selectivity as Well as Affinity. *JACS Au* **2021**, *1*, 2089–2099. [CrossRef]
125. Santos, A.; Bueno, P.R.; Davis, J.J. A Dual Marker Label Free Electrochemical Assay for Flavivirus Dengue Diagnosis. *Biosens. Bioelectron.* **2018**, *100*, 519–525. [CrossRef]
126. Åberg, V.; Fällman, E.; Axner, O.; Uhlin, B.E.; Hultgren, S.J.; Almqvist, F. Pilicides Regulate Pili Expression in E. Coli without Affecting the Functional Properties of the Pilus Rod. *Mol. Biosyst.* **2007**, *3*, 214–218. [CrossRef]
127. Schultz, K.M.; Klug, C.S. Characterization of and Lipopolysaccharide Binding to the E. Coli LptC Protein Dimer. *Protein Sci.* **2018**, *27*, 381–389. [CrossRef]
128. Ma, F.; Rehman, A.; Liu, H.; Zhang, J.; Zhu, S.; Zeng, X. Glycosylation of Quinone-Fused Polythiophene for Reagentless and Label-Free Detection of E. Coli. *Anal. Chem.* **2015**, *87*, 1560–1568. [CrossRef]
129. Dechtrirat, D.; Gajovic-Eichelmann, N.; Wojcik, F.; Hartmann, L.; Bier, F.F.; Scheller, F.W. Electrochemical Displacement Sensor Based on Ferrocene Boronic Acid Tracer and Immobilized Glycan for Saccharide Binding Proteins and E. Coli. *Biosens. Bioelectron.* **2014**, *58*, 1–8. [CrossRef]
130. Zhao, L.; Li, C.; Qi, H.; Gao, Q.; Zhang, C. Electrochemical Lectin-Based Biosensor Array for Detection and Discrimination of Carcinoembryonic Antigen Using Dual Amplification of Gold Nanoparticles and Horseradish Peroxidase. *Sens. Actuators B Chem.* **2016**, *235*, 575–582. [CrossRef]
131. Wang, L.; Shan, J.; Feng, F.; Ma, Z. Novel Redox Species Polyaniline Derivative-Au/Pt as Sensing Platform for Label-Free Electrochemical Immunoassay of Carbohydrate Antigen 199. *Anal. Chim. Acta* **2016**, *911*, 108–113. [CrossRef] [PubMed]
132. Thapa, A.; Soares, A.C.; Soares, J.C.; Awan, I.T.; Volpati, D.; Melendez, M.E.; Fregnani, J.H.T.G.; Carvalho, A.L.; Oliveira, O.N. Carbon Nanotube Matrix for Highly Sensitive Biosensors to Detect Pancreatic Cancer Biomarker CA19-9. *ACS Appl. Mater. Interfaces* **2017**, *9*, 25878–25886. [CrossRef] [PubMed]
133. Sletmoen, M.; Gerken, T.A.; Stokke, B.T.; Burchell, J.; Brewer, C.F. Tn and STn Are Members of a Family of Carbohydrate Tumor Antigens That Possess Carbohydrate–Carbohydrate Interactions. *Glycobiology* **2018**, *28*, 437–442. [CrossRef] [PubMed]

134. Silva, M.L.S.; Gutiérrez, E.; Rodríguez, J.A.; Gomes, C.; David, L. Construction and Validation of a Sambucus Nigra Biosensor for Cancer-Associated STn Antigen. *Biosens. Bioelectron.* **2014**, *57*, 254–261. [CrossRef]
135. Kveton, F.; Blsakova, A.; Lorencova, L.; Jerigova, M.; Velic, D.; Blixt, O.; Jansson, B.; Kasak, P.; Tkac, J. A Graphene-Based Glycan Biosensor for Electrochemical Label-Free Detection of a Tumor-Associated Antibody. *Sensors* **2019**, *19*, 5409. [CrossRef] [PubMed]
136. Patil, S.A.; Bshara, W.; Morrison, C.; Chandrasekaran, E.V.; Matta, K.L.; Neelamegham, S. Overexpression of α2,3sialyl T-Antigen in Breast Cancer Determined by Miniaturized Glycosyltransferase Assays and Confirmed Using Tissue Microarray Immunohistochemical Analysis. *Glycoconj. J.* **2014**, *31*, 509–521. [CrossRef]
137. Niu, Y.; He, J.; Li, Y.; Zhao, Y.; Xia, C.; Yuan, G.; Zhang, L.; Zhang, Y.; Yu, C. Multi-Purpose Electrochemical Biosensor Based on a "Green" Homobifunctional Cross-Linker Coupled with PAMAM Dendrimer Grafted p-MWCNTs as a Platform: Application to Detect α2,3-Sialylated Glycans and α2,6-Sialylated Glycans in Human Serum. *RSC Adv.* **2016**, *6*, 44865–44872. [CrossRef]
138. Niu, Y.; He, J.; Li, Y.; Zhao, Y.; Xia, C.; Yuan, G.; Zhang, L.; Zhang, Y.; Yu, C. Determination of α2,3-Sialylated Glycans in Human Serum Using a Glassy Carbon Electrode Modified with Carboxylated Multiwalled Carbon Nanotubes, a Polyamidoamine Dendrimer, and a Glycan-Recognizing Lectin from Maackia Amurensis. *Microchim. Acta* **2016**, *183*, 2337–2344. [CrossRef]
139. Yuan, Q.; He, J.; Niu, Y.; Chen, J.; Zhao, Y.; Zhang, Y.; Yu, C. Sandwich-Type Biosensor for the Detection of α2,3-Sialylated Glycans Based on Fullerene-Palladium-Platinum Alloy and 4-Mercaptophenylboronic Acid Nanoparticle Hybrids Coupled with Au-Methylene Blue-MAL Signal Amplification. *Biosens. Bioelectron.* **2018**, *102*, 321–327. [CrossRef]
140. Dall'olio, F.; Chiricolo, M.; Ceccarelli, C.; Minni, F.; Marrano, D.; Santini, D. β-Galactoside α2,6 Sialyltransferase in Human Colon Cancer: Contribution of Multiple Transcripts to Regulation of Enzyme Activity and Reactivity with Sambucus Nigra Agglutinin. *J. Cancer* **2000**, *88*, 58–65. [CrossRef]
141. Gao, L.; He, J.; Xu, W.; Zhang, J.; Hui, J.; Guo, Y.; Li, W.; Yu, C. Ultrasensitive Electrochemical Biosensor Based on Graphite Oxide, Prussian Blue, and PTC-NH$_2$ for the Detection of α2,6-Sialylated Glycans in Human Serum. *Biosens. Bioelectron.* **2014**, *62*, 79–83. [CrossRef] [PubMed]
142. Li, Y.; He, J.; Niu, Y.; Yu, C. Ultrasensitive Electrochemical Biosensor Based on Reduced Graphene Oxide-Tetraethylene Pentamine-BMIMPF6 Hybrids for the Detection of α2,6-Sialylated Glycans in Human Serum. *Biosens. Bioelectron.* **2015**, *74*, 953–959. [CrossRef] [PubMed]
143. Li, J.; He, J.; Zhang, C.; Chen, J.; Mao, W.; Yu, C. Dual-Type Responsive Electrochemical Biosensor for the Detection of α2,6-Sialylated Glycans Based on AuNRs-SA Coupled with c-SWCNHs/S-PtNC Nanocomposites Signal Amplification. *Biosens. Bioelectron.* **2019**, *130*, 166–173. [CrossRef] [PubMed]
144. Zhao, Y.; Chen, J.; Zhong, H.; Zhang, C.; Zhou, Y.; Mao, W.; Yu, C. Functionalized Ag/Fe-MOFs Nanocomposite as a Novel Endogenous Redox Mediator for Determination of α2,6-Sialylated Glycans in Serum. *Microchim. Acta* **2020**, *187*, 649. [CrossRef]
145. Stowell, S.R.; Ju, T.; Cummings, R.D. Protein Glycosylation in Cancer. *Annu. Rev. Pathol. Mech. Dis.* **2015**, *10*, 473–510. [CrossRef]
146. Xie, S.; Wang, F.; Wu, Z.; Joshi, L.; Liu, Y. A Sensitive Electrogenerated Chemiluminescence Biosensor for Galactosyltransferase Activity Analysis Based on a Graphitic Carbon Nitride Nanosheet Interface and Polystyrene Microsphere-Enhanced Responses. *RSC Adv.* **2016**, *6*, 32804–32810. [CrossRef]
147. Chen, Z.; Zhang, L.; Liu, Y.; Li, J. Highly Sensitive Electrogenerated Chemiluminescence Biosensor for Galactosyltransferase Activity and Inhibition Detection Using Gold Nanorod and Enzymatic Dual Signal Amplification. *J. Electroanal. Chem.* **2016**, *781*, 83–89. [CrossRef]
148. Bertok, T.; Lorencova, L.; Chocholova, E.; Jane, E.; Vikartovska, A.; Kasak, P.; Tkac, J. Electrochemical Impedance Spectroscopy Based Biosensors: Mechanistic Principles, Analytical Examples and Challenges towards Commercialization for Assays of Protein Cancer Biomarkers. *ChemElectroChem* **2019**, *6*, 989–1003. [CrossRef]
149. Garrote, B.L.; Santos, A.; Bueno, P.R. Label-Free Capacitive Assaying of Biomarkers for Molecular Diagnostics. *Nat. Protoc.* **2020**, *15*, 3879–3893. [CrossRef]
150. Kanyong, P.; Patil, A.V.; Davis, J.J. Annual Review of Analytical Chemistry Functional Molecular Interfaces for Impedance-Based Diagnostics. *Annu. Rev. Anal. Chem. 2020* **2020**, *13*, 183–200. [CrossRef]
151. Loyprasert, S.; Hedström, M.; Thavarungkul, P.; Kanatharana, P.; Mattiasson, B. Sub-Attomolar Detection of Cholera Toxin Using a Label-Free Capacitive Immunosensor. *Biosens. Bioelectron.* **2010**, *25*, 1977–1983. [CrossRef]
152. Bertok, T.; Sediva, A.; Katrlik, J.; Gemeiner, P.; Mikula, M.; Nosko, M.; Tkac, J. Label-Free Detection of Glycoproteins by the Lectin Biosensor down to Attomolar Level Using Gold Nanoparticles. *Talanta* **2013**, *108*, 11–18. [CrossRef]
153. Bertok, T.; Klukova, L.; Sediva, A.; Kasák, P.; Semak, V.; Micusik, M.; Omastova, M.; Chovanová, L.; Vlček, M.; Imrich, R.; et al. Ultrasensitive Impedimetric Lectin Biosensors with Efficient Antifouling Properties Applied in Glycoprofiling of Human Serum Samples. *Anal. Chem.* **2013**, *85*, 7324–7332. [CrossRef] [PubMed]
154. Pihíková, D.; Belicky, Š.; Kasák, P.; Bertok, T.; Tkac, J. Sensitive Detection and Glycoprofiling of a Prostate Specific Antigen Using Impedimetric Assays. *Analyst* **2016**, *141*, 1044–1051. [CrossRef] [PubMed]
155. Pihikova, D.; Kasak, P.; Kubanikova, P.; Sokol, R.; Tkac, J. Aberrant Sialylation of a Prostate-Specific Antigen: Electrochemical Label-Free Glycoprofiling in Prostate Cancer Serum Samples. *Anal. Chim. Acta* **2016**, *934*, 72–79. [CrossRef] [PubMed]
156. Chocholova, E.; Bertok, T.; Lorencova, L.; Holazova, A.; Farkas, P.; Vikartovska, A.; Bella, V.; Velicova, D.; Kasak, P.; Eckstein, A.A.; et al. Advanced Antifouling Zwitterionic Layer Based Impedimetric HER2 Biosensing in Human Serum: Glycoprofiling as a Novel Approach for Breast Cancer Diagnostics. *Sens. Actuators B Chem.* **2018**, *272*, 626–633. [CrossRef]

157. Wang, L.; Filer, J.E.; Lorenz, M.M.; Henry, C.S.; Dandy, D.S.; Geiss, B.J. An Ultra-Sensitive Capacitive Microwire Sensor for Pathogen-Specific Serum Antibody Responses. *Biosens. Bioelectron.* **2019**, *131*, 46–52. [CrossRef] [PubMed]
158. Oliveira, R.M.B.; Fernandes, F.C.B.; Bueno, P.R. Pseudocapacitance Phenomena and Applications in Biosensing Devices. *Electrochim. Acta* **2019**, *306*, 175–184. [CrossRef]
159. Land, K.J.; Boeras, D.I.; Chen, X.S.; Ramsay, A.R.; Peeling, R.W. REASSURED Diagnostics to Inform Disease Control Strategies, Strengthen Health Systems and Improve Patient Outcomes. *Nat. Microbiol.* **2018**, *4*, 46–54. [CrossRef] [PubMed]
160. Scholz, F. Voltammetric Techniques of Analysis: The Essentials. *ChemTexts* **2015**, *1*, 17. [CrossRef]
161. Tkac, J.; Davis, J.J. An Optimised Electrode Pre-Treatment for SAM Formation on Polycrystalline Gold. *J. Electroanal. Chem.* **2008**, *621*, 117–120. [CrossRef]
162. Cruz-Pacheco, A.F.; Quinchia, J.; Orozco, J. Cerium Oxide–Doped PEDOT Nanocomposite for Label-Free Electrochemical Immunosensing of Anti-P53 Autoantibodies. *Microchim. Acta* **2022**, *189*, 1–13. [CrossRef]

Review

Formation of miRNA Nanoprobes—Conjugation Approaches Leading to the Functionalization

Iveta Vilímová , Katel Hervé-Aubert * and Igor Chourpa

EA6295 Nanomédicaments et Nanosondes, Université de Tours, 37200 Tours, France
* Correspondence: katel.herve@univ-tours.fr; Tel.: +33-247-367-157

Abstract: Recently, microRNAs (miRNA) captured the interest as novel diagnostic and prognostic biomarkers, with their potential for early indication of numerous pathologies. Since miRNA is a short, non-coding RNA sequence, the sensitivity and selectivity of their detection remain a cornerstone of scientific research. As such, methods based on nanomaterials have emerged in hopes of developing fast and facile approaches. At the core of the detection method based on nanotechnology lie nanoprobes and other functionalized nanomaterials. Since miRNA sensing and detection are generally rooted in the capture of target miRNA with the complementary sequence of oligonucleotides, the sequence needs to be attached to the nanomaterial with a specific conjugation strategy. As each nanomaterial has its unique properties, and each conjugation approach presents its drawbacks and advantages, this review offers a condensed overview of the conjugation approaches in nanomaterial-based miRNA sensing. Starting with a brief recapitulation of specific properties and characteristics of nanomaterials that can be used as a substrate, the focus is then centered on covalent and non-covalent bonding chemistry, leading to the functionalization of the nanomaterials, which are the most commonly used in miRNA sensing methods.

Keywords: miRNA; conjugation strategy; covalent bonding; non-covalent bonding; nanomaterial

Citation: Vilímová, I.; Hervé-Aubert, K.; Chourpa, I. Formation of miRNA Nanoprobes—Conjugation Approaches Leading to the Functionalization. *Molecules* **2022**, *27*, 8428. https://doi.org/10.3390/molecules27238428

Academic Editor: Jahir Orozco Holguín

Received: 27 September 2022
Accepted: 25 November 2022
Published: 2 December 2022

Publisher's Note: MDPI stays neutral with regard to jurisdictional claims in published maps and institutional affiliations.

1. Introduction

The detection of miRNAs (short, non-coding RNA sequences of approximately 19 to 25 nucleotides) has tremendous potential for early diagnosis of varying disorders, most notably life-threatening and unpredictable diseases such as cancer. Their function as circulating biomarkers is however hindered by their sparseness in body fluids, small size, and minor differences between the types of miRNAs. Thus, their detection in a short time with high accuracy in biological samples remains a challenge. Body fluids affect large spectra of detecting systems due to their high viscosity, the pH influence, the presence of other interfering biomolecules, and as previously mentioned, low amounts of target miRNA [1].

The miRNA expression spectrum in respective body fluids can differ not only depending on the disease type and pathological conditions but also on other aspects related to the patient (e.g., medication, diet, age, etc.) The body fluids in which miRNAs are detected can be obtained by either invasive or non-invasive procedures. Currently, the research is more often focused on the detection in body fluids obtained by non-invasive or weakly invasive means, such as urine, whole blood, and serum. For the miRNA capture, the body fluid is typically mixed with nanoprobe suspension [1,2].

Methods and approaches of miRNA detection based on nanotechnologies offer an alternative path to a faster and less complicated detection process, nevertheless also with specific challenges, including reproducibility, standardization, optimization, normalization, and data processing. The techniques vary in utilized material (organic, inorganic, or hybrid), nanostructures (nanorods, nanowires, nanosheets, etc.), and the strategy of the complementary DNA application (hairpin conjugates, molecular beacons, catalytic self-assembly, spherical nucleic acids, etc.) [3–5].

As such, hybrid NPs are particularly interesting since they combine the unique physical properties of the inorganic core (especially useful for their detection/quantification) with the biocompatibility/biospecificity of the organic shell.

After the selection of the nanomaterial as the substrate, for example, for the nanoprobe, the next step is the functionalization via an appropriate conjugation strategy, generally in the form of the attachment of an oligonucleotide sequence. The cornerstone of miRNA targeting approaches is antisense technology based on hybridizing the target miRNA with a complementary oligonucleotide sequence, leading to the formation of a duplex with higher stability. Usually, a complementary miRNA of the same length as the target miRNA is used, since such molecules are widely commercially available along with a wide range of modifications. Other possibilities, such as single- or double-stranded DNA, various hairpin probes, and molecular beacons, depend on the chosen approach and subsequent detection technique. The form of the target miRNA depends on the chosen media of miRNA detection—for validation of a method in model solutions, it is possible to use commercially produced sequences mimicking the naturally occurring miRNA, whose detection in biofluids can be hindered by other biomolecules and proteins present [6].

The preparation of the functionalized material can be, at times, less scrutinized on the road to miRNA detection, the reports often understandably focused on achieving the lowest detection limits [7]. The review articles regarding miRNA detection methods analyze new developments and novel experimental techniques [5,8] or are devoted to some particular nanomaterial [9] or functionalized ligands [10]. However, to the best of our knowledge, none of those reviews is focused on the general strategy of functionalization and on its validation, in particular on the important details of the experimental protocol. In view of this situation, we suggest this review as an attempt to gain a deeper insight in the conjugation strategies and methods (Figure 1). In particular, this review is focused on the most common conjugation approaches applied to connect the nanomaterial and complementary sequences that are used to target specific miRNA. Since each nanomaterial presents its own unique properties and characteristics beyond those common to all, they are briefly outlined in the first part, along with the suggestion of potential characterization techniques. The following section describes the choice of conjugation approaches, divided into covalent and non-covalent ones, with drawbacks and advantages presented for each.

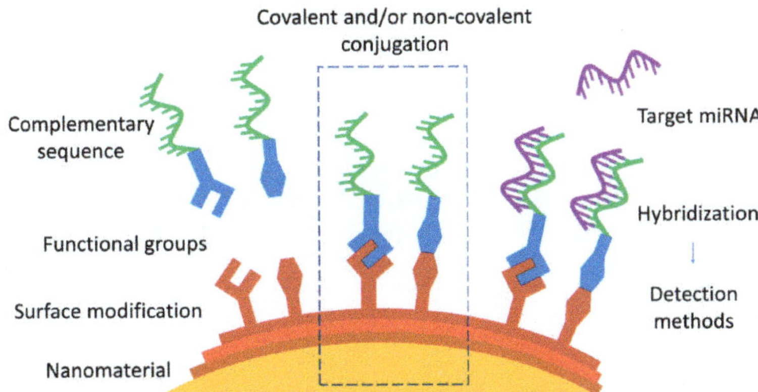

Figure 1. Position of the focus of this review, i.e., approaches of nanomaterial functionalization by covalent or non-covalent conjugation with complementary miRNA or DNA sequence as a part of the strategy of miRNA nanoprobe development.

2. Nanomaterials Used as Substrates

All nanomaterials mentioned in this brief overview have high surface energy and a high surface-to-volume ratio, leading to a higher catalytical activity. Their more specific properties vary depending on the nanomaterial, which oftentimes leads to a combination of two or more types of nanomaterials to take better advantage of their strengths (Table 1).

2.1. Gold Nanoparticles

Gold nanoparticles (AuNPs) have excellent catalytic and electrical properties and unique optical properties, which are extensively studied for signal amplification with the aim of reaching highly sensitive biosensing. Their optical properties are rooted in the localized surface plasmon resonance (LSPR) phenomenon, and additionally, the plasmonic surface can quench the fluorescence. With the increase of their size and/or upon aggregation of the AuNPs in aqueous suspensions (colloids), their LSPR band shifts are detected by means of a UV-Vis spectrometer or seen with a naked eye as a color change [11].

AuNPs are biocompatible due to their chemical stability in biological fluids. Possible toxicity concerns are connected to their long-term retention in the body. On the other hand, AuNPs can be easily modified with miRNA molecules, particularly due to the gold surface affinity to the thiol group [12]. Used as electrochemical or optical sensors in fluorescence, surface-enhanced Raman scattering (SERS), surface plasmon resonance (SPR), or colorimetry-based detection approaches, AuNPs can either be dispersed in the aqueous media [13] or immobilized on solid support [14]. The enhancement of AuNPs is widely used in the applications of biosensors in miRNA detection, achieving higher selectivity and sensitivity starting from sub-nanomolar levels [11].

2.2. Silver Nanoparticles

Compared with AuNPs, the LSPR band of silver nanoparticles (AgNPs) has a shorter position (approximately 400 nm instead of 535 nm). Nevertheless, both the monodispersity and reproducibility are more difficult to control with AgNPs compared with AuNPs. This leads to a frequent combination of AgNPs with other metals, such as gold, for example, in the form of core–shell nanoparticles [15–17].

AgNPs have found application as sensors in electrochemical and optical detection due to their high extinction coefficient, high scattering-to-extinction ratio, and high field enhancement. AgNPs show distinct amplified signals [18] along with strong Raman and fluorescence enhancement. Successful approaches are based on plasmonic properties of AgNPs implemented in platforms applying SERS [19], LSPR [20], or fluorescence readouts for detecting miRNA [21]. Similar to other metal nanoparticles, AgNPs also need a compatible organic coating before their use in biological systems to positively influence their stability and possible cytotoxicity [22,23].

2.3. Magnetic Nanoparticles

The term "magnetic nanoparticles" (MNPs) generally encompasses metal (e.g., iron, nickel platinum, and cobalt), metal oxide (e.g., iron oxides Fe_3O_4, γ-Fe_2O_3, and ferrites), and metal alloy NPs (e.g., FePt and FeCo). MNPs present several advantages enabling their application in biomedicine, since after proper modification, MNPs are non-toxic and biocompatible and can be utilized as nanovectors for specific targeting. Their magnetic properties allow the use of magnetic separation for the binding and detection of biomolecules [24,25]. MNPs are often combined with other nanomaterial types to unite the specific advantages [26].

Considering that miRNAs are generally present in very low concentrations in body fluids, the possibility to reconcentrate the miRNAs captured by MNPs via magnetic separation or extraction presents an enticing enhancement in the detection process. The miRNA extraction from a larger volume sample also represents insight into a larger population of the molecules and provides simpler handling procedures. However, the magnetic separation can be hampered if the viscous drag of body fluids overwhelms the magnetophoretic force [27].

2.4. Quantum Dots

Quantum dots (QDs) are luminescent semiconductor nanocrystals with the possibility to tune the maximum wavelength of their light emission spectra. Stable QDs can be obtained by an easy one-pot and/or one-step synthesis directly in a water medium [28]. Aqueous suspensions of QDs have a high photoluminescent quantum yield and resistance to photobleaching [29]. These qualities prompted their advancing application as optical, electrochemical, and chemiluminescent biosensors [4].

Due to high surface reactivity, QDs can serve as a nanoscale scaffold for further functionalization with common conjugation chemistry.

All in all, QDs are useful in miRNA detection due to their strong fluorescence provided by high quantum yield, narrow emission, and broad absorption spectra which provide multicolor labels with one light source and a strongly active surface for conjugations. Interactions on the surface of QDs lead to the activation or quenching of the fluorescence signal, allowing miRNA detection in complex media, such as body fluids.

2.5. Carbon Nanomaterials

Various carbon nanomaterials offer different characteristics (higher surface area, biocompatibility, and non-toxicity) useful for biosensing. Generally, carbon nanoparticles have strong and adjustable photoluminescence [30].

Carbon nanomaterials for miRNA detection can be in the form of carbon nanoparticles [31], carbon nanotubes [32], nanofibers [33], quantum dots [34], fullerenes [35], graphene nanosheets [36], and graphene oxide [37].

The good electrical conductivity and sensitivity of carbon nanomaterials are applied for the design and construction of electrochemical biosensors, particularly for electrode surface modification and for the preparation of modified electrodes [30,38–40].

Table 1. Types of nanomaterials used as substrates for miRNA nanoprobes.

Nanomaterial	Advantages	Disadvantages	References
Gold	<ul><li>easy synthesis and modification</li><li>chemical stability</li><li>affinity to a variety of ligands</li><li>plasmonic properties</li></ul>	<ul><li>poor degradability</li><li>possible long-term toxicity</li></ul>	[11,14,41]
Silver	<ul><li>electrochemical detection</li><li>plasmonic properties stronger than with gold</li></ul>	<ul><li>poor degradability</li><li>possible cytotoxicity</li></ul>	[18,20,42]
Magnetic	<ul><li>superparamagnetism</li><li>magnet-based purification andseparation</li></ul>	<ul><li>in larger size prone to magnetic aggregation</li></ul>	[1,43,44]
Quantum dots	<ul><li>strong and stable fluorescence</li><li>high surface reactivity</li><li>high electron transfer</li></ul>	<ul><li>stabilizers and surfactants needed on the surface</li></ul>	[28,45,46]
Carbon	<ul><li>widely available carbon sources</li><li>fluorescence properties</li><li>electrocatalytic activity</li></ul>	<ul><li>few in vivo studies performed</li></ul>	[30,37,47]

2.6. Characterization of Properties of the Nanomaterial before and after Conjugations

The determination of the physico-chemical properties of the chosen nanomaterial goes hand in hand with the need for monitoring their changes, both before and after the successful conjugation of a biomolecule. Some characterization techniques are more popular than others; however, as a rule, there is usually a combination of two or more methods used to cover possible inaccuracies and weak spots of the measurements.

Dynamic light scattering (DLS) measurements remain crucial in the confirmation of the size distribution of nanoparticles in aqueous dispersion and their colloidal stability [43]. The determination of the hydrodynamic diameter (DH) is usually joined by the evaluation of the zeta potential representative of the surface charges since the change of this parameter is often used to confirm the surface modification of the nanomaterial [48,49]. However, with small biomolecules such as miRNA, the changes in DH and zeta potential are generally too small to be reliable proof of conjugation. Although DLS and zetametry measurements are very common and can be found in the majority of performed studies, they are also necessarily joined by other complementary techniques which allow for more accurate confirmations [50].

Gel electrophoresis (of both the agarose and the polyacrylamide types) can be used to determine whether the conjugation between the nanomaterial and the ligand, or between two ligands, occurred. The migration of the unattached ligands is noticeably different compared with the migration of the conjugated ones, allowing for the separation of both. As a control, samples of NPs conjugated to a nucleotide sequence complementary to target miRNA, the non-conjugated NPs [44,51], and the solutions of free sequences are often used [52,53]. However, when it comes to confirmation of the binding of a large number of biomolecules, it is necessary to have a reliable purification technique to eliminate the excess of free ligands.

Structural and elemental analyses with X-ray diffraction (XRD) and X-ray photoelectron spectroscopy are often combined [38] and are generally used as a precise way to study the composition and bulk properties of a nanomaterial, including the presence of biomolecules on its surface. XRD provides information about the structural properties of a nanomaterial [44], including crystallinity and phase, while it can also give a rough idea of the average size of the NPs. XPS is the most sensitive spatially resolved technique enabling the determination of the bonding nature and elemental ratio in the nanomaterial, data from which the composition of the layers can be deduced [35]. Both techniques are less efficient with very small NPs, with smaller precision in structural measurements, and with amorphous NPs, where different atomic lengths can affect the measurement.

Thermogravimetric analysis (TGA) is centered on the weight loss of a sample as a function of increased temperature. Due to the organic content, the characteristic weight loss profile before and after the conjugation is investigated [12,14]. Generally, TGA is useful to confirm the presence of organic molecules conjugated to the inorganic nanomaterial, as is also the case with spectroscopic techniques.

Fourier transform infrared (FT-IR) spectroscopy is based on the detection of certain functional groups present on the surface of the nanomaterial in the event of successful conjugation and is considered acceptable for the detection of biological ligands onto inorganic nanomaterial [34,54,55]. Nevertheless, FT-IR is not always sensitive enough. Additionally, non-reacted components need to be removed with the appropriate purification method prior to FT-IR measurements.

Fluorescence spectroscopy is quite favored for confirmation of both the successful functionalization of the nanomaterial and the subsequent hybridization of target miRNA. The strategies are based on either the quenching of the fluorescence signal [56] or its increase after the binding event [34]. It can be the nanomaterial itself that has fluorescent properties, such as QDs, or a complementary sequence to the target miRNA modified with fluorescent dye. Therefore, in the case of using fluorescent components, the characterization of the nanoprobe and the detection of the target miRNA capture are closely entwined. Notably, it is important to have an efficient strategy of sample purification or a reliable way to distinguish the signals, as the non-conjugated fluorescence components can easily interfere with the accuracy of the measurements.

3. Conjugation Strategies Leading to Functionalization

The coating of the nanomaterial, more specifically of nanoparticles, improves the chemical and colloidal stability, provides surface modification available for further functionalization, and can alter the properties of NPs [57].

Selection of the conjugation method is influenced by: (i) available reactive groups on the surface of the nanomaterial and on the ligands intended to be conjugated; (ii) chemical and mechanical stability of both precursors and final product; and (iii) simplicity and reproducibility of the synthetic procedure, along with its cost-effectiveness [58].

Nanomaterials can be conjugated with various targeting biomolecules (such as proteins, nucleic acids, antibodies, and small drug molecules) via covalent or non-covalent assembly, which is more or less stable, respectively. Covalent conjugation is possible to achieve through the direct conjugation method, click chemistry, cross-linking strategies, etc. Non-covalent conjugation relies on physical interactions, such as hydrogen bonds, and interactions by electrostatic attraction or hydrophobic behavior. It should be noted that experimental approaches for formulating miRNA nanoprobes often combine covalent and non-covalent attachment to better exploit the advantages of both strategies, summarized in Table 2. Similarly, the most common approaches mentioned in the text are put into focus in Figure 2.

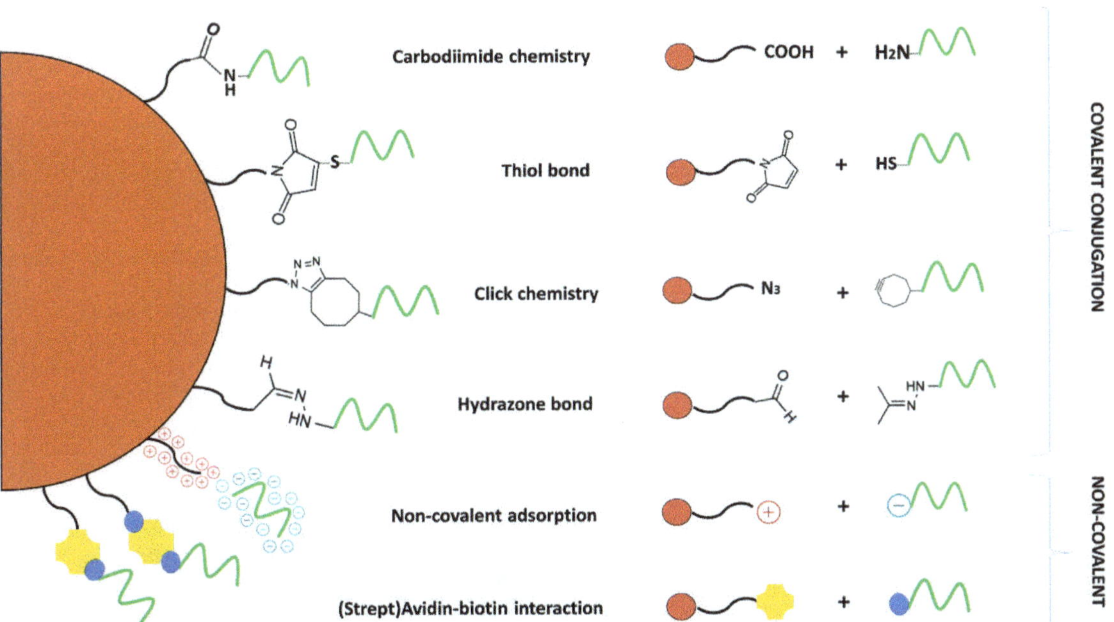

Figure 2. Schematic representations of covalent and non-covalent conjugation approaches.

3.1. Covalent Conjugation

Covalent conjugation offers high stability and reaction efficiency. Several functional groups can form a covalent bond on the surface of nanomaterial and ligand, such as carboxyl (-COOH), amine (-NH$_2$), aldehyde (-CHO), and thiol (-SH, sulfhydryl). In the case of the absence of a reactive group on some biomolecules and drugs, modification is possible before the conjugation [58]. The most popular covalent conjugations are carbodiimide chemistry and thiol bond, with widespread application.

3.1.1. Carboxylic Acid–Amine Bond

Amines, a general label for compounds with a nitrogen atom with a free electron pair, are reactive due to their nucleophilic ability (donating the free electron pair to form a bond). This is especially the case for primary amines, making them ideal participants in conjugation with reactive groups. Amines are also widely used for the modification of biomolecules. For example, nearly all proteins have free amine at the N-terminus. The ending parts of proteins (C-terminus) often contain carboxyl groups which are also found on the side chains of amino acids (e.g., glutamate and aspartate). Carboxylic acids are very potent for fast reactions with nucleophilic compounds, where the hydrogen from acid is removed to form an anion, which is unsusceptible to additional reactions with a second nucleophilic compound [59].

One of the most widely used types of links is the amide bonds formed by the reaction of the amine group with N-hydroxysuccinimidyl (NHS)-activated carboxylic compound. In this method, the carboxylic compound has a first reaction with 1-ethyl-3-(3-dimethylaminopropyl)carbodiimide (EDC) and NHS to form an acyl amino ester, followed by a subsequent reaction with an amine to create the amide bond of exceptional stability. EDC is a universal cross-linking agent; however, after the reaction with the carboxyl group, some unstable intermediate is formed with a very short life in aqueous solutions, hence, the addition of NHS acts as a stabilizing agent [9].

In miRNA detection, NHS can be used for activation of carboxyl-modified nanomaterials, such as magnetic beads [60,61], and subsequent attachment of NH_2-modified miRNA antisense strain. For example, the carboxylic acid group can be attached to the surface magnetic nanoparticles with silica coating [55]. Alternatively, a single-stranded DNA probe with COOH− group on one end can be covalently bound to amino-modified magnetic iron oxide nanoparticles [62].

A similar principle was utilized by Mahani et al., with a molecular beacon in a hairpin structure with a fluorescent quencher at one end and the NH_2 group at the other. Carbon QD with the carboxyl groups on the surface formed an amide link with a molecular beacon, and only weak fluorescence was observed for the whole complex. In the presence of target miRNA-21, the hybridization of the hairpin structure led to the opening of the molecular beacon loop. The change in the distance between the quencher and the QD led to a change and increase of the fluorescent signal [63].

Amination, a reductive reaction where the amine group is combined with an organic molecule by transforming a carbonyl group (such as the carboxylic group), was used by Salimi et al. for the functionalization of graphene oxide. Graphene oxide is suitable for reduction and functionalization, and after functionalization with amine, the resulting graphene presents acceptable material for the attachment of biomolecules. Salimi et al. used this fact in the fabrication of an electrochemical biosensor for the selective detection of miRNA. Amino-functionalized reduced graphene oxide was assembled on the surface of a glassy carbon electrode with glutaraldehyde as a cross-linking agent, and a special sequence of miRNA-containing terminal amino group was attached for the detection of target miRNA [64].

Horny et al. utilized magnetic hyperthermia, with superparamagnetic core–shell γ-Fe_2O_3–SiO_2 NPs working as nanoheaters. Maghemite cores were slightly pre-aggregated before coating with a silica shell, the elongated shape proposing suitable properties for magnetic hyperthermia. Since silica-coated core–shell nanoparticles with a maghemite γ-Fe_2O_3 core had -NH_2 groups present on the surface, it allowed for the attachment of carboxyl-modified DNA probes [65].

Gao et al. combined carbon dots with graphene oxide. Carbon dots were linked with the amine-modified complementary DNA, and this complex was adsorbed on the surface of graphene oxide, quenching the fluorescence signal of the carbon dots. In the presence of target miRNA, the DNA desorbed from the surface of graphene oxide, hybridizing with the target miRNA and renewing the fluorescence signal, in proportion to the concentration of target miRNA [34].

Notably, amine modification of complementary oligonucleotide sequences continues to be the popular approach. Yao et al. conjugated iodide-modified magnetic beads with carboxyl groups on the surface and amine-modified DNA probes [42]. Xu et al. linked an amine-modified DNA probe with core–shell Au–Ag NPs covered with 5,5′-dithiobis (2-nitrobenzoic acid) [53]. Yazdanparast et al. synthesized magnetic core–shell Fe_3O_4–Ag NPs with carboxylic groups on the surface and joined them with amine-modified C-miRNA, applying the complex on a magnetic bar carbon paste electrode [54].

Carbodiimide chemistry is one of the most common approaches to binding miRNA to the nanomaterial. The reaction can be performed in an aqueous medium with no need for a complex solvent system, leading to a widespread application. However, the best efficiency is achieved in acidic pH, and it is necessary to perform it in buffers without additional amine and carboxylic groups. Depending on the nanomaterial, the challenge is presented in the form of the alignment of ligands and their binding orientation on the surface of the nanomaterial with multiple amine groups present. Moreover, the presence of additional amine and carboxylic groups (in ligands, on the nanomaterial, and in biological fluids) can also lead to a lower selectivity of the reaction [58,66].

3.1.2. Thiol Bond

Thiol is a compound with the -SH pair linked to single carbon or a carbon-containing group of atoms. Thiols play a large role in biological systems, as they maintain suitable levels of oxidation and reduction state of cells, proteins, and organisms. It is also possible to link thiol and amine groups covalently (using a heterobifunctional coupling agent with one sulfhydryl-reactive and one amine-reactive group) or to link thiol and alcohol groups (similarly with an agent containing a hydroxyl reactive site). Alternatively, a biomolecule can be attached directly to the surface of the nanomaterial, using a dative (coordination) bond, generally formed by the biomolecule donating two electrons from a single atom. This usually happens with metal affinity coordination and thiol interactions. The formed bond is longer than the covalent bond, with higher energy and sensitivity to oxidation and change in pH. Metal-affinity bonds are formed between cationic metal on the surface of the nanomaterial and, for example, imidazole ring present at histidine residues. Affinity is increased with a larger number of histidine amino acids [9].

The high affinity of thiol groups for gold results in a self-assembled monolayer on the gold surface, or simply in the Au–S bond (which is more stable than a self-assembled monolayer), making it quite popular and gaining it widespread application. The strength of the bond between isolated thiols and gold depends on the properties of the gold surface, interaction time, and the pH of the solution. Coordinate Au–SH bond shifts to a covalent nature of the Au–S bond with higher pH. The reaction offers better efficiency at neutral pH and is suitable for nanomaterials or ligands that are less stable at acidic pH [67]. In many cases, thiolated capture probes with complementary sequences to target miRNA are used in some combination with AuNPs [21]. For example, Al Mubarak et al. covalently bound a hairpin probe of complementary DNA to the surface localized AuNPs on a gold substrate by using 1,4-benzenedithiol (a benzene core with two -SH groups in opposite places) [68].

M. Wang et al. obtained a strong electrochemical response of AgNPs in combination with AuNPs on a glassy carbon electrode for miRNA detection. AgNPs were adsorbed on the electrode subsequently after AuNPs with a clearly detectable silver peak. When AuNPs functionalized with thiolated complementary DNA were present on the electrode, steric hindrance and electrostatic repulsion prevented the adsorption of AgNPs, decreasing the signal response of AgNPs. In the presence of target miRNA, the complementary DNA was hybridized on the surface of AuNPs, forming a duplex. Duplex-specific nuclease, an enzyme able to disrupt the bond between the nucleotides in the duplex, leads to the elimination of the duplex from the surface of AuNPs. AgNPs could be then adsorbed on the electrode, providing detectable signal amplification and leading to the determination of the amount of target miRNA [39].

A similar approach of AuNPs (combined with polypyrrole-reduced graphene oxide) deposited on a glassy carbon electrode enabled the immobilization of thiolated capture probes, permitting the further formation of hairpin probes [52]. In the same vein, the self-assembly of thiolated capture probes and their subsequent immobilization on a pencil graphite electrode, modified with carbon black material and AuNPs, allowed the capture of the target miRNA [40].

A complementary DNA probe modified with SH can be covalently bonded to the AuNP surface [14,29,41], and similarly, the same type of modification can be used for a conjugation on AgNPs [18,69], core–shell Au–Ag NPs [26], Au–Ag nanocages [20], AuNP–peptide nanotube composites [12], Au-coated conglomerates of superparamagnetic NPs [43], citrate-capped AuNPs electrostatically adsorbed to Fe_3O_4 NPs [70], Ag nanoclusters on Au electrode [71], and hollow Au–Ag nanospheres [72].

The maleimide chemistry, along with the carbodiimide chemistry, is a very common strategy in miRNA adsorption to the nanomaterial. The functional groups of the reaction are less common in biological fluids, leading to better selectivity and ligand orientation. However, the thiolation of ligands can lead to a change in their chemical structure, negatively influencing the binding affinity. The presence of proteins containing thiol groups (such as in serums) can also lead to a less selective reaction [10].

3.1.3. Click Chemistry and Hydrazone Bond

As miRNA capture and detection can still be considered a new branch of scientific research, many of the covalent conjugation strategies that are regularly used for binding targeting biomolecules or ligands to nanomaterials [9,10] have not yet been applied for miRNA, or only rarely. Two such examples are presented here, namely click chemistry and hydrazone bond.

In the variety of conjugation strategies with biomolecules exists the problem of undesired side reactions, due to the vast number of reactive sites and functional groups. While this is partially managed with specific coupling agents, they also may react with other functional groups (one such example is NHS, able to react with sulfhydryl and hydroxyl groups). This led to the development of bio-orthogonal chemistry, where reagents are stable (not prone to oxidization or hydrolyzation process) and have a reactive site that reacts only with a specific functional group.

These types of reactions are often referred to as click chemistry with their concept lying in a reaction that occurs quickly, under moderate conditions, and leads to a specific product (possibly with an easily removed by-product). Leading advancements of click chemistry were the developments of copper-catalyzed click chemistry followed by the copper-free approach, Diels–Alder cycloadditions, or Thiol–Michael click reactions. Click chemistry is also very popular in gene therapeutics, for easy conjugation of RNA, mRNA, and siRNA with, for example, fluorescent molecules [9,73].

Lu et al. used aggregation of AuNPs amplified by click chemistry for the detection of plant miRNA by DLS. Two probes, probe A, modified with dibenzocyclooctyne, and probe B, modified with azide, were synthesized to form a complementary sequence to target miRNA together. Hybridization with target miRNA leads to the close vicinity of both functional groups, forming a complex. Similarly, two types of AuNPs, with two sequences complementary to probe A and probe B, led to the respective attachment of both probes, and when in the presence of target miRNA, allowed the formation of bigger aggregates of AuNPs [74].

L. Zhou et al. employed click chemistry, i.e., the copper (I)-catalyzed azide–alkyne cycloaddition to combine two nucleic acid strands containing G-quadruplex DNA sequences. Alkyne-modified hairpin strand and azide-modified strand were bound by click chemistry reaction, forming complementary DNA capture probes, which were deposited on a Au electrode modified with fullerene nanoparticles [35].

Another type of bio-orthogonal conjugation method has a basis in reductive amination reaction (amine group is combined with organic molecule), where a hydrazone bond is formed by a reaction between aldehydes and hydrazide groups. The synthetic procedure is straightforward with good reproducibility, and the unique advantage is the pH sensitivity of the resulting bond—the bond is stable at neutral pH but becomes easily affected at acidic pH [58].

Ye Wang applied hydrazone coupling chemistry catalyzed by aniline to conjugate a specific hexahistidine peptide to amine-functionalized DNA, which led to this complex being self-assembled onto the hydrophilic surface of fluorescent quantum dots [75].

These strategies are less commonly applied for the nano-conjugation of complementary miRNA sequences to the nanomaterial, due to a relatively recent focus on the conjugation via click chemistry, combined with even more recent interest in miRNA-related functionalization. Nevertheless, there is significant potential for more studies in this field.

3.2. Non-Covalent Conjugation

The formation of macromolecules through non-covalent bonding is very common in nature, using for example, hydrophobic effects, coordination chemistry, hydrogen bonding, and dipole interactions (electrostatic and weaker dispersion forces). Along with covalent attachment, thiolated molecules can be also used for non-covalent conjugation (generally depending on the pH of the solvent). If the biomolecule has available thiol residue, the sulfhydryl group can form a dative bond on the surface of the nanomaterial.

The method of non-covalent bonding presents several advantages, such as fast formulation, avoidance of complex bonding, and the possibility of a quick release in target sites while minimizing cytotoxicity. This results in simple and cost-effective synthesis requiring fewer chemical reagents, allowing for industrial production and clinical applications. Nevertheless, compared with the covalent bonds, non-covalent linking is weaker and more prone to dissolution out of specific conditions. Often the stability of such bonds is weak in biofluids, presenting a challenge to overcome for in vivo applications.

3.2.1. Non-Covalent Adsorption—Electrostatic and Hydrophobic/Hydrophilic Interactions

Electrostatic and hydrophobic/hydrophilic interactions are very well-known, their attractiveness being in the rapid binding process with no further chemical modification steps necessary. However, this type of passive adsorption results in a non-specific bond between the nanomaterial and the biomolecule, leading to reduced stability in biological fluids, and complicating the in vivo applications and the characterization of the functionalized nanomaterial.

Electrostatic interaction and subsequent adsorption on the surface of the nanomaterial originate from the attraction of two opposite charges. This interaction is non-specific, relying on the strength of the respective charges of the two interacting parts, and is susceptible to changes in experimental conditions (such as temperature, pH, and ionic strength). There is also a possibility of electrostatic complex dissociation due to the presence of other competing biomolecules and the following exchange [9].

As miRNA molecules are negatively charged, the high positive charge of polyethylenimine-capped AuNPs is suitable for the concentration of target miRNA-155. Hakimian et al. mixed the resulting complex with citrate-capped AuNPs covalently attached to a thiolated hairpin DNA probe. Both types of AuNPs formed cross-linking aggregates due to probe–target attachment, allowing detectable aggregation of AuNPs [49].

Ajgaonkar et al. used graphene quantum dots doped with nitrogen as a slightly positively charged nanoplatform, non-covalently bonding the single-stranded DNA capture probe. The electrostatic binding of the capture probe and the $\pi-\pi$ stacking onto the surface of the graphene-QDs altered their intrinsic fluorescence [76].

Despite the potential of non-covalent adsorption, the application remains limited due to the challenging and often weak stability presented by such bonds.

3.2.2. Streptavidin–Biotin

A well-known example of affinity interactions is the (strept)avidin–biotin interaction, which is irreversible and stable (unaffected by variations of temperature, pH, organic solvents, or denaturing agents). The natural high-affinity interaction occurring between (strept)avidin and biotin is similar to the interaction between enzyme and substrate (or receptor and ligand), and the resulting bond is approaching the strength of the covalent bonding.

Streptavidin is similar to avidin (glycoprotein consisting of four identical subunits), and as a biotin-binding protein, it possesses four binding sites to biotin, one for each subunit. Compared with avidin, it has a lower charge with a smaller possibility of electrostatic interaction with other biomolecules (or cell membranes), which also makes the protein less soluble in an aqueous medium. Furthermore, streptavidin is a non-glycosylated protein (without carbohydrate residues), making the likelihood of a non-specific bond to other molecules via carbohydrate receptors significantly smaller. It is possible to insert biotin (vitamin B7) into biomolecules without affecting their properties or activity. Carboxylic acid in this small molecule also allows covalent conjugation [77].

Both the nanomaterial and/or the biomolecule can be biotinylated, making it a very popular conjugation of choice. Likewise, streptavidin and biotin modifications are widely commercially available, allowing for easy use (one such example is pre-prepared streptavidin-modified magnetic nanoparticles, e.g., magnetic beads [78,79], QDs [75], biotin-modified RNA [13], or biotinylated DNA sequences [78]).

Streptavidin modification of NPs can be also performed with a cross-linking agent, such as presented by Chan et al. concerning silica-coated iron oxide NPs. After the amino-functionalization of the NPs, streptavidin was bonded to the surface with glutaraldehyde acting as a cross-linking agent. The biotinylated DNA was then linked to the streptavidin-coated NPs, and the resulting nanoprobes were used for the capture of a complex formed by target miRNA and molecular beacon containing target miRNA sequence [80].

In our previous publication, we modified the surface of PEGylated SPOINs with streptavidin. Biotin at the end of the PEG chain allowed close coverage of streptavidin, and the creation of the sufficient streptavidin layer of available biotin-binding spots enabled the functionalization with biotinylated complementary miRNA sequences, leading to the formation of nanoprobes with the ability to capture the target miRNA [50].

As was already mentioned, the combination of two or more conjugation approaches is very common. Cheng et al. deposited a thiolated hairpin probe on the surface of a gold electrode via the Au–S bond, and its loop structure was opened with the hybridization of target miRNA. The second hairpin probe modified with biotin lost the loop structure in the presence of the open first probe, leaving the biotinylated end to bind with AgNPs modified with streptavidin. The addition of more biotinylated AgNPs led to the aggregation of AgNPs and a stronger electrochemical signal [81].

A drawback in the extremely popular streptavidin–biotin conjugation is the large structure of the final complex, as streptavidin is a protein, and biotin itself is usually linked to the respective nanomaterial (or biomolecule) using a spacer (such as PEG chain) or cross-linking agent, adding to the final size and potentially affecting the binding rate [9].

Table 2. Covalent and non-covalent strategies for the functionalization of nanomaterials.

	Conjugation Strategy	Advantages	Disadvantages	References
Covalent conjugation	Carboxylic acid—amine bond	+ high stability; + high conjugation efficiency; + no chemical modification on the ligands	− low selectivity; − absence of control over ligand orientation; − reactions in acidic and water-free media	[42,54,65]
	Thiol bond	+ fast and selective reactions; + neutral pH media that avoid NP aggregation	− thiolation may hinder the biological activities of ligands; − non-selective in media containing thiol groups	[14,26,52]
	Click chemistry	+ site-specific conjugation; + higher yield; + control of ligand orientation; + mild conditions	− relatively new and more complicated than other methods; − possible toxicity with Cu(I) catalysts	[33,74]
	Hydrazone bond	+ stable at neutral pH	− easily affected by acidic pH	[75]
Non-covalent conjugation	Non-covalent adsorption: Electrostatic and hydrophobic/hydrophilic interactions	+ fast and easy formulation; + reduced toxicity due to high charges of NPs	− weak interactions and poor stability − harder to characterize	[49,76,82]
	(Strept)avidin–biotin	+ specific interaction + irreversible and stable + good orientation of conjugated ligands	− the large structure of the final complex − the potentially lower binding rate	[13,80,81]

4. Conclusions

Since the recognition of miRNAs as novel prognostic biomarkers, their accurate sensing and detection became crucial, especially in non-invasively obtained biofluids. As such, methods based on nanomaterials present several advantages stemming from unique properties (e.g., optical, magnetic, and electrochemical) depending on the chosen nanomaterial. Quite often, two or more types of nanomaterial can be combined in an effort to bring together several of their advantageous properties.

Covalent conjugations (carbodiimide bond, thiol bond, click chemistry, and hydrazone bond) form a stable covalent bond, while non-covalent conjugations (non-covalent adsorption, electrostatic, hydrophobic/hydrophobic interactions, and streptavidin–biotin interaction) rely on physical interactions between the ligand and the nanomaterial. Covalent conjugation often offers high stability, yield, and efficiency, at the cost of sensitivity to the medium of the reaction and pH, with, at times, lower selectivity. Similarly, non-covalent conjugation shows fast and easy formulation with the avoidance of complex chemical modification; however, the interactions are often weak and harder to characterize.

The mentioned limitations present interesting challenges in the functionalization of the nanomaterial, with the procedure generally expected to be simple, fast, and efficient. First, functionalization is often difficult to determine and requires a combination of two or more analytical techniques. Second, the overall stability and longevity of a nanoprobe or functionalized nanomaterial have to be controlled. Third, the binding affinity of the chosen nanoplatform connected to the subsequent miRNA capture, and its sensitivity to the media, pH, and composition needs to be better determined. On the other hand, the biomolecules naturally present in the biological samples can interfere with the functionalized nanoplatforms and lead to decreased selectivity and/or sensitivity. This issue depends on the experimental design of the capture and still needs to be studied in depth.

Considering the relatively young research field of miRNA detection, some of the conjugation approaches, which were successfully applied for ligands and biomolecules other than RNA sequences and their variations, have yet to be used. This fact leaves ample space for presenting novel applications on the road to functionalization with the goal of miRNA sensing.

Author Contributions: Conceptualization, K.H.-A., I.C. and I.V.; investigation, I.V.; data curation, I.V.; writing—original draft preparation, I.V.; writing—review and editing, K.H.-A. and I.C.; visualization, I.V.; supervision, K.H.-A. and I.C.; project administration, K.H.-A.; funding acquisition, K.H.-A. All authors have read and agreed to the published version of the manuscript.

Funding: This research was funded by "Ligue Nationale contre le Cancer" (the local committees 37, 53 and 56), in the project NANOmiR.

Institutional Review Board Statement: Not applicable.

Informed Consent Statement: Not applicable.

Data Availability Statement: No new data were created or analyzed in this study. Data sharing is not applicable to this article.

Conflicts of Interest: The authors declare no conflict of interest.

References

1. Gessner, I.; Fries, J.W.U.; Brune, V.; Mathur, S. Magnetic Nanoparticle-Based Amplification of MicroRNA Detection in Body Fluids for Early Disease Diagnosis. *J. Mater. Chem. B* **2021**, *9*, 9–22. [CrossRef] [PubMed]
2. Junqueira-Neto, S.; Batista, I.A.; Costa, J.L.; Melo, S.A. Liquid Biopsy beyond Circulating Tumor Cells and Cell-Free DNA. *Acta Cytol.* **2019**, *63*, 479–488. [CrossRef] [PubMed]
3. Chandrasekaran, A.R.; Punnoose, J.A.; Zhou, L.; Dey, P.; Dey, B.K.; Halvorsen, K. DNA Nanotechnology Approaches for MicroRNA Detection and Diagnosis. *Nucleic Acids Res.* **2019**, *47*, 10489–10505. [CrossRef]
4. Masud, M.K.; Umer, M.; Hossain, M.S.A.; Yamauchi, Y.; Nguyen, N.T.; Shiddiky, M.J.A. Nanoarchitecture Frameworks for Electrochemical MiRNA Detection. *Trends Biochem. Sci.* **2019**, *44*, 433–452. [CrossRef] [PubMed]
5. Jet, T.; Gines, G.; Rondelez, Y. Advances in Multiplexed Techniques for the Detection and Quantification of MicroRNAs. *Chem. Soc. Rev.* **2021**, *50*, 4141–4161. [CrossRef]
6. Grijalvo, S.; Alagia, A.; Jorge, A.F.; Eritja, R. Covalent Strategies for Targeting Messenger and Non-Coding RNAs: An Updated Review on SiRNA, MiRNA and AntimiR Conjugates. *Genes* **2018**, *9*, 74. [CrossRef]
7. Shen, Z.; He, L.; Wang, W.; Tan, L.; Gan, N. Highly Sensitive and Simultaneous Detection of MicroRNAs in Serum Using Stir-Bar Assisted Magnetic DNA Nanospheres-Encoded Probes. *Biosens. Bioelectron.* **2020**, *148*, 111831. [CrossRef]
8. Khashayar, P.; Al-Madhagi, S.; Azimzadeh, M.; Scognamiglio, V.; Arduini, F. New Frontiers in Microfluidics Devices for MiRNA Analysis. *TrAC Trends Anal. Chem.* **2022**, *156*, 116706. [CrossRef]
9. Pereira, G.; Monteiro, C.A.P.; Albuquerque, G.M.; Pereira, M.I.A.; Cabrera, M.P.; Cabral Filho, P.E.; Pereira, G.A.L.; Fontesa, A.; Santos, B.S. (Bio)Conjugation Strategies Applied to Fluorescent Semiconductor Quantum Dots. *J. Braz. Chem. Soc.* **2019**, *30*, 2536–2560. [CrossRef]
10. Nguyen, P.V.; Allard-vannier, E.; Chourpa, I.; Herv, K. Nanomedicines Functionalized with Anti-EGFR Ligands for Active Targeting in Cancer Therapy: Biological Strategy, Design and Quality Control. *Int. J. Pharm.* **2021**, *605*, 120795. [CrossRef]
11. Coutinho, C.; Somoza, Á. MicroRNA Sensors Based on Gold Nanoparticles. *Anal. Bioanal. Chem.* **2019**, *411*, 1807–1824. [CrossRef] [PubMed]
12. Yaman, Y.T.; Vural, O.A.; Bolat, G.; Abaci, S. One-Pot Synthesized Gold Nanoparticle-Peptide Nanotube Modified Disposable Sensor for Impedimetric Recognition of MiRNA 410. *Sens. Actuators B Chem.* **2020**, *320*, 128343. [CrossRef]
13. Lee, T.; Mohammadniaei, M.; Zhang, H.; Yoon, J.; Choi, H.K.; Guo, S.; Guo, P.; Choi, J.W. Single Functionalized PRNA/Gold Nanoparticle for Ultrasensitive MicroRNA Detection Using Electrochemical Surface-Enhanced Raman Spectroscopy. *Adv. Sci.* **2020**, *7*, 1902477. [CrossRef] [PubMed]
14. Akbal Vural, O.; Yaman, Y.T.; Bolat, G.; Abaci, S. Human Serum Albumin−Gold Nanoparticle Based Impedimetric Sensor for Sensitive Detection of MiRNA-200c. *Electroanalysis* **2021**, *33*, 925–935. [CrossRef]
15. González-Rubio, G.; De Oliveira, T.M.; Altantzis, T.; La Porta, A.; Guerrero-Martínez, A.; Bals, S.; Scarabelli, L.; Liz-Marzán, L.M. Disentangling the Effect of Seed Size and Crystal Habit on Gold Nanoparticle Seeded Growth. *Chem. Commun.* **2017**, *53*, 11360–11363. [CrossRef] [PubMed]
16. Polte, J. Fundamental Growth Principles of Colloidal Metal Nanoparticles—A New Perspective. *CrystEngComm* **2015**, *17*, 6809–6830. [CrossRef]

17. Fischer, A.; Thuenemann, A.F.; Emmerling, F.; Rademann, K.; Kraehnert, R.; Polte, J.; Tuaev, X.; Wuithschick, M. Formation Mechanism of Colloidal Silver Nanoparticles: Analogies and Differences to the Growth of Gold Nanoparticles. *ACS Nano* **2012**, *6*, 5791–5802. [CrossRef]

18. Li, H.; Cai, Q.; Yan, X.; Jie, G.; Jie, G. Ratiometric Electrochemical Biosensor Based on Silver Nanoparticles Coupled with Walker Amplification for Sensitive Detection of MicroRNA. *Sens. Actuators B Chem.* **2022**, *353*, 131115. [CrossRef]

19. Pang, Y.; Wang, C.; Wang, J.; Sun, Z.; Xiao, R.; Wang, S. Fe3O4@Ag Magnetic Nanoparticles for MicroRNA Capture and Duplex-Specific Nuclease Signal Amplification Based SERS Detection in Cancer Cells. *Biosens. Bioelectron.* **2016**, *79*, 574–580. [CrossRef]

20. Li, M.; Cheng, J.; Yuan, Z.; Shen, Q.; Fan, Q. DNAzyme-Catalyzed Etching Process of Au/Ag Nanocages Visualized via Dark-Field Imaging with Time Elapse for Ultrasensitive Detection of MicroRNA. *Sens. Actuators B Chem.* **2021**, *330*, 129347. [CrossRef]

21. Zhou, W.; Tian, Y.F.; Yin, B.C.; Ye, B.C. Simultaneous Surface-Enhanced Raman Spectroscopy Detection of Multiplexed MicroRNA Biomarkers. *Anal. Chem.* **2017**, *89*, 6120–6128. [CrossRef] [PubMed]

22. Sharma, V.K.; Siskova, K.M.; Zboril, R.; Gardea-Torresdey, J.L. Organic-Coated Silver Nanoparticles in Biological and Environmental Conditions: Fate, Stability and Toxicity. *Adv. Colloid Interface Sci.* **2014**, *204*, 15–34. [CrossRef] [PubMed]

23. Panáček, A.; Kvítek, L.; Smékalová, M.; Večeřová, R.; Kolář, M.; Röderová, M.; Dyčka, F.; Šebela, M.; Prucek, R.; Tomanec, O.; et al. Bacterial Resistance to Silver Nanoparticles and How to Overcome It. *Nat. Nanotechnol.* **2018**, *13*, 65–71. [CrossRef] [PubMed]

24. Knežević, N.; Gadjanski, I.; Durand, J.O. Magnetic Nanoarchitectures for Cancer Sensing, Imaging and Therapy. *J. Mater. Chem. B* **2019**, *7*, 9–23. [CrossRef]

25. Janko, C.; Ratschker, T.; Nguyen, K.; Zschiesche, L.; Tietze, R.; Lyer, S.; Alexiou, C. Functionalized Superparamagnetic Iron Oxide Nanoparticles (SPIONs) as Platform for the Targeted Multimodal Tumor Therapy. *Front. Oncol.* **2019**, *9*, 59. [CrossRef]

26. Li, M.; Li, J.; Zhang, X.; Yao, M.; Li, P.; Xu, W. Simultaneous Detection of Tumor-Related MRNA and MiRNA in Cancer Cells with Magnetic SERS Nanotags. *Talanta* **2021**, *232*, 122432. [CrossRef]

27. Leong, S.S.; Yeap, S.P.; Lim, J.K. Working Principle and Application of Magnetic Separation for Biomedical Diagnostic at High- and Low-Field Gradients. *Interface Focus* **2016**, *6*, 20160048. [CrossRef]

28. Goryacheva, O.A.; Novikova, A.S.; Drozd, D.D.; Pidenko, P.S.; Ponomaryeva, T.S.; Bakal, A.A.; Mishra, P.K.; Beloglazova, N.V.; Goryacheva, I.Y. Water-Dispersed Luminescent Quantum Dots for MiRNA Detection. *TrAC—Trends Anal. Chem.* **2019**, *111*, 197–205. [CrossRef]

29. Song, W.; Zhang, F.; Song, P.; Zhang, Z.; He, P.; Li, Y.; Zhang, X. Untrasensitive Photoelectrochemical Sensor for MicroRNA Detection with DNA Walker Amplification and Cation Exchange Reaction. *Sens. Actuators B Chem.* **2021**, *327*, 128900. [CrossRef]

30. Asadian, E.; Ghalkhani, M.; Shahrokhian, S. Electrochemical Sensing Based on Carbon Nanoparticles: A Review. *Sens. Actuators B Chem.* **2019**, *293*, 183–209. [CrossRef]

31. Li, H.; Li, Y.; Li, W.; Cui, L.; Huang, G.; Huang, J. A Carbon Nanoparticle and DNase I-Assisted Amplified Fluorescent Biosensor for MiRNA Analysis. *Talanta* **2020**, *213*, 120816. [CrossRef] [PubMed]

32. Liu, Q.; Bai, W.; Guo, Z.; Zheng, X. Enhanced Electrochemiluminescence of Ru(Bpy)32+-Doped Silica Nanoparticles by Chitosan/Nafion Shell@carbon Nanotube Core-Modified Electrode. *Luminescence* **2021**, *36*, 642–650. [CrossRef] [PubMed]

33. Sahtani, K.; Aykut, Y.; Tanik, N.A. Lawsone Assisted Preparation of Carbon Nanofibers for the Selective Detection of MiRNA Molecules. *J. Chem. Technol. Biotechnol.* **2022**, *97*, 254–269. [CrossRef]

34. Gao, Y.; Yu, H.; Tian, J.; Xiao, B. Nonenzymatic DNA-Based Fluorescence Biosensor Combining Carbon Dots and Graphene Oxide with Target-Induced DNA Strand Displacement for MicroRNA Detection. *Nanomaterials* **2021**, *11*, 2608. [CrossRef]

35. Zhou, L.; Wang, T.; Bai, Y.; Li, Y.; Qiu, J.; Yu, W.; Zhang, S. Dual-Amplified Strategy for Ultrasensitive Electrochemical Biosensor Based on Click Chemistry-Mediated Enzyme-Assisted Target Recycling and Functionalized Fullerene Nanoparticles in the Detection of MicroRNA-141. *Biosens. Bioelectron.* **2020**, *150*, 111964. [CrossRef]

36. Wang, Y.; Li, M.; Zhang, Y. Electrochemical Detection of MicroRNA-21 Based on a Au Nanoparticle Functionalized g-C3N4 Nanosheet Nanohybrid as a Sensing Platform and a Hybridization Chain Reaction Amplification Strategy. *Analyst* **2021**, *146*, 2886–2893. [CrossRef] [PubMed]

37. Islam, M.N.; Gorgannezhad, L.; Masud, M.K.; Tanaka, S.; Hossain, M.S.A.; Yamauchi, Y.; Nguyen, N.T.; Shiddiky, M.J.A. Graphene-Oxide-Loaded Superparamagnetic Iron Oxide Nanoparticles for Ultrasensitive Electrocatalytic Detection of MicroRNA. *ChemElectroChem* **2018**, *5*, 2488–2495. [CrossRef]

38. Liu, L.; Wei, Y.; Jiao, S.; Zhu, S.; Liu, X. A Novel Label-Free Strategy for the Ultrasensitive MiRNA-182 Detection Based on MoS2/Ti3C2 Nanohybrids. *Biosens. Bioelectron.* **2019**, *137*, 45–51. [CrossRef]

39. Wang, M.; Chen, W.; Tang, L.; Yan, R.; Miao, P. Duplex-Specific Nuclease Assisted MiRNA Assay Based on Gold and Silver Nanoparticles Co-Decorated on Electrode Interface. *Anal. Chim. Acta* **2020**, *1107*, 23–29. [CrossRef]

40. Yammouri, G.; Mohammadi, H.; Amine, A. A Highly Sensitive Electrochemical Biosensor Based on Carbon Black and Gold Nanoparticles Modified Pencil Graphite Electrode for MicroRNA-21 Detection. *Chem. Afr.* **2019**, *2*, 291–300. [CrossRef]

41. Zhu, R.; Feng, H.; Li, Q.; Su, L.; Fu, Q.; Li, J.; Song, J.; Yang, H. Asymmetric Core–Shell Gold Nanoparticles and Controllable Assemblies for SERS Ratiometric Detection of MicroRNA. *Angew. Chem.* **2021**, *133*, 12668–12676. [CrossRef]

42. Yao, Y.; Zhang, H.; Tian, T.; Liu, Y.; Zhu, R.; Ji, J.; Liu, B. Iodide-Modified Ag Nanoparticles Coupled with DSN-Assisted Cycling Amplification for Label-Free and Ultrasensitive SERS Detection of MicroRNA-21. *Talanta* **2021**, *235*, 122728. [CrossRef]

43. Mehdipour, M.; Gloag, L.; Bennett, D.T.; Hoque, S.; Pardehkhorram, R.; Bakthavathsalam, P.; Gonçales, V.R.; Tilley, R.D.; Gooding, J.J. Synthesis of Gold-Coated Magnetic Conglomerate Nanoparticles with a Fast Magnetic Response for Bio-Sensing. *J. Mater. Chem. C* **2021**, *9*, 1034–1043. [CrossRef]

44. Wang, H.; Tang, H.; Yang, C.; Li, Y. Selective Single Molecule Nanopore Sensing of MicroRNA Using PNA Functionalized Magnetic Core-Shell Fe_3O_4-Au Nanoparticles. *Anal. Chem.* **2019**, *91*, 7965–7970. [CrossRef] [PubMed]

45. Borghei, Y.S.; Hosseini, M.; Ganjali, M.R. A Label-Free Luminescent Light Switching System for MiRNA Detection Based on Two Color Quantum Dots. *J. Photochem. Photobiol. A Chem.* **2020**, *391*, 112351. [CrossRef]

46. Niu, X.; Lu, C.; Su, D.; Wang, F.; Tan, W.; Qu, F. Construction of a Polarity-Switchable Photoelectrochemical Biosensor for Ultrasensitive Detection of MiRNA-141. *Anal. Chem.* **2021**, *93*, 13727–13733. [CrossRef]

47. Pothipor, C.; Jakmunee, J.; Bamrungsap, S.; Ounnunkad, K. An Electrochemical Biosensor for Simultaneous Detection of Breast Cancer Clinically Related MicroRNAs Based on a Gold Nanoparticles/Graphene Quantum Dots/Graphene Oxide Film. *Analyst* **2021**, *146*, 4000–4009. [CrossRef]

48. Wang, W. Label-Free Electrochemical Sensor for MicroRNA Detection Based on a Gold Nanoparticles@Poly(Methylene Blue)-Modified Electrode and a Target Cyclic Amplification Strategy. *Int. J. Electrochem. Sci.* **2020**, *15*, 4631–4639. [CrossRef]

49. Hakimian, F.; Ghourchian, H.; Hashemi, A.S.; Arastoo, M.R.; Behnam Rad, M. Ultrasensitive Optical Biosensor for Detection of MiRNA-155 Using Positively Charged Au Nanoparticles. *Sci. Rep.* **2018**, *8*, 2943. [CrossRef]

50. Vilímová, I.; Chourpa, I.; David, S.; Soucé, M.; Hervé-Aubert, K. Two-Step Formulation of Magnetic Nanoprobes for MicroRNA Capture. *RSC Adv.* **2022**, *12*, 7179–7188. [CrossRef]

51. Afzalinia, A.; Mirzaee, M. Ultrasensitive Fluorescent MiRNA Biosensor Based on a "Sandwich" Oligonucleotide Hybridization and Fluorescence Resonance Energy Transfer Process Using an Ln(III)-MOF and Ag Nanoparticles for Early Cancer Diagnosis: Application of Central Composite Design. *ACS Appl. Mater. Interfaces* **2020**, *12*, 16076–16087. [CrossRef]

52. Bao, J.; Hou, C.; Zhao, Y.; Geng, X.; Samalo, M.; Yang, H.; Bian, M.; Huo, D. An Enzyme-Free Sensitive Electrochemical MicroRNA-16 Biosensor by Applying a Multiple Signal Amplification Strategy Based on Au/PPy–RGO Nanocomposite as a Substrate. *Talanta* **2019**, *196*, 329–336. [CrossRef] [PubMed]

53. Xu, W.; Zhao, A.; Zuo, F.; Khan, R.; Hussain, H.M.J.; Chang, J. Au@Ag Core-Shell Nanoparticles for MicroRNA-21 Determination Based on Duplex-Specific Nuclease Signal Amplification and Surface-Enhanced Raman Scattering. *Microchim. Acta* **2020**, *187*, 384. [CrossRef] [PubMed]

54. Yazdanparast, S.; Benvidi, A.; Azimzadeh, M.; Tezerjani, M.D.; Ghaani, M.R. Experimental and Theoretical Study for MiR-155 Detection through Resveratrol Interaction with Nucleic Acids Using Magnetic Core-Shell Nanoparticles. *Microchim. Acta* **2020**, *187*, 479. [CrossRef]

55. Gessner, I.; Yu, X.; Jüngst, C.; Klimpel, A.; Wang, L.; Fischer, T.; Neundorf, I.; Schauss, A.C.; Odenthal, M.; Mathur, S. Selective Capture and Purification of MicroRNAs and Intracellular Proteins through Antisense-Vectorized Magnetic Nanobeads. *Sci. Rep.* **2019**, *9*, 2069. [CrossRef]

56. Borghei, Y.S.; Hosseini, M. A New Eye Dual-Readout Method for MiRNA Detection Based on Dissolution of Gold Nanoparticles via LSPR by CdTe QDs Photoinduction. *Sci. Rep.* **2019**, *9*, 5453. [CrossRef] [PubMed]

57. Hermanson, G.T. *Bioconjugate Techniques*, 3rd ed.; Audet, J., Preap, M., Eds.; Elsevier: London, UK, 2013; ISBN 9780123822390.

58. Amaral, G.; Bushee, J.; Cordani, U.G.; Kawashita, K.; Reynolds, J.H.; Almeida, F.F.M.D.E.; de Almeida, F.F.M.; Hasui, Y.; de Brito Neves, B.B.; Fuck, R.A.; et al. *Applications of Nanocomposite Materials in Drug Delivery*; Woodhead Publishing: Cambridge, UK, 2013; Volume 369, ISBN 9789896540821.

59. Sunasee, R.; Narain, R. *Covalent and Noncovalent Bioconjugation Strategies*, 1st ed.; Narain, R., Ed.; Wiley: Hoboken, NJ, USA, 2014; ISBN 9781118359143.

60. Xu, Z.-H.; Wang, H.; Wang, J.; Zhao, W.; Xu, J.-J.; Chen, H.-Y. Bidirectional Electrochemiluminescent Sensing: An Application in Detecting MiRNA-141. *Anal. Chem.* **2019**, *91*, 12000–12005. [CrossRef]

61. Chen, W.Y.; Wang, C.H.; Wang, K.H.; Chen, Y.L.; Chau, L.K.; Wang, S.C. Development of Microfluidic Concentrator Using Ion Concentration Polarization Mechanism to Assist Trapping Magnetic Nanoparticle-Bound MiRNA to Detect with Raman Tags. *Biomicrofluidics* **2020**, *14*, 014102. [CrossRef]

62. Gu, T.; Li, Z.; Ren, Z.; Li, X.; Han, G. Rare-Earth-Doped Upconversion Nanocrystals Embedded Mesoporous Silica Nanoparticles for Multiple MicroRNA Detection. *Chem. Eng. J.* **2019**, *374*, 863–869. [CrossRef]

63. Mahani, M.; Mousapour, Z.; Divsar, F.; Nomani, A.; Ju, H. A Carbon Dot and Molecular Beacon Based Fluorometric Sensor for the Cancer Marker MicroRNA-21. *Microchim. Acta* **2019**, *186*, 132. [CrossRef]

64. Salimi, A.; Kavosi, B.; Navaee, A. Amine-Functionalized Graphene as an Effective Electrochemical Platform toward Easily MiRNA Hybridization Detection. *Meas. J. Int. Meas. Confed.* **2019**, *143*, 191–198. [CrossRef]

65. Horny, M.C.; Gamby, J.; Dupuis, V.; Siaugue, J.M. Magnetic Hyperthermia on γ-Fe_2O_3@SiO_2 Core-Shell Nanoparticles for Mi-RNA 122 Detection. *Nanomaterials* **2021**, *11*, 149. [CrossRef] [PubMed]

66. Lillo, A.M.; Lopez, C.L.; Rajale, T.; Yen, H.J.; Magurudeniya, H.D.; Phipps, M.L.; Balog, E.R.M.; Sanchez, T.C.; Iyer, S.; Wang, H.L.; et al. Conjugation of Amphiphilic Proteins to Hydrophobic Ligands in Organic Solvent. *Bioconjug. Chem.* **2018**, *29*, 2654–2664. [CrossRef]

67. Xue, Y.; Li, X.; Li, H.; Zhang, W. Quantifying Thiol-Gold Interactions towards the Efficient Strength Control. *Nat. Commun.* **2014**, *5*, 4348. [CrossRef] [PubMed]

68. Al Mubarak, Z.H.; Premaratne, G.; Dharmaratne, A.; Mohammadparast, F.; Andiappan, M.; Krishnan, S. Plasmonic Nucleotide Hybridization Chip for Attomolar Detection: Localized Gold and Tagged Core/Shell Nanomaterials. *Lab Chip* **2020**, *20*, 717–721. [CrossRef] [PubMed]
69. Zhu, Q.; Li, H.; Xu, D. Sensitive and Enzyme-Free Fluorescence Polarization Detection for MiRNA-21 Based on Decahedral Sliver Nanoparticles and Strand Displacement Reaction. *RSC Adv.* **2020**, *10*, 17037–17044. [CrossRef]
70. Ma, X.; Ma, Y.; Ejeromedoghene, O.; Kandawa-schulz, M.; Song, W.; Wang, Y. Dual-Mode Corroborative Detection of MiRNA-21 Based on Catalytic Hairpin Assembly Reaction and Oxygen Reduction Reaction Combined Strategy for the Signal Amplification. *Sens. Actuators B. Chem.* **2022**, *364*, 131924. [CrossRef]
71. Zhao, Y.; Lu, C.; Zhao, X.; Kong, W.; Zhu, S.; Qu, F. A T-Rich Nucleic Acid-Enhanced Electrochemical Platform Based on Electroactive Silver Nanoclusters for MiRNA Detection. *Biosens. Bioelectron.* **2022**, *208*, 114215. [CrossRef]
72. Sun, Y.; Li, T. Composition-Tunable Hollow Au/Ag SERS Nanoprobes Coupled with Target-Catalyzed Hairpin Assembly for Triple-Amplification Detection of MiRNA. *Anal. Chem.* **2018**, *90*, 11614–11621. [CrossRef]
73. Astakhova, K.; Ray, R.; Taskova, M.; Uhd, J.; Carstens, A.; Morris, K. "clicking" Gene Therapeutics: A Successful Union of Chemistry and Biomedicine for New Solutions. *Mol. Pharm.* **2018**, *15*, 2892–2899. [CrossRef]
74. Lu, X.; Fan, W.; Feng, Q.; Qi, Y.; Liu, C.; Li, Z. A Versatile Dynamic Light Scattering Strategy for the Sensitive Detection of Plant MicroRNAs Based on Click-Chemistry-Amplified Aggregation of Gold Nanoparticles. *Chem.—Eur. J.* **2019**, *25*, 1701–1705. [CrossRef] [PubMed]
75. Wang, Y.; Howes, P.D.; Kim, E.; Spicer, C.D.; Thomas, M.R.; Lin, Y.; Crowder, S.W.; Pence, I.J.; Stevens, M.M. Duplex-Specific Nuclease-Amplified Detection of MicroRNA Using Compact Quantum Dot-DNA Conjugates. *ACS Appl. Mater. Interfaces* **2018**, *10*, 28290–28300. [CrossRef] [PubMed]
76. Ajgaonkar, R.; Lee, B.; Valimukhametova, A.; Nguyen, S.; Gonzalez-rodriguez, R.; Coffer, J.; Akkaraju, G.R.; Naumov, A. V Detection of Pancreatic Cancer MiRNA with Biocompatible Nitrogen-Doped Graphene Quantum Dots. *Materials* **2022**, *15*, 5760. [CrossRef] [PubMed]
77. Su, X.; Wu, Y.J.; Robelek, R.; Knoll, W. Surface Plasmon Resonance Spectroscopy and Quartz Crystal Microbalance Study of Streptavidin Film Structure Effects on Biotinylated DNA Assembly and Target DNA Hybridization. *Langmuir* **2005**, *21*, 348–353. [CrossRef] [PubMed]
78. Tian, B.; Han, Y.; Wetterskog, E.; Donolato, M.; Hansen, M.F.; Svedlindh, P.; Strömberg, M. MicroRNA Detection through DNAzyme-Mediated Disintegration of Magnetic Nanoparticle Assemblies. *ACS Sens.* **2018**, *3*, 1884–1891. [CrossRef]
79. Djebbi, K.; Shi, B.; Weng, T.; Bahri, M.; Elaguech, M.A.; Liu, J.; Tlili, C.; Wang, D. Highly Sensitive Fluorescence Assay for MiRNA Detection: Investigation of the DNA Spacer E Ff Ect on the DSN Enzyme Activity toward Magnetic-Bead-Tethered Probes. *ASC Omega* **2022**, *1*, 2224–2233. [CrossRef]
80. Chan, H.N.; Ho, S.L.; He, D.; Li, H.W. Direct and Sensitive Detection of Circulating MiRNA in Human Serum by Ligase-Mediated Amplification. *Talanta* **2020**, *206*, 120217. [CrossRef]
81. Cheng, W.; Ma, J.; Cao, P.; Zhang, Y.; Xu, C.; Yi, Y.; Li, J. Enzyme-Free Electrochemical Biosensor Based on Double Signal Amplification Strategy for the Ultra-Sensitive Detection of Exosomal MicroRNAs in Biological Samples. *Talanta* **2020**, *219*, 121242. [CrossRef]
82. Segal, M.; Biscans, A.; Gilles, M.E.; Anastasiadou, E.; De Luca, R.; Lim, J.; Khvorova, A.; Slack, F.J. Hydrophobically Modified Let-7b MiRNA Enhances Biodistribution to NSCLC and Downregulates HMGA2 In Vivo. *Mol. Ther.—Nucleic Acids* **2020**, *19*, 267–277. [CrossRef]

molecules

Article

Site-Directed Mutants of Parasporin PS2Aa1 with Enhanced Cytotoxic Activity in Colorectal Cancer Cell Lines

Miguel O. Suárez-Barrera [1,2,3], Lydia Visser [1], Efraín H. Pinzón-Reyes [2], Paola Rondón Villarreal [2], Juan S. Alarcón-Aldana [2] and Nohora Juliana Rueda-Forero [2,*]

[1] Department of Pathology and Medical Biology, University Medical Center Groningen, University of Groningen, 9700 AB Groningen, The Netherlands
[2] Facultad de Ciencias Médicas y de la Salud, Instituto de Investigación MASIRA, Universidad de Santander, Bucaramanga 680002, Colombia
[3] Max Planck Tandem Group in Nanobioengineering, Institute of Chemistry, Faculty of Natural and Exacts Sciences, University of Antioquia, Medellín 050015, Colombia
* Correspondence: juliana.forero@udes.edu.co

Abstract: Parasporin 2 has cytotoxic effects against numerous colon cancer cell lines, making it a viable alternative to traditional treatments. However, its mechanism of action and receptors remain unknown. In this study, site-directed mutagenesis was used to obtain PS2Aa1 mutants with variation in domain I at positions 256 and 257. Variants 015, 002, 3-3, 3-35, and 3-45 presented G256A, G256E, G257A, G257V, and G257E substitutions, respectively. Cytotoxicity tests were performed for the cell viability of cell lines SW480, SW620, and CaCo-2. Mutants 3-3, 3-35, and 3-45 efficiently killed the cell lines. It was found that the activated forms of caspase-3 and PARP were in higher abundance as well as increased production of γH2AX when 3-35 was used to treat CaCo-2 and SW480. To assess possible membrane-binding receptors involved in the interaction, an APN receptor blocking assay showed reduced activity of some parasporins. Hence, we performed molecular docking and molecular dynamics simulations to analyze the stability of possible interactions and identify the residues that could be involved in the protein–protein interaction of PS2Aa1 and APN. We found that residues 256 and 257 facilitate the interaction. Parasporin 3-35 is promising because it has higher cytotoxicity than PS2Aa1.

Keywords: anticancer; parasporin; site-directed mutagenesis; apoptosis; APN receptor

Citation: Suárez-Barrera, M.O.; Visser, L.; Pinzón-Reyes, E.H.; Rondón Villarreal, P.; Alarcón-Aldana, J.S.; Rueda-Forero, N.J. Site-Directed Mutants of Parasporin PS2Aa1 with Enhanced Cytotoxic Activity in Colorectal Cancer Cell Lines. *Molecules* **2022**, *27*, 7262. https://doi.org/10.3390/molecules27217262

Academic Editor: Jahir Orozco Holguín

Received: 28 September 2022
Accepted: 24 October 2022
Published: 26 October 2022

Publisher's Note: MDPI stays neutral with regard to jurisdictional claims in published maps and institutional affiliations.

1. Introduction

The use of bacteria and their byproducts, i.e., attenuated or genetically modified, has begun to increase in recent decades based on their ability to recognize specific characteristics of cancer cells. They can act directly on mechanisms involved in the proliferation and growth of tumor cells [1]. Additionally, their use as antitumor agents represents a promising strategy because of the ease of their genetic manipulation, which allows biomolecules with improved toxic activity and specificity to be obtained [1–4]. Mizuki et al., in the first report on Parasporin (PS) in 2000 [5], described its potential as an anticancer molecule, specifically in leukemia. Subsequently, most studies have focused on the characterization and screening of new proteins identified from *Bacillus thuringiensis* (*Bt*) strains and isolates [6–10]. Akiba and Okumura proposed in 2017 that the mechanism of action of this protein is in inducing apoptosis [11], but many unknowns persist up to now, especially regarding the identity of the major PS receptor, with N-aminopeptidase (APN) and GPI-anchored proteins having been proposed as candidates [6,7,12,13]. Parasporin 2Aa1 (PS2Aa1) has shown cytotoxic effects against several human cancer cells, being highly specific for human liver and colon cancer cells [7–9], thus representing an additional treatment alternative for colorectal cancer. This protein is a polypeptide that requires proteases such as proteinase K to switch from a pro-toxin of 37 kDa to its active form of 30 kDa, which is shown to be highly cytotoxic and selective toward different cancer cell lines [10]. Furthermore, this protein is believed to act specifically on the membrane of target cells through domain I, which indicates the presence of a specific PS2Aa1 receptor whose identity is so far unknown [12]. Once PS2Aa1 is bound to the receptor, through domains II and III, it forms oligomers that permeabilize the membrane and lead to the formation of pores, which induces structural cytoskeletal alterations, organelle fragmentation, alterations in cell morphology, and finally lysis of susceptible cells [8]. Considering this scenario, our research group is studying the structure–function relationship of Parasporin 2, while exploring its possible use as a therapeutic alternative. This is how we recently designed and prepared peptides in which loop 1 of domain I was mutated, and the selection of this site in the design was supported by in silico modeling with APN. In that study, the Loop1–PS2Aa and P264–G274, peptides exhibited stronger anticancer activity than the wild type against SW480 and SW620, respectively, and demonstrated high effectiveness and selectivity; hence, they were proposed as possible alternative therapeutic agents for the treatment of colon cancer. In this sense, we decided to take domain I of PS2Aa1, mutate the residues that presented differentiated activities in the reported peptides, and consider additional positions for the modification of these peptides to generate the libraries of PS2Aa1 mutants, which we describe below [6]. The goal of this work was to obtain a characterized set of PS2Aa1 mutants with modifications in domain I and assess its cytotoxicity in colon cancer cells. Moreover, we used molecular docking and molecular dynamics simulations to analyze the possible interactions between PS2Aa1 and APN receptor, which have been proposed as a possible interacting protein. Additionally, our work contributes to the understanding of the relation between the structure and function of these proteins and their possibilities of being used as new treatments in the future.

2. Results

2.1. Cloning and Obtention of PS2Aa1 Mutants

The gene *ps2Aa1* (NCBI accession number AB099515.1) was cloned in the plasmid pET30a to generate the construct *pET-30a/ps2Aa1*, which was corroborated by restriction assays with a band of ~1 kb for PS2Aa1 and a band of ~5.4 kb for pET30 (Figure 1). Then, the construct was transformed into *E. coli* strain DE3BL21 for the production of plasmid and subsequent site-direct mutagenesis to create four libraries, one for each primer set.

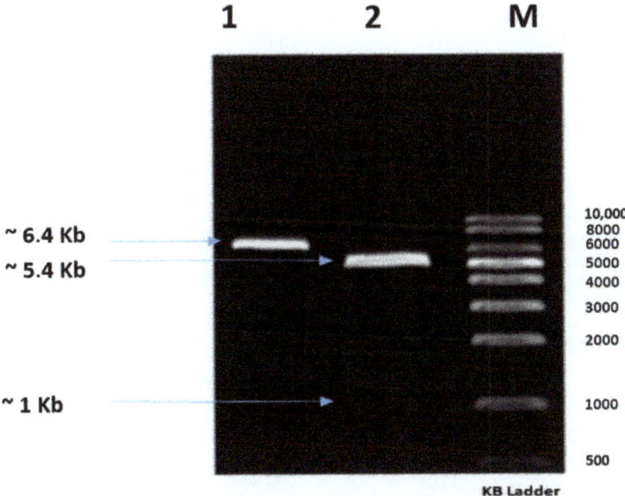

Figure 1. Agarose gel showing the construct *pET-30a/ps2Aa1* treated with the restriction enzymes *Kpn*I and *Hind*III; Lane 1: construct pET-30a/ps2Aa1; Lane 2: construct *pET-30a/ps2Aa1* treated with the restriction enzymes *Kpn*I and *Hind*III; Lane M: molecular weight marker.

A low mutation rate was obtained, whereby only 12 of the 71 selected mutants had mutations in their sequences. Therefore, from the 12 PS2Aa1 mutants, 6 were chosen for cytotoxicity assays. Mutant 0-2 presented an amino acid change from glycine (G) to aspartic acid (D) at position 256 of its amino acid chain corresponding to position 1000 of the nucleotide sequence; mutant 0-6 had an amino acid change from glycine to valine (V) in the same position; mutant 0-15, like the previous proteins, presented an amino acid change at position 256 to become alanine (A); therefore, these three mutants were obtained from the primer set flanking nucleotide position 1000. On the other hand, mutants 3-35 and 3-45 presented amino acid changes in their sequences at position 257 (position 1003 in the nucleotide chain), changing from glycine (G) to valine (V) and glutamic acid (E), respectively. It is, therefore, assumed that these mutants were obtained from the primer set targeting position 1003.

2.2. Description of Parasporins Obtained from PS2Aa1 Using Site-Directed Mutagenesis

From the libraries in *E. coli*, 103 colonies were sequenced. Mutagenesis was successful, although not in all the colonies because of the randomness attributed to the reaction by adding equimolar conditions of adenine, guanine, cytosine, or thymine.

The sequence analysis showed 99.9% identity with parasporin 2 of *Bt*, determined by BLASTn searching of the contigs for each variant, and MatGat, similarity/identity matrices for DNA [14]. The contigs were translated to their respective amino acid sequence and a multiple sequence alignment was performed comparing each of these sequences with the native parasporin reported in the PDB (code: 2ztb). The substitutions obtained at residues 256 and 257 were mostly glycine for alanine (A), aspartic acid (D), valine (V), and glutamic acid (E), as described in Table 1.

Table 1. Description of parasporins obtained from PS2Aa1 using site-directed mutagenesis.

ID (Library)	Type/Site of the Mutation	Mutant Name	Amino Acid Sequencing
Wt.	PS2Aa1 (PDB 2ZTB)	-	251-KRVGP**GG**HYF-260
02	SUS/256	G256D	251-KRVGPDGHYF-260
015	SUS/256	G256A	251-KRVGPAGHYF-260
3-3	SUS/257	G257A	251-KRVGPGAHYF-260
3-35	SUS/257	G257V	251-KRVGPGVHYF-260
3-45	SUS/257	G257E	251-KRVGPGEHYF-260

2.3. Cytocidal Activities of PS2Aa1 Variants in Human Colorectal Cancer Cells

We first compared the native *Bt* protein 4R2 with the recombinant PS2Aa1 and found that the activities were comparable in SW480 and SW620, but PS2Aa1 was more effective than 4R2 in CaCo-2 (Figure S1). Cell line NCM460 was used as a control. In this normal colorectal cell line, none of the parasporins showed any relevant effect, consistent with previous reports of specificity of the parasporin to cancer cells (Figure S1). The variants showed different patterns in the cell lines. In SW480, the protein with the highest cytotoxic activity was 3-35 with an IC_{50} of 0.32 µg/mL, more than three times that of PS2Aa1 (IC_{50} = 1 µg/mL). All of the variants were effective, with 002 having the lowest cytotoxicity (IC_{50} = 4.68 µg/mL). In SW620, 3-35 was also the most cytotoxic (IC_{50} 2.06 µg/mL), while 3.45 was the least (IC_{50} 17.77 µg/mL). Against CaCo-2 cells, the mutant 3-35 was the most cytotoxic (IC_{50} 0.96 µg/mL), with similar results observed for 3-3 and 3.45, whereas 0015 and 002 were not effective (Table 2).

Table 2. Cytocidal activities of PS2Aa1 variants in human colorectal cancer cells.

Parasporin	IC50 Parasporin µg·mL^{-1} (95% CI)		
	SW480	**SW620**	**CaCo-2**
0015	2.42 (1.59–4.70)	ND	ND
002	4.68 (2.57–17.74)	ND	ND
3-3	2.02 (1.51–2.96)	ND	0.98 (0.83–1.16)
3-35	0.32 (0.22–0.43)	2.46 (2.07–3.00)	0.88 (0.81–0.95)
3-45	0.78 (0.65–0.92)	ND	0.96 (0.84–1.09)
P2SAa1	1.00 (0.65–1.55)	5.70 (4.10–9.50)	2.57 (1.84–4.15)
4R2	0.62 (0.41–0.86)	ND	ND

ND: Non-detectable.

Comparisons with PS2Aa1 were based on ordinary one-way ANOVA employing Tukey's multiple comparison test for SW480 and CaCo-2. All the comparisons were significantly different (**, ***), except between the activity of PS2Aa1 and the native strain 4R2 in SW480 (Figure 2).

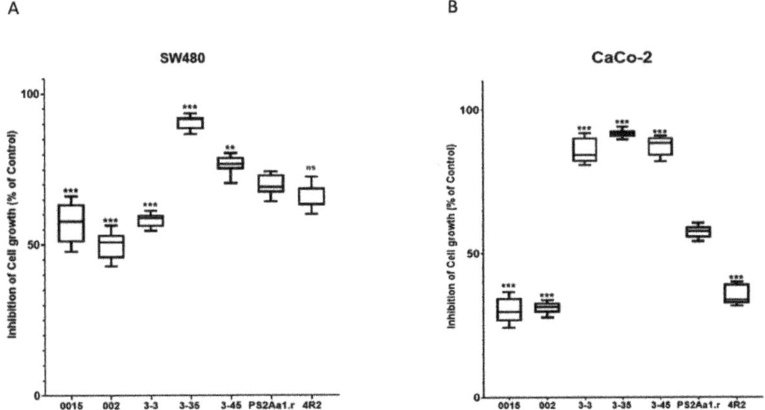

Figure 2. Tukey's multiple comparison test for (**A**) SW480 and (**B**) CaCo-2. The results for the percentage of cell growth inhibition correspond to 5 µg·mL^{-1} for each parasporin. PS2Aa1 was used as a control. ** $p \leq 0.01$, and *** $p \leq 0.001$, ns (non-significant).

2.4. Variants of PS2Aa1 Induce Apoptosis in Human Colorectal Cancer Cells

To expand our results, we performed an analysis with annexin V (AV) and propidium iodide (PI) to characterize the percentage of cells in apoptosis and death. For this part of the study, the cell lines SW480 and CaCo-2 and the parasporins 015, 3-35, and PS2Aa1 were selected, owing to the observation of relevant cytocidal activity.

After the treatment of CaCo-2 and SW480 with 5 µg/mL of each parasporin, higher percentages of apoptotic cells, 25% and 37%, respectively, were observed for both cell lines when treated with parasporin 3-35 followed by PS2Aa1 and, lastly, 015 (Figure 3A,B), which is similar to the cytotoxicity results, where 3-35 also excels. The differences between PS2Aa1 and 015 were not remarkable, and no differences in dead cells were detected in all of the treatments, suggesting the late activation of apoptosis in the cell lines by parasporin 2Aa1 and its variants as an effect of the pore-forming action of these proteins, which was previously suggested to be the mechanism of action [13,15–17]. HSD Tukey testing was performed with the results of annexin V-positive cells, showing that there are significant differences between PS2Aa1.r and 3-35 in the treatment of CaCo-2 (*). For SW480, it was shown that there are differences between PS2Aa1.r and 3-35 (**) and Ps2Aa1.r and the control (*); $p \leq 0.05$ (Figure 3B).

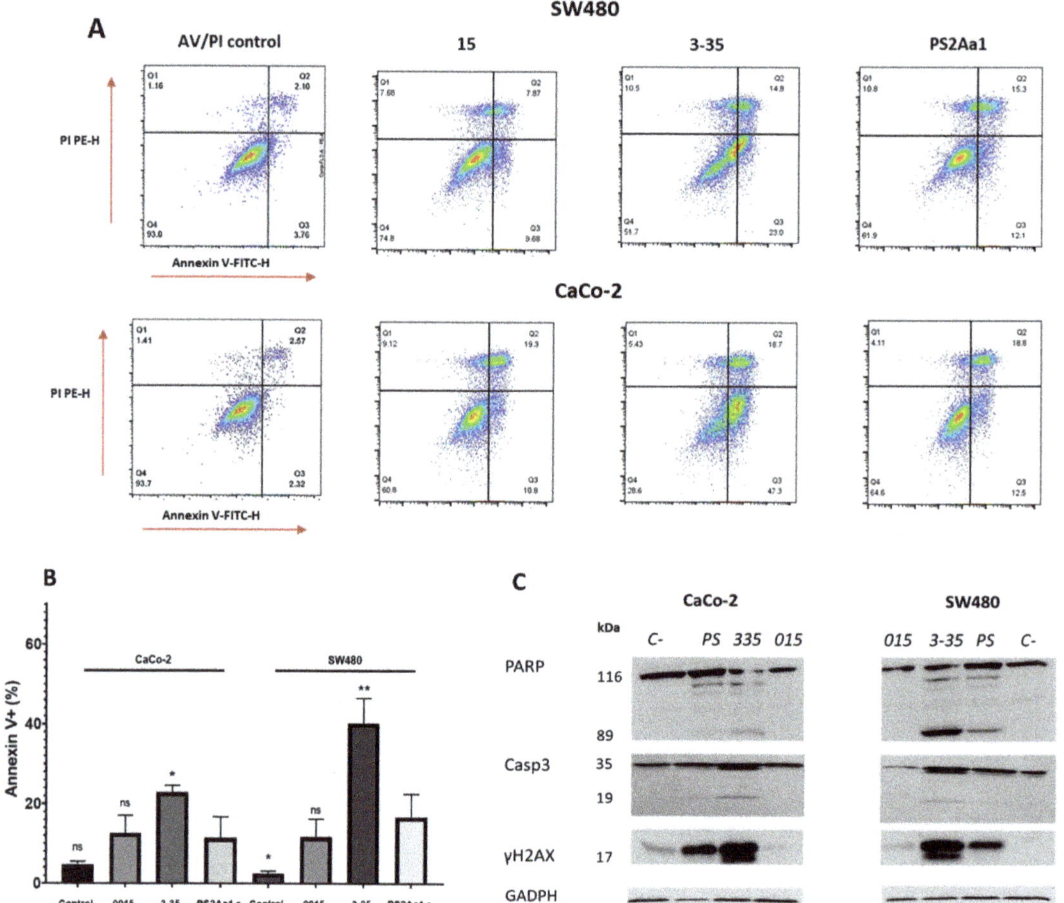

Figure 3. Cytotoxic effects of parasporins to CaCo-2 and SW480. Detection of annexin V/PI (**A**), statistical analysis of induction of apoptosis by parasporins (**B**). Western blot of PARP, caspase-3 (Casp3), and γH2AX in SW480 and CaCO-2 after treatment with the indicated parasporins (**C**). GAPDH was used as a loading control. The amount of toxins used for each treatment was 5 ug/mL over 48 h. * $p \leq 0.05$, ** $p \leq 0.01$, ns (non-significant).

To further investigate the induction of apoptosis, we also measured different markers using Western blot analysis, such as cleaved caspase-3, cleaved poly (ADP-ribose) polymerase-1 (PARP), and Histone 2 family member, phosphorylated on serine 139 (γ-H2AX). For both CaCo-2 and SW480, activation of caspase-3 and PARP was more induced by the treatment with parasporin 3-35, with multiple bands obtained for both proteins. PARP cleavage is more pronounced in SW480 and corresponds with the lowest IC_{50} (0.32 µg·mL^{-1}) in the cytotoxicity assay (Table 2). We did not see differences between the results of caspase-3 and γH2AX in the cell lines.

In the case of γH2AX, the band intensity in the treatment with 3-35 is the highest. The presence of a slightly higher molecular weight band at 17 kDa indicates the phosphorylated form of this modified histone involved in the repair of DNA damage (Figure 3C).

2.5. Cytotoxic Activity of PS2Aa1 Variants Is Affected by APN Receptor Inhibition

We next checked for the presence of APN as a possible receptor for PS2Aa1. The highest level of expression of the APN receptor was obtained in the positive control cell line HL60, and the protein was not present in negative controls MCF-7 and U2932 (Figure 4A). In the case of colorectal cancer cells, the amount of APN is higher in CaCo-2 compared with SW480. We next blocked APN with an inhibitor and tested the effect of the different parasporins at 5 μg·mL^{-1}. The effect of 3-35 and PS2Aa1 on SW480 and Caco-2 was diminished to the point of no cytotoxicity at the lowest concentration of APN inhibitor (5 μM) (Figure 4B,C), which strongly suggest that the effect of 3-35 and PS2Aa1 is presumably dependent on binding to APN. To further understand these results, in silico analysis was performed via docking and molecular dynamics simulations.

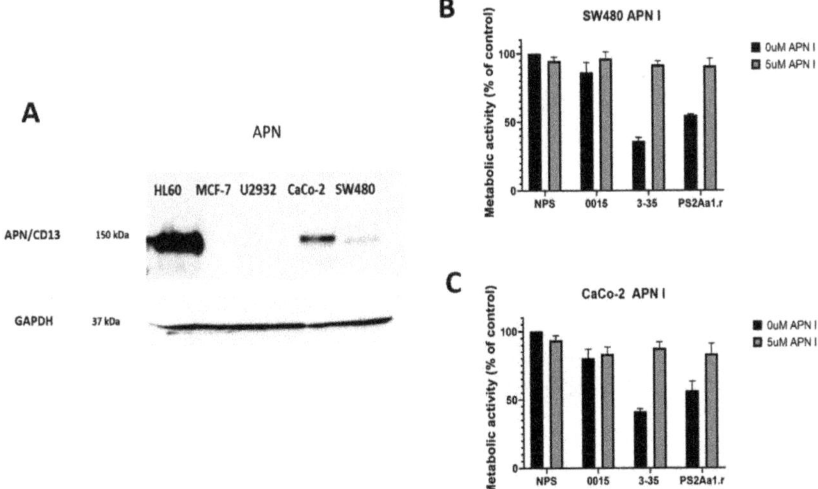

Figure 4. (**A**) Detection of APN receptor in different cell lines HL60, MCF-7, U2932, CaCo-2, and SW480. Metabolic activity of (**B**) SW480 and (**C**) CaCO-2 without parasporin (NP), parasporin 0015, 3-35, and the recombinant of PS2Aa1 (PS2Aa1.r) at 5 μg·mL^{-1} in the presence or absence of the APN inhibitor.

2.6. Molecular Docking and Molecular Dynamics Analysis Highlight Residues 256 and 257 of Domain I of PS2Aa1

The predicted models were ranked based on the number of hydrogen bonds in the interface, with the top 10 shown in Table 3. After visual inspection, model number 560 was selected to perform the molecular dynamics simulations. Figure 5 shows the selected complex between PS2Aa1 and APN with the relevant residues in the interaction.

The 560-model selected by molecular docking analysis was subjected to molecular dynamics computational simulations to identify possible residues of wild-type PS2Aa1, which showed preference in protein–protein interaction (PS2Aa1–APN). As shown in Table 4, after three molecular dynamics simulation replicates, some of the PS2Aa1 residues were in contact with APN for the longest simulation time and at a smaller average distance from the center of mass of the PSAa1 residue and the closest APN residues. Table 4 records the wild-type PS2Aa1 residues, for each of the three simulation replicates, that were in contact for more than 80% of the simulation time and that also maintained inter-action distances of less than 5 Å. These two conditions suggest in silico that these residues could be of interest in the PS2Aa1–APN interaction.

Table 3. Top 10 models according to the number of hydrogen bonds in the interface.

Top	Model	Number of Hydrogen Bonds in the Interface
1	560	9
2	3413	9
3	105	9
4	819	8
5	994	8
6	2079	8
7	3521	8
8	1015	8
9	1742	8
10	708	8

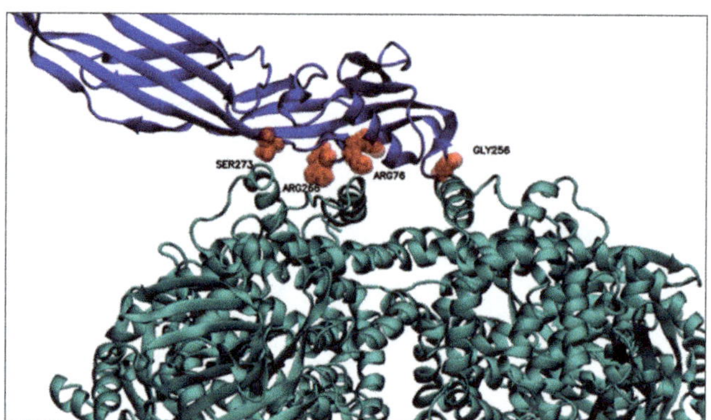

Figure 5. Relevant residues for PS2Aa1–APN interaction. Relevant residues for interacting with the APN receptor are displayed. GLY256, ARG76, ARG266, and SER273. PS2Aa1 and APN are colored blue and teal, respectively.

Table 4. Residues in contact with APN receptors.

MD	Contact Residues (>80%) *
Replicate 1	PRO255, GLY256
Replicate 2	ARG76, PRO238, ILE239, THR240, VAL241, ARG266, GLY256, GLY257, THR272, SER273, GLY274
Replicate 3	ARG76, PRO238, ILE239, THR240, VAL241, ASP242, PRO255, GLY256, GLY257, ARG266, ASP267, ASN270, THR272, SER273, GLY274, THR275

* The residues in contact with APN receptors at 80% frequency are shown for each independent MD.

Table 5 below presents a review of these residues from their prevalence in the three replicates, that is, the frequency with which these residues were identified as preferential for interaction in the molecular dynamic simulations (Table 4). There, GLY256 seems to be the residue of conspicuous importance, as it is the only residue recorded in all three replicas of molecular dynamics; the importance of this residue could be explained based on its position, because GLY256 is in a PS2Aa1 loop (Figure 5), a mobile and flexible part of the protein. Moreover, GLY256 presents stable dynamics (SD 0.21 Å), with an average minimum distance of 4.44 Å to APN (Figure 6), and is the closest residue.

Table 5. Prevalence of PS2Aa1 residues among MD replicates.

MD	PS2Aa1 Residues
Replicate 3	GLY256
Replicate 2	ARG76 *, PRO238, ILE239, THR240, VAL241, PRO255, GLY257, ARG266 *, THR272, SER273 *, GLY274
Replicate 1	ASP267, ASN270, THR275

* Residues with the highest frequency in the MD replicates.

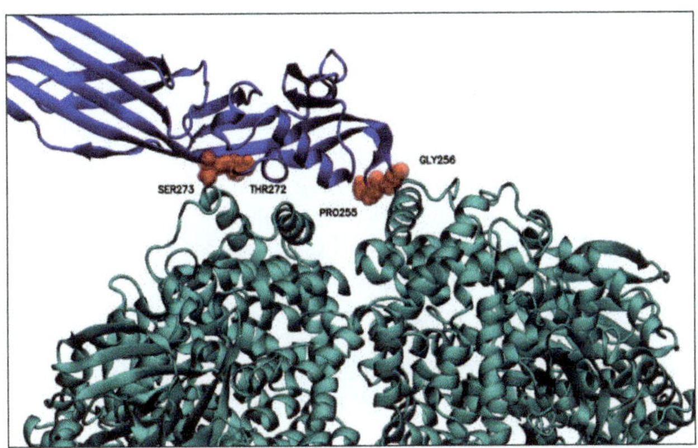

Figure 6. Residues relevant for the interaction with APN at distances less than 6 Å between the center of mass of the relevant PS2Aa1 residues and the nearest APN residue. Above, PS2Aa1 and APN are colored blue and teal, respectively.

Residues ARG76, PRO238, ILE239, THR240, VAL241, PRO255, GLY257, ARG266, THR272, SER273, and GLY274 had high prevalence in two MD replicates (Table 6), highlighting the significant prevalence (frequency) of residues ARG76, ARG266, and SER273 that were close to the APN receptor. These three residues are found within the beta-sheet secondary structure of PS2Aa1 (Figure 5), which could limit their interacting with APN.

Table 6. Distances of the residues of PS2Aa1 relevant to interacting with APN.

Comparison	Mean	SD
GLY256	4.44	0.21
THR272	5.53	0.61
SER273	5.58	0.09
PRO255	5.77	0.26
THR240	6.04	0.27
ARG266	6.29	0.70
PRO238	6.30	0.58
GLY274	6.33	0.23
ASP267	6.49	0.44
ARG76	6.58	0.55
GLY257	6.64	0.69
ASN270	7.04	0.79
THR275	7.59	0.45
VAL241	7.64	0.75
ILE239	7.74	0.25

There are also some residues that could have had a modest role in the PS2Aa1–APN interaction because they appeared in only one MD replicate (ASP267, ASN270, and THR275) and were the farthest residues from APN. Likewise, the previous group of residues were located in the beta-sheet secondary structure of PS2Aa1, which reduced their chance to interact with APN.

Lastly, some high-prevalence residues have an average distance of less than 6 Å (Table 6). These were PRO255, GLY256, THR272, and SER273, with GLY256 having the shortest distance and significant relevance in all the simulations (Table 6, Figure 6).

These results also suggest the participation of amino acids PRO255 and GLY256 as part of the loop of domain I of PS2Aa1 in the interactions with the APN receptor (Figure 6), being within the top 4 of the residues with the shortest interaction distance (Table 6).

3. Discussion

In this study, genetic modification was used to obtain mutants with substitutions in residues 256 and 257 of PS2Aa1. These modifications allowed us to build a library of new parasporins with different activities against colorectal cancer cell lines. Site-directed mutagenesis was performed considering the results of Cruz et al. [6], where peptides from loop 1 of PS2Aa1 had remarkable activity and adherence with SW480. It was established that the oligonucleotides with the mutations incorporated for site-directed mutagenesis should be present at the N-terminal end (variable region), specifically in domain I of PS2Aa1 because, based on previous molecular docking studies, this region is the one presumably responsible for specific binding to membrane receptors [14]. The search for conserved domains found that PS2Aa1 shares conserved domains with aerolysin-type βPFT proteins, which comprises a highly conserved region corresponding to the chain of the C-terminal end and a highly variable region at the N-terminal end of the protein, which usually contains recognition signals [11] and corresponds to domain I of the βPFT proteins. Its high variability means that this family of proteins has various binding receptors that are highly specific.

For the mutants, we obtained a selectivity index of 18.9 and 17.5 for PS2Aa1 against SW480 and SW620, respectively, and 46.4 and 34.4 for variant 3-35 against SW480 and SW620, respectively (data not shown). PS2Aa1 is described as an aerolysin and beta-pore-forming protein type because of their shared homology [15]. Parasporins and aerolysins are anticancer proteins [16] and the cytogenetic effects of aerolysin produced by *Aeromonas hydrophila* on normal and tumor cells have been studied; however, the latter present a toxic effect upon normal cell lines, unlike parasporins, which have an undetectable or only slight effect on normal cell lines [7,8] and a greater effect on several cancer cell lines.

Concerning the IC$_{50}$, 40.15 µg·mL^{-1} PS2Aa1 and 39.93 µg·mL^{-1} 3-35 were noted for CHOK-1 and >100 µg·mL^{-1} for both against NCM460 presenting the parasporins and their variants as potential candidates for cancer treatment in the future. Moreover, in this study, we decided to analyze the possible interaction between PS2Aa1 and the designed mutants with the membrane receptor APN. Our results were like those found by Periyasamy et al. (2016) [7], where the use of an APN inhibitor reduced the cytotoxic activity of PS2Aa1. We found that the lowest concentration of APN inhibitor led to the loss of the cytotoxic activity of 3-35 and PS2Aa1 with CaCo-2 and SW480. Although not conclusive, our results showed that variant 3-35 requires, to some extent, APN receptor for its cytotoxic activity against Caco-2 and SW480. However, it is interesting to note that the mutant 3-35 achieved the highest cytotoxicity with cell line SW480, despite the CaCo-2 cell line having higher amounts of APN. Hence, it is likely that APN is not the only receptor directly involved in the mechanism of action of these toxins [6,17]. As previously mentioned, GPI-anchored proteins play a crucial role in the interaction with the cell membrane. These receptors seem to be more relevant in the activity of the PS in cancer cell lines because cell lines such as MCF-7 lack APN, but PS2Aa1 has strong activity against this line [18]. It has been suggested that APN could be a receptor in the mechanism of action of PS2Aa1 because parasporin activity decreased substantially in HCT116 cells, a colorectal cancer cell line, when treated with an APN inhibitor [7]. In addition, it is known that PS2Aa1 is related to the Cry proteins, which use APN as a receptor in the midgut of insects [17,19–23]; in this scenario, APN receptor is important for parasporin activity. The data indicate that, although there was a decrease in cytotoxic activity, it was not completely lost, which leads to the assumption that APN is important, but it is not the only receptor involved. Considering another binding receptor for PS2Aa1, the GPI-anchored proteins have also been suggested as relevant to the binding to the cell membrane. It was determined that the glycan core, which is part of CD59, is essential for the recognition of the aerolysin-type parasporin [6,7,12].

PS2Aa1 in its native form can induce apoptosis in mammalian cancer cells [24] like PS produced by *B. thuringiensis* A1519, which induces the mitochondrial apoptosis pathway in Jurkat cells via caspase-3 and -9 cleavage followed by the release of cytochrome C [25]. In this study, we showed through cytotoxicity assays that the mutant 3-35 had a more remarkable effect on cancer cell lines, and the amount of cleaved caspase-3 and PARP and the production of γH2AX were reported for the first time. We also presented the effect of this mutant in early and late apoptosis with respect to recombinant PS2Aa1; these results suggest that the point variation G257V on Loop 1 can alter the behavior of the protein measured by its differentiated cytotoxicity and apoptosis induction when compared with the native PS2Aa1.

The molecular docking and dynamics simulations showed that it is likely that a stable interaction between domain I of PS2Aa1 and APN receptor could take place. Moreover, the mutation of mutant 3-35 was performed in the region of amino acids involved in interactions during the molecular dynamics simulations (Figure 7). Residue 256 was present in all three replicates, with the smallest interaction distance between its center of mass and the APN contact surface (Figure 7). The experimental results showed that its mutation reduced the interaction and cytotoxicity, so the G in position 256 seems relevant for the cytotoxic effects of PS2Aa1. In contrast, the mutation of position 257, which was involved in two replicates of molecular dynamic simulation, led to an improved interaction and cytotoxicity, i.e., the change of glycine for valine improved the cytocidal activity of the designed mutant 3-35 by creating a more nonpolar amino acid, suggesting that it feasibly conferred more stability on the protein. This mutant outperformed native Ps2Aa1 and 4R2 in terms of cytotoxicity in all three tested cell lines: SW480, SW620, and CaCo-2.

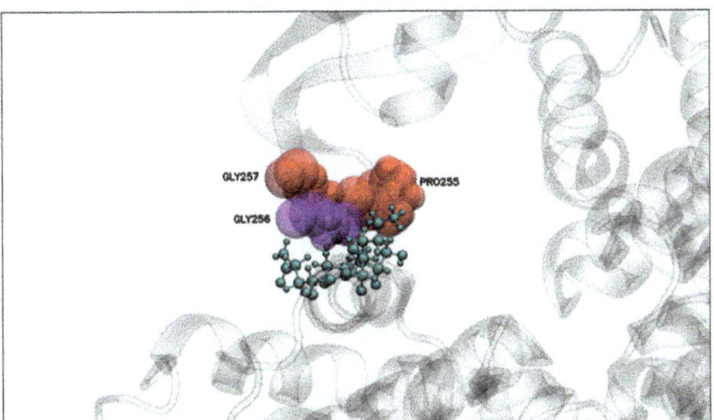

Figure 7. PS2Aa1–APN interaction. Relevant residues for interaction with APN are analyzed in Figure 6. Distances were computed between the center of mass of the relevant residues of PS2Aa1 and the closest residue from APN.

Finally, the use of site-directed mutagenesis might helpful to obtain a clearer understanding of how parasporin 2 works and to shed some light on understanding the action mechanism while generating proteins with enhanced activity like variant 3-35. This study reveals interesting features of domain I and highlights the relevance of residues 256 and 257 in the interaction with possible receptors that might be interacting with PS2Aa1, such as APN, and membrane-anchored receptors or GPI-anchored proteins [12].

4. Materials and Methods

4.1. Bacterial Strains and Culture Conditions

E. coli strain DE3BL21 harboring the construct pET30a + PS2Aa1 and *Bacillus thuringiensis* (*Bt*) BMB171 were cultured in Luria Bertani (LB) broth incubated at 37 °C with constant agitation for 24 h.

4.2. Parasporin Site-Directed Mutagenesis

Mutagenesis assays were carried out using the Gene-Art Site-Directed Mutagenesis Plus kit according to the manufacturer's instructions and four sets of designed primers (Table 7). Reaction and amplification conditions were developed according to the kit specifications.

Table 7. Primers designed for site-directed mutagenesis of PS2Aa1.

Primer Id.	Sequence
-Pet100Ps2a_994_FW	5′-TTC CCA AAA GAG TAG NGC CAG GTG GGC ATT A-3′
-Pet100Ps2a_994_RV	5′-TAA TGC CCA CCT GGC NCT ACT CTT TTG GGA A-3′
-Pet100Ps2a_997_FW	5′-CCA AAA GAG TAG GGC NAG GTG GGC ATT ATT T-3′
-Pet100Ps2a_997_RV	5′-AAA TAA TGC CCA CCT NGC CCT ACT CTT TTG G-3′
-Pet100Ps2a_1000_FW	5′-AAA GAG TAG GGC CAG NTG GGC ATT ATT TTT G-3′
-Pet100Ps2a_1000_RV	5′-CAA AAA TAA TGC CCA NCT GGC CCT ACT CTT T-3′
-Pet100Ps2a_1003_FW	5′-GAG TAG GGC CAG GTG NGC ATT ATT GGT T-3′
-Pet100Ps2a_1003_RV	5′-AAC CAA AAA TAA TGC NCA CCT GGC CCT ACT C-3′

Each of the mutation reactions was directly transformed into chemically competent TOP10F and DH5αT1 cells, 70 μL of the transformed cells was then inoculated on plates with LB + kanamycin agar (25 μg/mL) + X-gal + Isopropyl β-D-1-thiogalactopyranoside (IPTG) and incubated for 18 h at 37 °C, and white/blue cell screening was performed.

4.3. Library Verification

The positive clones (blue colonies) were cultured in LB plates with kanamycin (25 μg/mL). A single colony was selected and cultured in 5 mL LB broth at 37 °C overnight with agitation at 200 rpm. The plasmid pET-30PS2-Variant was isolated using the Wizard Plus SV Minipreps DNA purification system (Promega®, Madison, WA, USA) according to the manufacturer's instructions. Sanger sequencing of the desired fragments using universal T7 primers was performed by Macrogen® (Seoul, Korea). The ~1000 bp contig was assembled using DNA Baser version 5.15 (Sequence assembly software) and used in BLASTn search (National Center for Biotechnology Information; www.ncbi.nim.nih.gov accessed on 19 September 2022). The protein sequence was deduced using the Translate tool available at Expasy (https://web.expasy.org/translate/, accessed on 20 September 2022) followed by multiple alignment sequence analysis of the variants in comparison with native PS2Aa1 using Bio-Edit Software [26]. Finally, variants with non-silent mutations were transformed into *Bt* BMB171 via electroporation using two consecutive electric pulses of 1.5 kV during 4.5 ms. Following this procedure, four PS2Aa1 mutant libraries were created.

4.4. Preparation of Activated Parasporin Proteins

Bt BMB171 variants were cultivated in LB broth and incubated for 5 days at 30 °C. The cells were then harvested by centrifugation at 10,000 rpm for 10 min; the pellet containing the precipitated parasporin proteins was solubilized in 5 mL of solubilization buffer (56 mM Na_2CO_3 (pH: 11.4) and 11 mM of dithiothreitol (DTT) (pH: 11.4)) for 2 h at 37 °C. Insoluble material was removed by centrifugation at 10,000 rpm for 10 min and the supernatant was passed through a 0.22 μm membrane filter. The filtrate (70 mL) pH was adjusted to 8 using 1 M Tris-HCl (pH 8).

The solubilized proteins were digested using proteinase K (final concentration of 185 µg/mL) for 1 h at 37 °C. Phenylmethylsulfonyl fluoride (PMSF) was added (final concentration 1 mM) to stop proteolytic processing. To confirm the presence of the parasporin proteins, SDS–PAGE analysis was performed as previously described [19]. The protein concentration was determined using the Bio-Rad Protein Assay (Bio-Rad Laboratories, Mississauga, ON, Canada).

4.5. Colon Cancer Cell Lines

Colon cancer cell lines were obtained from the American Type Culture Collection (ATCC). Lines SW-480 and CaCo-2 were cultured in Dulbecco's modified Eagle's medium (DMEM) with 25 mM glucose and 2 mM L-glutamine, supplemented with 10% fetal bovine serum (FBS), 10,000 µg/mL penicillin and streptomycin, and 1% non-essential amino acids; SW620 and NCM460 were cultured in RPMI supplemented with 10% FBS. The cultures were incubated in a humidified incubator at 37 °C under a 5% CO_2 atmosphere.

4.6. Cytotoxicity Assays of PS2Aa1 Mutants in Colon Cancer Cells

To characterize the anti-proliferative activity of the toxins PS2Aa1 and mutants, concentrations ranging from 0.25 to 5 µg/mL of the activated protein were prepared. The cells were incubated with parasporin for 72 h, and anticancer activity was assessed on the basis of cell viability following incubation with alamar blue (BioRad, Hercules, CA, USA) for 5 h at 37 °C under a 5% CO_2 atmosphere. Emission at 560 nm and excitation at 590 nm were measured using a CLARIOstar reader. IC_{50} concentrations were calculated from these data.

4.7. APN Detection and Blocking Assay

Western blot detection of APN was carried out using antibody CD13/APN (Cell Signaling Technology®, Beverly, MA, USA) and GAPDH was used as a loading control. The proteins extracted from HL60 and MCF-7 were used as the positive and negative control, respectively. The blocking assay was performed as reported by Periyasamy et al. 2016 [7]. Briefly, CaCo-2 and SW480 cells were used, but, because of the presence of APN receptors, the concentration of the APN blocker Dinitroflavone (Santa Cruz Biotechnology, Inc., Dallas, TX, USA) was diluted from 5 to 50 µM and added to the cells, with triplication of samples. This was followed by incubation for 72 h at 37 °C under a 5% CO_2 atmosphere using 3 µg of parasporin. The results of metabolic activity were analyzed using the alamar blue assay, as previously described [7].

4.8. Molecular Docking and Molecular Dynamics Analysis

Molecular docking analysis was performed to determine possible interactions between PS2Aa1 and the APN receptor. The 3D structures of PS2Aa1 and APN were downloaded from the Protein Data Bank with PDB ID 2ZTB and 6ATK, respectively. The simulations were performed using the protein–protein global docking protocol of Rosetta [27] with flags construct-5000, -spin, -randomize1, and -randomize2, and simulations were run in the software version 3.9. Later, the Interface Analyzer protocol of Rosetta [28] was used to analyze the interaction between PS2Aa1 and APN. Next, the top 10 models were visually inspected for the selection of a suitable model in which the amino acids corresponding to domain I of PS2Aa1 were interacting with a valid region of APN. Subsequently, the selected model was used to perform molecular dynamics (MD) simulations, in triplicate, for PS2Aa1–APN complexes. The MD simulation protocol included solvation with TIP3P water molecules, and Na^+ and Cl^- ions were added to ensure the neutrality of the system at an ion concentration of 150 mM. The Amber ff19SB force field for proteins was implemented for all systems.

The system was first minimized and equilibrated using Amber18 software. For the parasporin–APN complexes, stepped minimization, heating, and balancing were performed. Complexes were first minimized for 5000 (steepest descent) and 10,000 (conjugate gradient) followed by heating for 1 ns and then in an NVT to 300 K. For all simulations, an 8 Å cutoff was used for unbounded Coulombic and Lennard–Jones interactions and for periodic boundary conditions with a particle mesh Ewald treatment of long-range Coulombic interactions. A 2 fs time-step was employed by the SHAKE algorithm and the production steps were 100 ns for both systems.

4.9. Statistical Analysis

Type 1 ANOVA and Tukey's test was performed using Graphad Prism 8. Statistical significance was indicated as * $p \leq 0.05$, ** $p \leq 0.01$, and *** $p \leq 0.001$, with an IC of 95%.

Supplementary Materials: The following supporting information can be downloaded at https://www.mdpi.com/article/10.3390/molecules27217262/s1. Figure S1: Cytocidal activities of variants of Parasporin 2 (PS2Aa1) obtained with site-directed mutagenesis to CaCo2 cell lines; Figure S2: Cytocidal activities of variants of Parasporin 2 (PS2Aa1) obtained with site-directed mutagenesis to SW620 cell lines; Figure S3: Cytocidal activities of variants of Parasporin 2 (PS2Aa1) obtained with site-directed mutagenesis to SW480; Figure S4: Cytocidal activities of variants of Parasporin 2 (PS2Aa1) obtained with site-directed mutagenesis to NCM460 cell lines.

Author Contributions: Conceptualization, M.O.S.-B., L.V., E.H.P.-R., P.R.V. and N.J.R.-F.; Investigation, M.O.S.-B., L.V., E.H.P.-R., P.R.V., J.S.A.-A. and N.J.R.-F.; Methodology, M.O.S.-B., L.V., E.H.P.-R., P.R.V. and N.J.R.-F.; Project administration, N.J.R.-F.; Writing—original draft, M.O.S.-B., L.V., E.H.P.-R., P.R.V., J.S.A.-A. and N.J.R.-F.; Writing—review and editing, M.O.S.-B., L.V. and N.J.R.-F. All authors have read and agreed to the published version of the manuscript.

Funding: The work has been funded by MINCIENCIAS, MINEDUCACIÓN, MINCIT, and ICETEX, through the Program Ecosistema Científico Cod. FP44842-211-2018 Project number, 58668.

Institutional Review Board Statement: Not applicable.

Informed Consent Statement: Not applicable.

Data Availability Statement: Not applicable.

Conflicts of Interest: The authors declare no conflict of interest. The funders had no role in the design of the study; in the collection, analyses, or interpretation of data; in the writing of the manuscript; or in the decision to publish the results.

Sample Availability: Not available.

References

1. Laliani, G.; Ghasemian Sorboni, S.; Lari, R.; Yaghoubi, A.; Soleimanpour, S.; Khazaei, M.; Hasanian, S.M.; Avan, A. Bacteria and cancer: Different sides of the same coin. *Life Sci.* **2020**, *246*, 11739. [CrossRef] [PubMed]
2. Desbats, M.A.; Giacomini, I.; Prayer-Galetti, T.; Montopoli, M. Metabolic Plasticity in Chemotherapy Resistance. *Front. Oncol.* **2020**, *10*, 281. [CrossRef]
3. Shriwas, O.; Mohapatra, P.; Mohanty, S.; Dash, R. The Impact of m6A RNA Modification in Therapy Resistance of Cancer: Implication in Chemotherapy, Radiotherapy, and Immunotherapy. *Front. Oncol.* **2021**, *10*, 3220. [CrossRef]
4. Baindara, P.; Mandal, S.M. Bacteria and bacterial anticancer agents as a promising alternative for cancer therapeutics. *Biochimie* **2020**, *177*, 164–189. [CrossRef]
5. Mizuki, E.; Park, Y.S.; Saitoh, H.; Yamashita, S.; Akao, T.; Higuchi, K.; Ohba, M. Parasporin, a Human Leukemic Cell-Recognizing Parasporal Protein of *Bacillus thuringiensis*. *Clin. Vaccine Immunol.* **2000**, *7*, 625–634. [CrossRef] [PubMed]
6. Cruz, J.; Suárez-Barrera, M.O.; Rondón-Villarreal, P.; Olarte-Diaz, A.; Guzmán, F.; Visser, L.; Rueda-Forero, N.J. Computational study, synthesis and evaluation of active peptides derived from Parasporin-2 and spike protein from Alphacoronavirus against colorectal cancer cells. *Biosci. Rep.* **2021**, *41*, 1964. [CrossRef]
7. Periyasamy, A.; Kkani, P.; Chandrasekaran, B.; Ponnusamy, S.; Viswanathan, S.; Selvanayagam, P.; Rajaiah, S. Screening and characterization of a non-insecticidal *Bacillus thuringiensis* strain producing parasporal protein with selective toxicity against human colon cancer cell lines. *Ann. Microbiol.* **2016**, *66*, 1167–1178. [CrossRef]
8. Abe, Y.; Shimada, H.; Kitada, S. Raft-targeting and oligomerization of parasporin-2, a *Bacillus thuringiensis* crystal protein with anti-tumour activity. *J. Biochem.* **2008**, *143*, 269–275. [CrossRef]
9. Ito, A.; Sasaguri, Y.; Kitada, S.; Kusaka, Y.; Kuwano, K.; Masutomi, K.; Mizuki, E.; Akao, T.; Ohba, M. A *Bacillus thuringiensis* crystal protein with selective cytocidal action to human cells. *J. Biol. Chem.* **2004**, *279*, 21282–21286. [CrossRef] [PubMed]

10. Mendoza-Almanza, G.; Esparza-Ibarra, E.L.; Ayala-Luján, J.L.; Mercado-Reyes, M.; Godina-González, S.; Hernández-Barrales, M.; Olmos-Soto, J. The cytocidal spectrum of *Bacillus thuringiensis* toxins: From insects to human cancer cells. *Toxins* **2020**, *12*, 301. [CrossRef] [PubMed]

11. Akiba, T.; Okumura, S. Parasporins 1 and 2: Their structure and activity. *J. Invertebr. Pathol.* **2017**, *142*, 1–6. [CrossRef]

12. Abe, Y.; Inoue, H.; Ashida, H.; Maeda, Y.; Kinoshita, T.; Kitada, S. Glycan region of GPI anchored-protein is required for cytocidal oligomerization of an anticancer parasporin-2, Cry46Aa1 protein, from *Bacillus thuringiensis* strain A1547. *J. Invertebr. Pathol.* **2017**, *142*, 71–81. [CrossRef]

13. Soberón, M.; Portugal, L.; Garcia-Gómez, B.I.; Sánchez, J.; Onofre, J.; Gómez, I.; Pacheco, S.; Bravo, A. Cell lines as models for the study of Cry toxins from *Bacillus thuringiensis*. *Insect Biochem. Mol. Biol.* **2018**, *93*, 66–78. [CrossRef] [PubMed]

14. Al-Baidani, S.J.; Al-Faisal, A.H.M.; Ali, N.A. Study of some cytogenetic effects of Aerolysin Produced by Aeromonas hydrophila on normal and tumor cells in vitro. *J. Biotechnol. Res. Cent.* **2010**, *4*, 11–16. [CrossRef]

15. Ohba, M.; Mizuki, E.; Uemori, A. Parasporin, a new anticancer protein group from *Bacillus thuringiensis*. *Anticancer Res.* **2009**, *29*, 427–433.

16. Katayama, H.; Kusaka, Y.; Yokota, H.; Akao, T.; Kojima, M.; Nakamura, O.; Mekada, E.; Mizuki, E. Parasporin-1, a novel cytotoxic protein from *Bacillus thuringiensis*, induces Ca^{2+} influx and a sustained elevation of the cytoplasmic Ca^{2+} concentration in toxin-sensitive cells. *J. Biol. Chem.* **2007**, *282*, 7742–7752. [CrossRef] [PubMed]

17. Borin, D.B.; Castrejón-Arroyo, K.; Cruz-Nolasco, A.; Peña-Rico, M.; Rorato, M.S.; Santos, R.C.V.; de Baco, L.S.; Pérez-Picaso, L.; Camacho, L.; Navarro-Mtz, A.K. Parasporin a13-2 of *Bacillus thuringiensis* isolates from the papaloapan region (Mexico) induce a cytotoxic effect by late apoptosis against breast cancer cells. *Toxins* **2021**, *13*, 476. [CrossRef]

18. Akiba, T.; Abe, Y.; Kitada, S.; Kusaka, Y.; Ito, A.; Ichimatsu, T.; Katayama, H.; Akao, T.; Higuchi, K.; Mizuki, E.; et al. Crystal structure of the parasporin-2 *Bacillus thuringiensis* toxin that recognizes cancer cells. *J. Mol. Biol.* **2009**, *386*, 121–133. [CrossRef] [PubMed]

19. Brasseur, K.; Auger, P.; Asselin, E.; Parent, S.; Cote, J.C.; Sirois, M. Parasporin-2 from a New *Bacillus thuringiensis* 4R2 Strain Induces Caspases Activation and Apoptosis in Human Cancer Cells. *PLoS ONE* **2015**, *10*, e0135106. [CrossRef]

20. Inagaki, Y.; Tang, W.; Zhang, L.; Du, G.; Xu, W.; Kokudo, N. Invasion of Hepatocellular Carcinoma Cells as Well as Angiogenesis. *Biosci. Trends* **2010**, *4*, 56–60. [PubMed]

21. Liao, C.; Jin, M.; Cheng, Y.; Yang, Y.; Soberón, M.; Bravo, A.; Liu, K.; Xiao, Y. *Bacillus thuringiensis* Cry1Ac Protoxin and Activated Toxin Exert Differential Toxicity Due to a Synergistic Interplay of Cadherin with ABCC Transporters in the Cotton Bollworm. *Appl. Environ. Microbiol.* **2022**, *88*, e02505-21. [CrossRef]

22. Karlova, R.; Weemen-Hendriks, M.; Naimov, S.; Ceron, J.; Dukiandjiev, S.; de Maagd, R.A. *Bacillus thuringiensis* delta-endotoxin Cry1Ac domain III enhances activity against Heliothis virescens in some, but not all Cry1-Cry1Ac hybrids. *J. Invertebr. Pathol.* **2005**, *88*, 169–172. [CrossRef] [PubMed]

23. Okumura, S.; Ohba, M.; Mizuki, E.; Crickmore, N.; Côté, J.-C.; Nagamatsu, Y.; Kitada, S.; Sakai, H.; Harata, K.; Shin, T. Parasporin Classification and Nomenclature. Available online: https://www.fitc.pref.fukuoka.jp/parasporin/intro.html (accessed on 15 October 2022).

24. Melo, A.L.D.A.; Soccol, V.T.; Soccol, C.R. *Bacillus thuringiensis*: Mechanism of action, resistance, and new applications: A review. *Crit. Rev. Biotechnol.* **2016**, *36*, 317–326. [CrossRef] [PubMed]

25. Kitada, S.; Abe, Y.; Maeda, T.; Shimada, H. Parasporin-2 requires GPI-anchored proteins for the efficient cytocidal action to human hepatoma cells. *Toxicology* **2009**, *264*, 80–88. [CrossRef] [PubMed]

26. Hall, T.A. BIOEDIT: A user-friendly biological sequence alignment editor and analysis program for Windows 95/98/NT. *Nucleic Acids Symp. Ser.* **1999**, *41*, 95–98.

27. Ehrlich, L.P.; Wade, R.C. Protein-protein docking. *Rev. Comput. Chem.* **2001**, *17*, 61–97. [CrossRef]

28. Lewis, S.; Stranges, P.B.; Jared, A.-B. InterfaceAnalyzer. Available online: https://www.rosettacommons.org/docs/latest/application_documentation/analysis/interface-analyzer (accessed on 15 October 2022).

Article

Development of Genistein Drug Delivery Systems Based on Bacterial Nanocellulose for Potential Colorectal Cancer Chemoprevention: Effect of Nanocellulose Surface Modification on Genistein Adsorption

Melissa Castaño [1], Estefanía Martínez [1], Marlon Osorio [1,2] and Cristina Castro [1,*]

[1] School of Engineering, Universidad Pontificia Bolivariana, Circular 1#70-01, Medellín 050031, Colombia
[2] School of Health Science, Biology Systems Research Group, Universidad Pontificia Bolivariana, Calle 78b #72a-159, Medellín 050031, Colombia
*   Correspondence: cristina.castro@upb.edu.co

Abstract: Genistein is an isoflavone with antioxidant, anti-inflammatory, and anticancer properties. That said, its use in the industry is limited by its low solubility in aqueous systems. In this work, bacterial nanocellulose (BNC) and BNC modified with cetyltrimethylammonium (BNC-CTAB) were evaluated as genistein-encapsulating materials for their controlled release in cancer chemoprevention. Thin films were obtained and characterized by contact angle, AFM, TEM, UV–Vis spectroscopy FTIR, and TGA techniques to verify surface modification and genistein encapsulation. The results show a decrease in hydrophilization degree and an increase in diameter after BNC modification. Furthermore, the affinity of genistein with the encapsulating materials was determined in the context of monolayer and multilayer isotherms, thermodynamic parameters and adsorption kinetics. Spontaneous, endothermic and reversible adsorption processes were found for BNC-GEN and BNC-CTAB-GEN. After two hours, the maximum adsorption capacity corresponded to 4.59 mg GEN·g^{-1} BNC and 6.10 mg GEN·g^{-1} BNC-CTAB; the latter was a more stable system. Additionally, in vitro release assays performed with simulated gastrointestinal fluids indicated controlled and continuous desorption in gastric and colon fluids, with a release of around 5% and 85%, respectively, for either system. Finally, the IC50 tests made it possible to determine the amounts of films required to achieve therapeutic concentrations for SW480 and SW620 cell lines.

Keywords: bacterial nanocellulose; surface modification; genistein; controlled drug delivery system; colorectal cancer

Citation: Castaño, M.; Martínez, E.; Osorio, M.; Castro, C. Development of Genistein Drug Delivery Systems Based on Bacterial Nanocellulose for Potential Colorectal Cancer Chemoprevention: Effect of Nanocellulose Surface Modification on Genistein Adsorption. *Molecules* **2022**, *27*, 7201. https://doi.org/10.3390/molecules27217201

Academic Editor: Jahir Orozco Holguín

Received: 30 July 2022
Accepted: 7 October 2022
Published: 24 October 2022

Publisher's Note: MDPI stays neutral with regard to jurisdictional claims in published maps and institutional affiliations.

1. Introduction

Annually, there are 9.0 million cancer deaths worldwide [1], and colorectal cancer (CRC) ranks third in the world and fourth with new cases reported [2,3]. The treatments for CRC include endoscopic resection, polypectomy, a partial colectomy, chemotherapy, and radiation therapy. All of these are invasive and are related to side effects [4,5]. Consequently, researchers seek to develop new strategies that can increase the specificity of treatments, reduce side effects, and prevent the formation and progression of cancer [6,7]. In recent years, novel treatments have focused on natural compounds as an alternative for cancer prevention and treatments due to their nontoxic and selective character [8].

Genistein, or 4',5,7-trihydroxyisoflavone, is a polyphenolic compound member of the isoflavones family that is derived from different food sources such as soybeans, legumes, broad beans, and lupins [9]. Soy isoflavones are known as phytoestrogens, due to the structural similarity between them and 17β-estradiol; in the case of genistein, carbons 4 and 7 on the phenol rings are similar to the OH groups of estradiol and form bonds with estrogen receptor (ER) residues. Carbon 7 binds to His475, and carbon 4 binds with Arg346 and

Glu305; these connections allow GEN to bind with isoforms α and β of ER and stimulate estrogen [10,11], making genistein a chemopreventive agent against cancer, especially in breast, prostate, and CRC [10]. It also influences inflammation, cell proliferation and the modulation of epigenetic changes, and prevents angiogenesis and has antioxidant activity toward cancer cells [9,10,12]. However, the pharmaceutical use of genistein is limited by its low solubility [13]. Encapsulation technology may provide a means of increasing the biodisponibility of compounds, protecting the bioactive compounds from oxidation, and enhancing their miscibility and absorption in aqueous systems [14].

Cellulose is the most abundant polymer on Earth; it is present in algae, wood, cotton, bacteria, and fungi [15]. Bacterial nanocellulose (BNC) is obtained from different bacteria genera such as *Acetobacter*, *Sarcina*, and *Komagateibacter* [16]. BNC is composed of β-1,4-D(+) glucose units, and it is lignin- and hemicellulose-free, which is an advantage for avoiding chemical isolation treatments [15,16]. Besides this, BNC is made up of a three-dimensional network of nanoribbons with diameters at the nanoscale [15]. Its structure is assembled from inter- and intramolecular hydrogen bonds between nanoribbons, allowing for the formation of structures with a large surface area, open porosity, and good tensile strength [17,18]. Likewise, BNC is biocompatible and nontoxic, and has a high potential in medical applications, such as in the treatment of wounds, the development of artificial skin, tissue regeneration, the manufacture of dental and artery implants, as well as in protein immobilization and controlled drug release [16,19–21]. In drug delivery systems, BNC has been used to encapsulate different compounds. For instance, Subtaweesin et al. (2018) loaded curcumin on BNC, and the results showed anticancer activity against A375 melanoma cancer cells and no significant cytotoxic activity against human keratinocytes and human dermal fibroblast [22]. However, some authors have modified the hydrophilic nature of BNC to improve the adsorption of nonpolar bioactive compounds [23]. S. Akagi et al. (2021) prepared cellulose in a culture supplemented with carboxymethylcellulose and hydroxypropylcellulose to carry hydrophobic cancer drugs such as paclitaxel (PTX) formulations, which reduced the side effects of PTX and toxicity and increased the therapeutic efficacy of hydrophobic compounds [24]. M. L. Cacicedo et al. (2016) modified BNC by an in-situ method with alginate addition to load doxorubicin (DOX); the results showed decreased HT-29 viability compared to free DOX and stability in time (14 days) [21]. Finally, Wang et al. (2019) coated plant-derived cellulose nanocrystals with cetyltrimethylammonium bromide (CTAB) to improve the dissolution rate and antioxidant activity of genistein in aqueous systems, with results showing a decrease in genistein crystalline structure, and an increase in the dissolution rate to 72–92% compared with the original genistein, as well as enhanced antioxidant activity [12]. Additionally, Qu et al. (2019) investigated CTAB modification to increase the hydrophobicity and enhance the dispersibility of fibrillated BNC for further applications. The results show that CTAB modification improved the thermal stability and hydrophobicity of fibrillated cellulose [25]. Nevertheless, the literature lacks information on the effect of this surface modification on genistein loaded on BNC, and its application for oral drug delivery in cancer chemoprevention. Therefore, this paper aims to develop drug delivery systems (DDS) based on BNC and BNC-CTAB for genistein, along with the mathematical model of the systems, material characterization, and in vitro studies.

2. Results

2.1. Calculation of Inhibitory Concentration (IC50) for Free Genistein

The half maximal inhibitory concentration (IC50) is a measure of the potency or drug efficacy; this indicates the amount of a bioactive compound that is necessary to inhibit a biological process by half. Genistein's ability to inhibit cancer cell growth was evaluated. These values are key to any subsequent DDS design. Figure 1 shows the results of the inhibitory effects of genistein on SW480, SW620, and HaCaT cell lines. The results show that cell inhibition is time- and dose-dependent, suggesting increased inhibition in all cell lines at 48 h. Previous results in the literature indicate IC50 values of 39.43, 50 and 52.62 μM against

CRC cells such as HCT-116, SW620, HT29 and SW480 cells, respectively [26–29]. Similarly, IC50 values of 127.60 and 140.30 µM demonstrate genistein's anticancer activities against MCF-7 and Hep3B cells, corresponding to hepatocellular and breast carcinomas [30,31]. Figure 1d summarizes the IC50 values of SW40, SW620, and HaCaT at 24 and 48 h in this study. Our results indicate that genistein has selectivity and anticancer activity against CRC cell lines. In this work, the IC50 values for free genistein were used for the development of oral DDS for the chemoprevention of CRC.

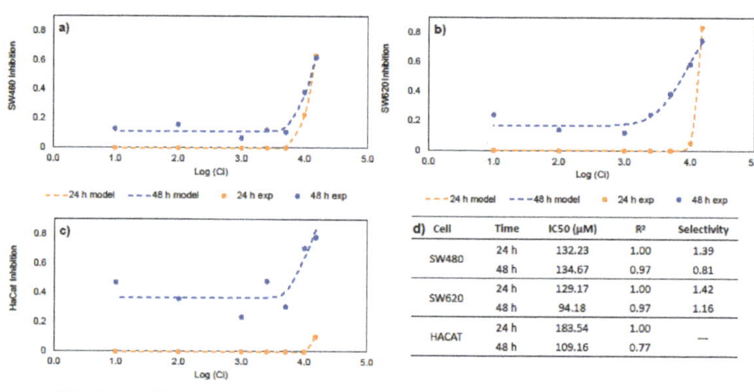

Figure 1. In vitro studies for IC50 and selectivity; (**a**) Hill equation for SW480 cells; (**b**) Hill equation for SW620 cells; (**c**) Hill equation for HaCaT cells; (**d**) IC50 for free genistein and selectivity.

2.2. Development of BNC and BNC-CTAB Materials

To improve the adsorption of nonpolar compounds such as genistein, the surface modification of BNC has emerged as a possibility to enhance the solubility and biodisponibilty in aqueous systems. Figure 2 shows the TEM micrographs of BNC before and after CTAB modification. A 3D network with an entangled structure of randomly oriented nanoribbons can be observed. For BNC nanoribbons (Figure 2a,b), the diameters ranged between c.a. 10 and 60 nm (Figure 2e). For BNC-CTAB nanoribbons (Figure 2c,d), these values were between 12 and 85 nm. This increase in diameter is not statistically significant. However, the BNC and BNC-CTAB diameter medians were 28.74 ± 14.72 and 34.47 ± 18.68 nm, respectively.

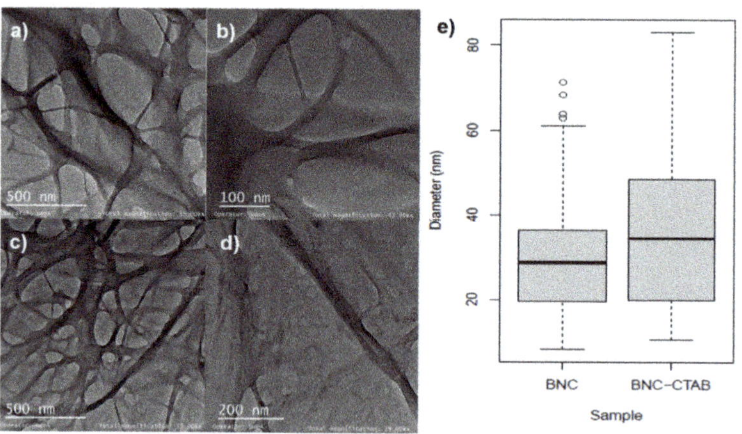

Figure 2. Morphological analysis (**a**) TEM micrographs of BNC at 15 kX; (**b**) TEM micrographs of BNC at 43 kX; (**c**) TEM micrographs of BNC-CTAB at 15 kX; (**d**) TEM micrographs of BNC-CTAB at 29 kX; (**e**) diameter distribution graphic.

The addition of cetyl trimethyl ammonium to BNC implies a hydrophobization of nanoribbons [32]. The contact angles formed between deionized water and dry films of BNC and BNC-CTAB correspond to 33.92° and 60.66°, respectively (Figure 3); the increase was statistically significant and corresponded to c.a. two-fold. The contact angle measurements indicate a decrease in hydrophilicity on the BNC-CTAB surface [33,34].

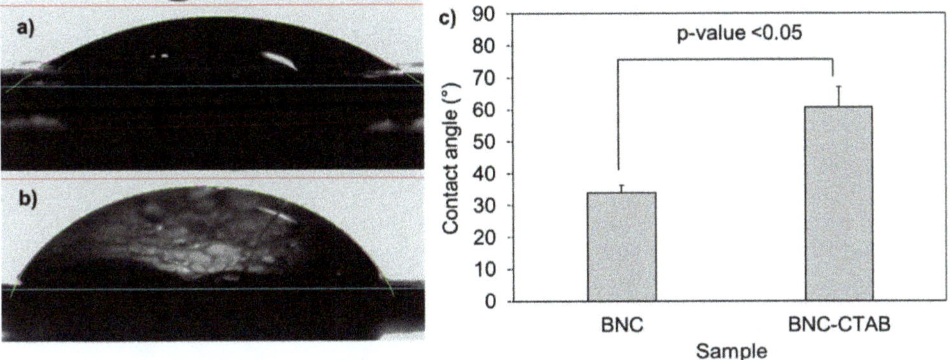

Figure 3. Contact angle experiments: (**a**) drop of deionized water in BNC; (**b**) drop of deionized water in BNC-CTAB; (**c**) average of contact angles. The sample groups were statistically different with *p* values < 0.05.

The FTIR spectra of BNC, BNC-CTAB, and CTAB are shown in Figure 4. For BNC, bands at 3330 and 2890 cm^{-1} are characteristic of O-H stretching and C-H stretching [12,21]. The band at 1640 cm^{-1} is related to residual water, and the peak at 1370 cm^{-1} represents C-H bending [12,35]. In addition, bands at 1430, 1056, and 898 cm^{-1} represent the CH$_2$ symmetrical bending, C-O vibration, and C-O-C stretching of the sugar ring, respectively [35,36]. For CTAB, characteristic bands were observed between 2925 and 2850 cm^{-1}, related to the symmetric and asymmetric stretching of the C-H bond of the alkyl chain [32]. The BNC-CTAB spectra show new peaks at 2915, 2840 and 1599 cm^{-1}; the first two are related to the alkyl chain of CTAB, and the last peak is related to carboxyl groups presented due to (2,2,6,6-Tetramethylpiperidin-1-yl)oxyl (TEMPO) pretreatment.

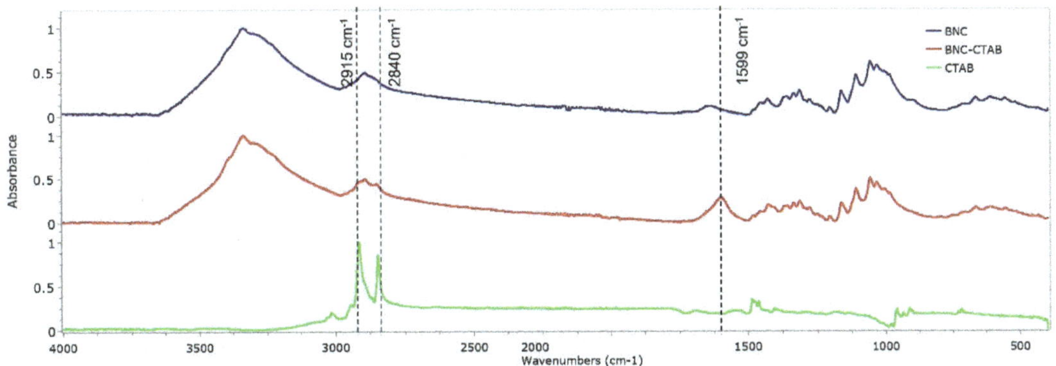

Figure 4. FTIR spectrum of BNC, BNC-CTAB, and free CTAB.

Figure 5 shows the results of the thermal analysis of BNC, BNC-CTAB ribbons, and free CTAB. The samples presented a thermal degradation temperature of 265 °C for BNC and 210 °C for BNC-CTAB and CTAB. Surface modification decreased the thermal degradation temperature of cellulose. Moreover, the BNC-CTAB ribbons contained less water than BNC, which is related to the larger amount of residue at higher temperatures in the BNC-CTAB sample. The DTG decomposition curves of BNC and CTAB (Figure 5b) demonstrate the appearance of one peak at 345 °C for BNC and 274 °C for CTAB, similar to what was reported in the literature [32,37]. In contrast, the DTG curve of BNC-CTAB shows two peaks; the first at 254 °C is related to the early degradation of alkyl chains and carboxylic groups, while the latter is found in samples due to TEMPO pretreatment, and the second peak at 337 °C is related to the unmodified parts of cellulose [32]. Finally, the previous results indicate the successful modification of BNC with CTAB.

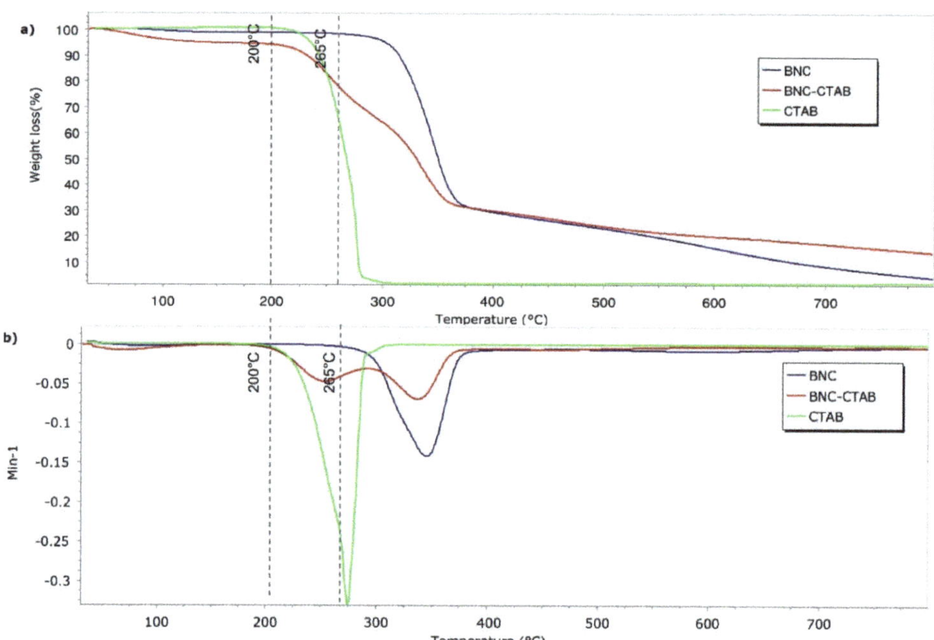

Figure 5. Thermogravimetric analysis of BNC, BNC-CTAB, and free CTAB: (**a**) TGA thermogram; (**b**) DTG curves.

2.3. Adsorption Studies

2.3.1. Adsorption Isotherms

Figure 6 shows the experimental data and their adjustment to monolayer models at 0, 23, and 40 °C, and Table 1 presents the adjustment and parameters of the Langmuir, Freundlich, Sips, and Toth models. The experimental data did not fit to multilayer models and can be seen in Appendix A. The BNC experimental data showed an increase in adsorption capacity with an increase in the genistein concentration, and subsequently, a plateau was reached when the concentration in equilibrium (C_e) was 17.91, 13.90, and 15.45 mg·L^{-1} for BNC at 0, 23, and 40 °C, respectively. Additionally, the experimental data show the adjustment ($R^2 > 0.95$) to the models of Langmuir, Freundlich, Sips, and Toth at the three temperatures; hence, these models adequately describe the adsorption process and indicate monolayer adsorption. Otherwise, the BNC-CTAB experimental data show an increase in adsorption capacity with an increase in the genistein concentration, followed by a plateau when the C_e was 10.68 and 10.81 mg·L^{-1} at 0 and 23 °C; at 40 °C, unstable adsorption was presented, while an increase in genistein adsorption was followed by a decrease in adsorption capacity, and standard deviation growth was presented. Therefore, it was found that the Langmuir, Freundlich, Sips, and Toth models are appropriately adjusted to temperatures of 0 and 23 °C ($R^2 > 0.95$).

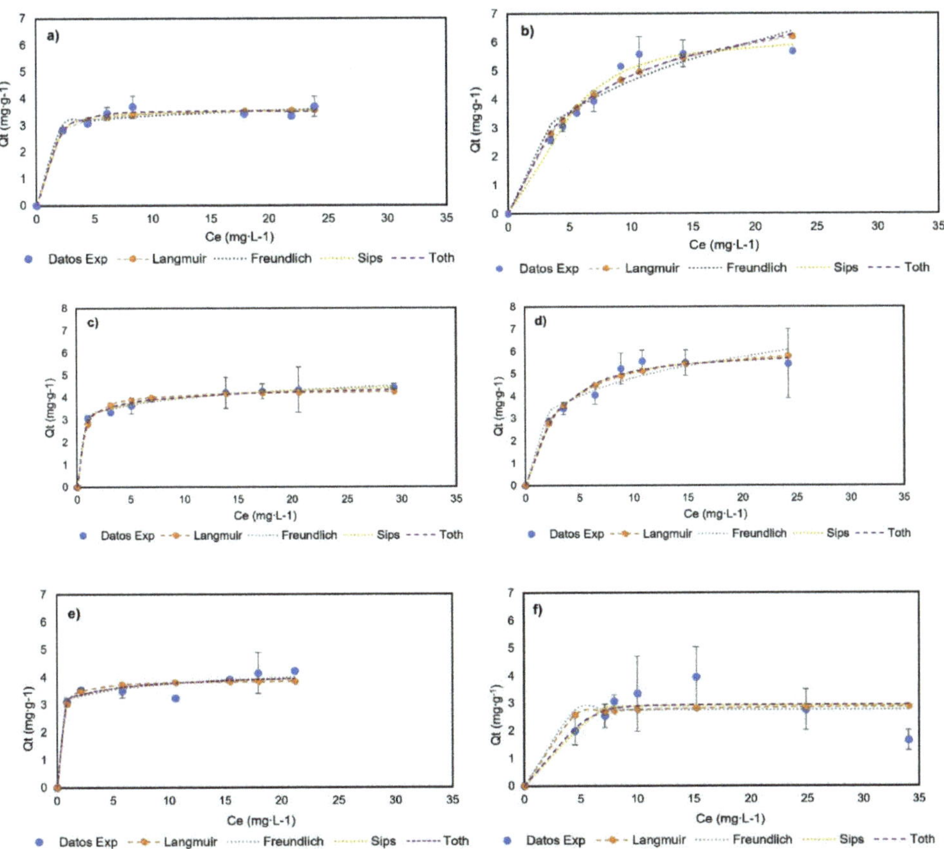

Figure 6. Adsorption isotherms of (**a**) BNC-GEN at 0 °C; (**b**) BNC-CTAB-GEN at 0 °C; (**c**) BNC-GEN at 23 °C; (**d**) BNC-CTAB-GEN at 23 °C; (**e**) BNC-GEN at 40 °C; (**f**) BNC-CTAB-GEN at 40 °C.

Table 1. Isotherms models for genistein adsorption in BNC and BNC-CTAB.

Isotherm Models		BNC-GEN			BNC-CTAB-GEN		
		0 °C	23 °C	40 °C	0 °C	23 °C	40 °C
Langmuir	Q_m (mg·g^{-1})	3.67	4.34	3.90	8.12	6.47	2.92
	K_L (L·mg^{-1})	1.58	1.80	3.95	0.15	0.35	1.68
	R^2	0.98	0.98	0.95	0.96	0.98	0.65
Freundlich	K_F	8.12	3.04	3.11	1.88	2.63	2.75
	n_f	0.15	8.44	12.13	2.53	3.81	4.30×10^7
	R^2	0.96	0.99	0.97	0.92	0.95	0.64
Sips	Q_{mS} (mg·g^{-1})	3.56	5.56	4.63	6.24	6.12	2.91
	a_S (L·mg^{-1})	0.96	1.12	2.07	0.05	0.31	0.00
	B_S	1.65	0.37	0.33	2.00	1.20	5.87
	R^2	0.98	0.99	0.96	0.98	0.98	0.71
Toth	Q_{mT} (mg·g^{-1})	4.47	4.84	4.99	6.55	5.99	2.93
	A_T (L·mg^{-1})	0.65	3.48	8.49	0.24	0.17	0.00
	z	1.09	0.50	0.24	0.92	1.38	5.14
	R^2	0.99	0.99	0.96	0.95	0.98	0.71

The adsorption isotherms of both systems indicate that the adsorption capacity of adsorbents does not depend on the initial concentration, but rather the available sites for the adsorption of genistein [38]. Likewise, the increment in the Langmuir constant (K_L) indicates the relation between the adsorption capacity and temperature. Moreover, the Freundlich constant provides information on the heterogeneity and irreversibility of the system—when $1/n_f < 0.1$ approaches zero, it implies an increase in the surface heterogeneity, while if $1/n_f < 0.1$ the process is irreversible, and $1/n_f > 1$ indicates an unfavorable adsorption isotherm [38]. Regarding reversibility, the results show that the adsorption of genistein in BNC at 23 °C and BNC-CTAB at 0 and 23 °C is reversible, while at 40 °C, it is irreversible for both systems. This means that later difficulties could arise for genistein release at 40 °C; in addition, the reversibility of the system depends on the temperature. For heterogeneity, an increase in temperature causes changes in the surface of the adsorbents, and as a consequence, the heterogeneity increases. The Freundlich isotherm showed that at 23 °C, BNC-GEN has a more heterogeneous surface than BNC-CTAB-GEN, while for the other temperatures, the results are not comparable. Furthermore, the Sips models indicate that at 23 °C, BNC-GEN shows a heterogenous surface, and BNC-CTAB-GEN a homogeneous surface. The Toth isotherm shows, similarly to the Freundlich, that heterogeneity increases with increasing temperature. At 0 °C, both systems are homogeneous, while at 23 °C, BNC-GEN is more heterogenous.

2.3.2. Determination of Thermodynamic Parameters

The thermodynamic results (Table 2) were obtained from K_L and the Van't Hoff equation (Equation (7)). For BNC-GEN, it was found that as the temperature increases, the spontaneity of the adsorption process increases; the positive value of ΔH indicates an endothermic process; likewise, a $\Delta H < 40$ kJ·mol^{-1} indicates physisorption, where the related molecular interactions correspond to hydrogen bonds [39]. Similarly, the parameters for BNC-CTAB-GEN indicate spontaneous adsorption processes, and endothermic processes for BNC-CTAB (40 kJ·mol^{-1} < ΔH < 80 kJ·mol^{-1}), meaning that the adsorption of genistein on BNC is due to a complex reaction, where physisorption predominates, as described by Lyubchik et al. (2020) and Scheufele et al. (2016) for liquid–solid systems [40,41]. In this case, molecular interactions correspond to electrostatic and hydrophobic interactions [41]. For both systems, the ΔS value shows an increase in the affinity of genistein molecules for BNC-CTAB and an increase in randomness in the liquid–solid interface of BNC-CTAB-GEN compared to BNC-GEN [41,42].

Table 2. Thermodynamic results for BNC-GEN and BNC-CTAB-GEN.

BNC-GEN			
T (°C)	**ΔG (kJ·mol^{-1})**	**ΔH (kJ·mol^{-1})**	**ΔS (J·mol^{-1}·K^{-1})**
0	−29.45		
23	−32.25	15.21	162.58
40	−36.14		
BNC-CTAB-GEN			
T (°C)	**ΔG (kJ·mol^{-1})**	**ΔH (kJ·mol^{-1})**	**ΔS (J·mol^{-1}·K^{-1})**
0	−24.099		
23	−28.225	41.36	238.33
40	−33.913		

According to the isotherm models, a greater adsorption capacity (Q_m) for BNC-GEN was reached at 23 °C, while the thermodynamic parameters indicate spontaneous and endothermic adsorption processes at this temperature. For BNC-CTAB-GEN, a greater adsorption capacity was reached at 0 °C; however, the affinity constant decreased at this temperature. Moreover, the thermodynamic parameters indicate spontaneous and endothermic adsorption processes. Hence, for subsequent assays, the selected temperature was 23 °C.

2.3.3. Adsorption Kinetics

For the adsorption kinetics, the temperature was 23 °C, with a genistein concentration of 42.28 mg·L^{-1}, to ensure the formation of the monolayer. Figure 7 and Table 3 show the experimental data and the model fit. For BNC-GEN, the results indicate an increase in adsorption capacity in the first 20 min and a desorption event at 60 min. After 100 min, the results showed the equilibrium. The models showed a low adjustment to the experimental data with R^2 < 0.800; therefore, these models are not suitable for explaining the adsorption of genistein on BNC over time. The low stability of the compound could be related to the hydrophilic nature of BNC and the formation of hydrogen bonds between water and genistein.

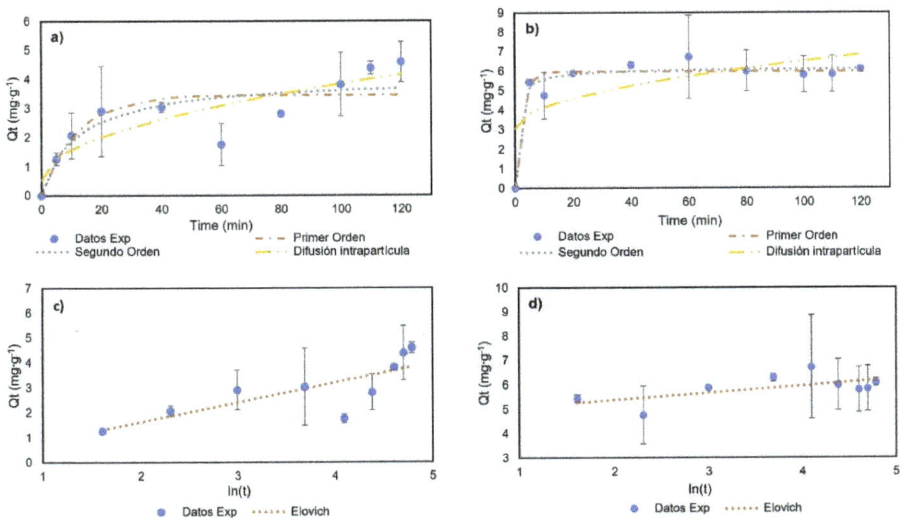

Figure 7. Adsorption kinetics results: (**a**) BNC-GEN; (**b**) BNC-CTAB-GEN; (**c**) BNC-GEN adjustment to Elovich model; (**d**) BNC-CTAB-GEN adjustment to Elovich model.

Table 3. Adsorption kinetics models.

	Pseudo-First-Order			Pseudo-Second-Order	
Parameters	**BNC-GEN**	**BNC-CTAB-GEN**	**Parameters**	**BNC-GEN**	**BNC-CTAB-GEN**
Q_e (mg·g^{-1})	3.46	5.98	Q_e (mg·g^{-1})	4.05	6.16
K_1 (min^{-1})	0.08	0.39	K_2 (g·mg^{-1}·min^{-1})	0.02	0.15
R^2	0.70	0.94	R^2	0.75	0.96
	Elovich			Intra-Particle Diffusion	
Parameters	**BNC-GEN**	**BNC-CTAB-GEN**	**Parameters**	**BNC-GEN**	**BNC-CTAB-GEN**
α (mg·g^{-1}·min^{-1})	0.845	4.15 × 10^6	C (mg·L^{-1})	0.53	3.07
β (g·mg^{-1})	1.27	3.431	K_3 (mg·g^{-1}·min$^{-1/2}$)	0.33	0.35
R^2	0.62	0.364	R^2	0.78	0.48

Furthermore, for BNC-CTAB-GEN, rapid adsorption was observed in the first 5 min followed by a desorption event at 10 min, and finally, equilibrium was reached at 20 min. The experimental data show a good adjustment (R^2 > 0.93) to the pseudo-first- (PFO) and pseudo-second-order (PSO) models, with a larger R^2 being shown by the PSO model. The CTAB surface modification gives genistein greater stability compared to BNC without modification.

Differences between both kinetics are related to the adsorption mechanism. The BNC-GEN adsorption mechanism is characterized by physisorption, in which adsorption and desorption events take place, while BNC-CTAB-GEN is characterized by a complex physical reaction, which helps the modified system to reach equilibrium faster. Finally, the good adjustment to the pseudo-second-order kinetics model indicates, similar to the Langmuir constants, that the adsorption process is not related to adsorbate concentration, but rather to the available sites on the adsorbent (BNC) [43].

2.4. Development of Thin Film Drug Delivery System and Characterization

To analyze the textures of the films, the BNC and BNC-CTAB samples were subjected to AFM. The parameters calculated for the surface roughness analysis of BNC and BNC-CTAB were arithmetical mean height or Sa, and root means square height or Sq (Figure 8). The results show an Sa of 64.25 ± 5.26 and 29.20 ± 3.81 nm, and an Sq of 81.69 ± 4.66 and 36.59 ± 2.46 nm, for BNC and BNC-CTAB, respectively. Therefore, BNC is statically rougher than BNC-CTAB. The incorporation of hydrophobic groups into modified cellulose causes the fibers to be homogeneously distributed on the surface after oven drying, avoiding nanoribbon agglomeration. Therefore, BNC is rougher than BNC-CTAB, and the difference is statistically significant.

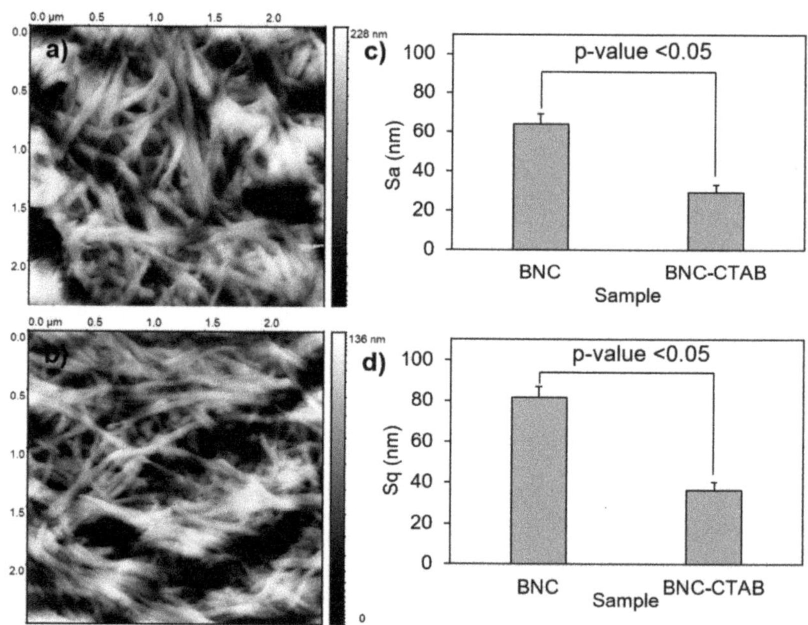

Figure 8. Roughness analysis; (**a**) AFM micrographs of BNC; (**b**) AFM micrographs of BNC-CTAB; (**c**) average of Sa; (**d**) average of Sq. The sample groups were statistically different with *p* values < 0.05.

To verify the adsorption of genistein in BNC and BNC-CTAB, Figure 9 shows the spectra of free genistein (GEN), BNC, BNC-GEN, and BNC-CTAB-GEN samples. For genistein, characteristic bands are observed at 3404, 1652, 1517, 1308, 1273, and 840–810 cm^{-1}. Due to the low proportion of genistein to the amount of cellulose in the films, the FTIR spectra for BNC-GEN do not show differences from the BNC spectra, which is attributed to the masking of genistein bands by the bands of BNC. However, in BNC-CTAB-GEN, there are changes in the heights and ratios of the bands, mainly at 1630 cm^{-1}, and between 1070 and 1040 cm^{-1}.

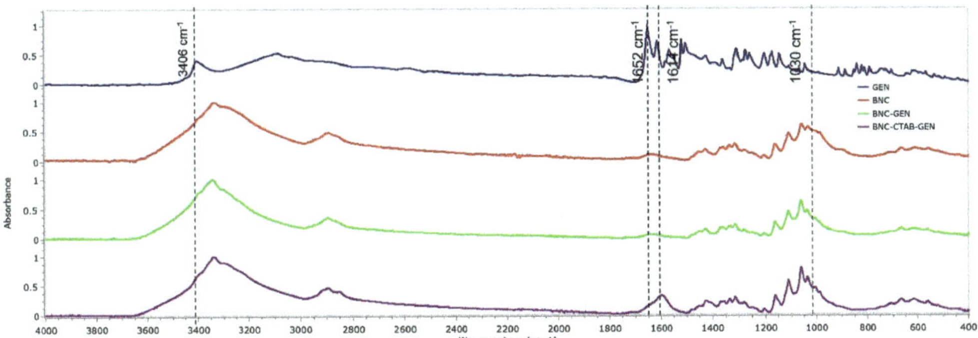

Figure 9. FTIR spectrum of GEN, BNC, BNC-GEN, and BNC-CTAB-GEN.

Figure 10 shows the thermal analysis of BNC, GEN, BNC-GEN, and BNC-CTAB-GEN. The samples showed a thermal degradation temperature of 265 °C for GEN and BNC-GEN, and 200 °C for BNC-CTAB-GEN. The DTG decomposition curves of BNC-GEN and GEN showed peaks at 350 and 375 °C, while the slight increase in the peak of DTG compared to that in BNC (Figure 10b) indicates an interaction between BNC and genistein. For BNC-CTAB-GEN, the DTG curve showed two peaks at 254 and 337 °C, the former being related to the early degradation of alkyl chains of CTAB and carboxylic groups of TEMPO pretreatment, and the latter being related to unmodified parts of cellulose, as described before in BNC-CTAB (Figure 10b). Both curves indicate an interaction between compound and adsorbent, but despite this, the interaction between BNC-CTAB and genistein is stronger than that between BNC and genistein, due to the appearance of a shoulder on the DTG curve of BNC-GEN at 321 °C.

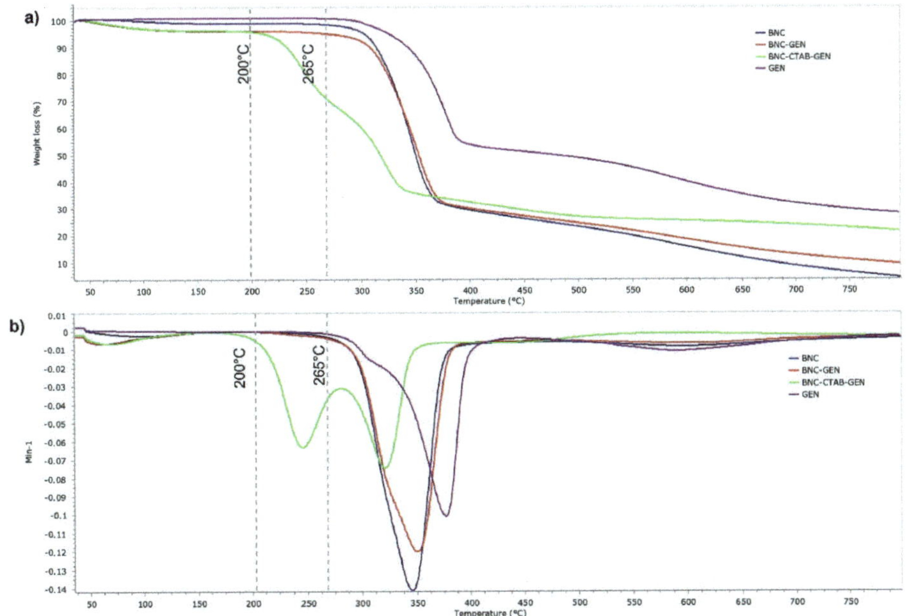

Figure 10. Thermogravimetric analysis of BNC, GEN, BNC-GEN, and BNC-CTAB-GEN; (**a**) TGA thermogram; (**b**) DTG curves.

2.5. In Vitro Release Study in Gastrointestinal Fluids

The maximum desorption capacities for BNC-GEN and BNC-CTAB-GEN were 46.6843 and 66.4316 mg·g^{-1}, respectively. BNC-CTAB was found to release 1.42 times more genistein than BNC. Table 4 shows the adjustment of experimental data to kinetic models. For both systems, a good fit to the second linear order model was observed (R^2 > 0.93) in the three simulated fluids. In stomach and small intestine fluids (pH 1.2 and 6.0, respectively), BNC-CTAB-GEN shows a higher desorption rate (h), while in colon fluid (pH 7.4), the BNC-GEN system shows higher h.

Table 4. Release kinetics models.

	Models	Stomach Fluid		Small Intestine Fluid		Colon Fluid	
		BNC-GEN	**BNC-CTAB-GEN**	**BNC-GEN**	**BNC-CTAB-GEN**	**BNC-GEN**	**BNC-CTAB-GEN**
PFO	Q_d (mg·g^{-1})	1.62	2.63	10.68	16.05	42.10	50.46
	K_1 (min^{-1})	3.29	12.00	1.78	1.78	0.16	12.00
	R^2	0.68	0.75	0.93	0.91	0.98	0.94
PSO	Q_d (mg·g^{-1})	2.19	3.64	11.68	17.32	44.46	63.70
	K_2 (g·mg^{-1}·min^{-1})	6.59×10^2	4.11×10^{-2}	6.18×10^{-4}	4.88×10^{-4}	4.28×10^{-4}	9.38×10^{-5}
	h (mg·g^{-1}·min^{-1})	0.32	0.54	0.08	0.15	0.85	0.38
	R^2	0.98	0.99	0.93	0.95	0.99	0.96

Moreover, the genistein release percentage from dried films of BNC-GEN and BNC-CTAB-GEN was determined and is shown in Figure 11. The results show that in the simulated stomach fluid, there is a sustained release of genistein that reaches almost 5% for both samples, while in small intestine fluid, approximately 20% of genistein is released in both samples; however, the release occurs in the first 5 min, and subsequently remains stable. Finally, the release reaches approximately 85% after 72 h in simulated colon fluid, but the release profile in colon fluid indicates a faster release in the BNC-GEN system and a controlled release for BNC-CTAB-GEN [44].

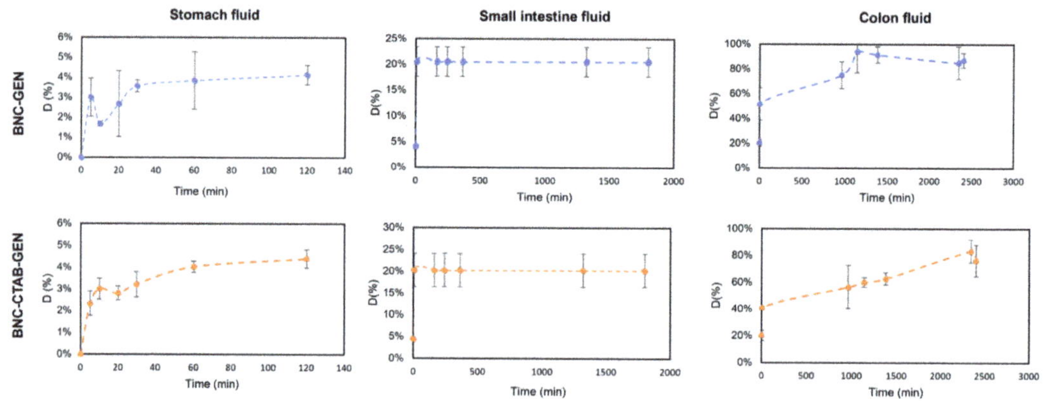

Figure 11. The release profile of genistein in gastrointestinal fluids.

The low release rate in stomach fluid indicates that BNC and BNC-CTAB protect genistein from acidic conditions while genistein encapsulation protects its bioactivity during its transit through stomach fluids [14]. As both systems contain cellulose, they release a small amount of active compound in acid conditions; when the pH is 7, the release of genistein is favored by the swelling of the cellulose.

Previous results indicate that BNC and BNC-CTAB could act as protective agents of genistein in the stomach, aiding its subsequent release, where cells would be responsible for metabolizing genistein [10]; thus, BNC and BNC-CTAB can be considered for use as genistein nanocarriers.

3. Discussion

Colorectal cancer is one of the most malign and deadly carcinomas worldwide [3,7]. As a result, we are seeking to develop new strategies that inhibit the formation and growth of cancer [6]. Recently, the use of natural compounds has increased due to their selective and nontoxic characteristics [8]. Genistein is a promising chemopreventive agent of cancer, with different effects, including influence on inflammation, cell proliferation, angiogenesis inhibition, and antioxidant activity [27], related to the regulation of ER. Toxicity tests should be performed to determine the concentration at which active compounds affect cancer cells and their mechanisms of action. Previous studies by Wang et al. (2012) have demonstrated that genistein treatment induces G2 phase detention and the inhibition of cell proliferation in SW40 cells, because of an increase in DKK1 expression [45]. Additionally, the results presented by Sun et al. (2022) suggest that genistein induced SW620 cell cycle arrest in the G2/M phase by targeting mutant p53 [46]. In the study carried out by Quin et al. (2015), it was suggested that the downregulation of miR-95 and SGK1, and Akt phosphorylation, could be related to the antitumor effects of genistein in CRC [26]. According to the results obtained in this work, genistein presented similar IC50 values compared to those in the literature and greater selectivity to SW480 and SW620 CRC cell lines compared to healthy keratocytes. The results obtained in cellular assays determine the subsequent design of DDS; hence, genistein was encapsulated in BNC materials for its potential application in CRC chemoprevention.

BNC has been widely used in drug delivery systems to encapsulate active compounds [22] because its large surface area allows for the adsorption and further desorption of different compounds. However, surface modification is proposed in the literature to increase the adsorption of nonpolar compounds such as genistein, as the higher degree of sample hydrophobicity is related to an increase in the stability and biodisponibility of active compounds in aqueous systems [14,23]. The diameters found for BNC nanoribbons correspond to the typical morphology of BNC reported previously by Castro et al. (2012) [47]. For BNC after CTAB modification, the diameter increases slightly; this corresponds to the presence of CTAB molecules and was previously reported by K. Syverud et al. (2011) and N. Zainuddin et al. (2017) upon the surface modification of cellulose nanofibers and nanocrystals, respectively [37,48]. However, the structure of BNC was not altered, and a high surface area after modification was still available for the adsorption and subsequent release of genistein, which is ideal for the development of drug delivery systems. Additionally, the decrease in the hydrophilicity of the BNC-CTAB surface is attributed to the adsorption of CTAB on the surface of BNC-TEMPO by ionic bonding; the COO^- groups act as active sites for the adsorption of CTA^+ polar heads [32].

Adsorption studies have suggested that BNC-CTAB possesses more binding sites that are available for genistein adsorption compared to BNC. However, isotherm models for BNC-CTAB showed a decrease in adsorption capacity when the temperature was increasing, and this could be attributed to the modified BNC properties; temperature reductions could increase sample wetting, and as a result, the increase in solute transfer from the liquid phase to the adsorbent surface occurs [49]. Moreover, thermodynamic parameters showed a slight decrease in terms of the ΔG values for BNC-CTAB-GEN compared to BNC-GEN, due to the energy of the hydrophobic interactions, which was lower than the hydrogen bond energy, and the increase in ΔH occurred because the BNC-CTAB-GEN system needs higher energy given that the intermolecular interactions are weaker.

The period of release assay was prolonged by 72 h because of its application in delivery systems for CRC, which are desired considering that the transit through the gastrointestinal tract could take time. The BNC and BNC-CTAB films do not show a burst release of genistein during transit through stomach and colon fluids. Additionally, the results show an increase in genistein desorption with increasing pH, until reaching a plateau after

48 h in the BNC-GEN system. In light of previous results, BNC and BNC-CTAB films acted as a carrier for genistein until it reached the intestinal fluids, where genistein is metabolized by cells [10]. Considering the IC50 values, 78.73 and 24.13 mg of BNC-GEN films, and 74.24 and 32.90 mg of BNC-CTAB-GEN are needed to reach a therapeutic concentration for SW480 at 24 and 48 h, respectively. For SW620, the amounts necessary are 76.91 and 43.71 mg of BNC-GEN films and 72.52 and 59.60 mg of BNC-CTAB-GEN films.

Finally, the modification of BNC with CTAB improves the adsorption profile of genistein, increases the loading capacity of the compound in the nanostructure from 53.91% to 71.73%, and provides stability during desorption, indicating a controlled and sustained release over time. Therefore, BNC surface modification could lead to favorable changes in the bioavailability of genistein, positively impacting the chemopreventive effect of the compound.

4. Materials and Methods

4.1. Materials

For the inoculum, glucose, peptone, yeast, sodium dihydrogen phosphate (NaH_2PO_4), potassium dihydrogen phosphate (KH_2PO_4), magnesium sulfate ($MgSO_4$), and citric acid (*Komagataeibacter medellinensis* strain) were used, all of analytical grade. For the commercial medium of BNC, raw cane sugar and acetic acid in food grade were used. In addition, cetyl trimethyl ammonium bromide (CTAB) and ethanol 96% were used for the synthesis and modification of BNC; all of these were of analytical grade. Genistein from Shanghai Yingrui Biopharma Co (CAS 446-72-0) was used with a purity $\geq$ 98%. Finally, for the in vitro studies, the reagents used were sodium taurocholate, pepsin, sodium chloride (NaCl), maleic acid, sodium hydroxide (NaOH), potassium chloride (KCl), and disodium phosphate (Na_2HPO_4) of analytical grade, and soy lecithin of food grade. Trichloroacetic acid (TCA) at 50 wt.%, sulforhodamine B, glacial acetic acid, and Tris base were employed (of cell culture grade), and SW480 (CCL-228), SW620 (CCL-227), and HaCaT (PCS-200-011) lines were used.

4.2. Calculation of Inhibitory Concentration (IC50) for Free Genistein

Cell studies were used to prove the inhibitory effect of free genistein. The cells tested in the experiment were the following: SW480 (adenocarcinoma colorectal cancer cells), SW620 (metastatic colorectal cancer cells) and HaCaT (non-malignant human Keratinocytes). The experiments were conducted under the protocols of Agudelo et al. (2017) [50]. Briefly, cells were seeded in 96-well culture plates at a concentration of 20000 cells per well for cancer cells and 15000 for HaCaT and cultured at 37 °C in 5 vol.% CO_2. After 24 h seeding, the cells were exposed to 7 different concentrations starting at 150 μM of genistein and incubated for 24 and 48 h. Cell monolayers were fixed to the well bottoms by adding 50 μL of 50 wt.% trichloroacetic acid (TCA) in each well, and the plates were incubated at room temperature for 1 h. The wells were then drained, rinsed twice with distilled water, and air dried. Sulforhodamine B (SRB) (0.4% w/v in 1 vol.% glacial acetic acid) was then added (100 μL/well), and the plates were incubated for 30 min. Unbound dye was drained and removed by washing 5 times with 1 vol.% glacial acetic acid. After air-drying the plate overnight, the dye was solubilized by adding 100 μL/well of 10 mM Tris base and stirred for 30 min at 37 °C. Absorbance at 490 nm was measured. All experiments were performed in quintuplicate. The absorbance of the control group (non-treated cells) was considered as 100% viability [50].

The percent inhibition was calculated using the following equation:

$$Inhibittion(\%) = \left[1 - \frac{OD_T}{OD_c}\right] * 100 \tag{1}$$

where OD_T is the optical density (OD) of treated cells and OD_c is that for control (non-treated cells). The concentration able to inhibit 50% of cells (IC50) was calculated using the 4-parameter Hill equation and nonlinear regression. As proposed by Bertrand et al.

(1992), Hill coefficients were found after plotting dose–response curves on the logarithmic scale [51]. Selectivity was calculated according to Equation (2).

$$S = \frac{IC50_{Non-cancer\ cells}}{IC50_{Cancer\ cells}} \tag{2}$$

4.3. Development of BNC and BNC-CTAB Materials

4.3.1. BNC Synthesis

BNC was synthesized from a non-commercial culture medium with raw cane sugar at 13 wt.% at pH 3.6, to which was added bacterial inoculum of *Komagataeibacter medellinensis* in plastic containers of 500 mL. This was incubated at room temperature for 15 days, and then, membranes were treated with 5 wt.% KOH for 14 h and washed to reach neutral pH. BNC was processed using a Super Masscolloider (MKCA 6-2) for the individualization of nanoribbons, and finally sterilized for subsequent use; the final concentration of the cellulose suspension was 1.37 wt.%.

4.3.2. BNC Surface Modification

A preliminary modification of BNC was performed with TEMPO, as described by Cañas-Gutiérrez et al. (2020) [52]; 1 g of dry BNC was suspended in 500 mL of distilled water with 17 mg of TEMPO and 170 mg of NaBr to acquire TEMPO-oxidized BNC. To initiate the oxidation reaction, a NaClO solution (10 mmol·g^{-1} cellulose) was added slowly to the TEMPO-oxidized BNC under stirring. The pH value was kept constant at 10.0 by adding NaOH 0.5M. Finally, the reaction was ceased by ethanol addition to the suspension. Products were washed with distilled water until neutral pH was reached. For the final modification, a CTAB solution (5mM) was prepared by diluting CTAB in deionized water. Later, 83 mL of the CTAB solution was added dropwise into 130 mL of TEMPO-oxidized BNC at 1.92% under magnetic stirring at room temperature. The mixture was heated to 60 °C for 30 min and then cooled down to room temperature. Modified cellulose (BNC-CTAB) was washed in dialysis membranes of 12–14 kD several times until unbound CTAB was removed and neutral pH reached. Therefore, no toxic leachate was generated.

4.3.3. Morphological Analysis

The fibrillar structure and morphology of BNC and BNC-CTAB were analyzed by Transmission Electron Microscopy (TEM). A 10 µL drop was deposited on a Formvar/Carbon 200 Mesh Copper grid and stained with 2% uranyl acetate. The images were taken with a TECNAI FEI microscope operating at 80 kV at magnifications of 7.00 kX to 43.00 kX. Finally, the software ImageJ was used to measure the number of pixels and calculate the nanoribbons' diameters; 60 measurements were taken per sample, and the results were compared using one-way analysis of variance (ANOVA). The statistical analysis was performed using RStudio.

4.3.4. Contact Angle Measurements

To evaluate the surface hydrophobicity of the dry films of BNC and BNC-CTAB, contact angle measurements were made. For this, a flat surface of the material was placed on a goniometer coupled to a Dataphysics OCA 15EC camera. The system was calibrated using the ASTM D7490 08 standard; after that, a deionized 8 µL water drop was deposited, and with the help of the software, the contact angle was measured. The test was performed five times in different areas of the material.

4.3.5. Chemical Analysis

To evaluate surface chemical changes, the infrared spectra of BNC and BNC-CTAB were obtained using a Nicolet 6700 spectrophotometer in ATR mode on a type IIA diamond crystal. The sample area was 0.5 mm^2, and constant pressure was applied to each sample. Infrared spectra were collected between 4000 and 400 cm^{-1} with a resolution of 4 cm^{-1}.

4.3.6. Thermal Analysis

Thermal degradation of the samples was evaluated using a thermogravimetric analyzer (Mettler Toledo TGA/SDTA 851E). A total of 8 mg of the dried sample was weighed before and after the surface modification and heated in a nitrogen atmosphere from 30 to 800 °C, with a heating rate of 10 °C·min^{-1}.

4.4. Adsorption Studies

4.4.1. Genistein Quantification

Successive solutions were prepared from a genistein stock solution (25.00 mg·L^{-1}). Absorbance values were obtained by UV–Vis spectroscopy at a wavelength of 260 nm on a UV–Vis Evolution 600 spectrophotometer. A curve with 6 points was made, and using the linear regression function, slope and intercept values were found. The equation was y = 0.127x + 0.032 with a correlation coefficient of 0.999. These values allowed us to determine the concentration of genistein in a subsequent test.

4.4.2. Adsorption Isotherms

A genistein stock solution (200 µg GEN·mL^{-1}ethanol) was initially prepared. Different mixtures of 10 mL were prepared as described in Table 5 and then placed in water baths at temperatures of 0, 23, and 40 °C.

Table 5. Concentrations of genistein for adsorption isotherms assays.

Sample	Genistein Concentration (mg·L^{-1})	Adsorbent Concentration (%)
GE$_9$	65.34	0.5
GE$_8$	51.48	0.5
GE$_7$	42.28	0.5
GE$_6$	38.64	0.5
GE$_5$	35.00	0.5
GE$_4$	26.72	0.5
GE$_3$	23.26	0.5
GE$_2$	19.80	0.5
GE$_1$	16.34	0.5
GE$_0$	0.00	0.5

After 2 h, samples were vacuum-filtered; 600 µL of this solution was diluted in 10 mL of ethanol to determine genistein concentration by UV–Vis spectroscopy. The experimental adsorption capacity Q_t (mg·g^{-1}) of genistein was calculated by the following equation [53]:

$$Q_t = \frac{(C_0 - C_t)V_i}{W} \tag{3}$$

where C_0 is the initial concentration of genistein in the solution (mg·L^{-1}), C_t is the genistein concentration at instant t (mg·L^{-1}), V_i is the volume of solution during adsorption isotherms assay (L), and W is the weight of BNC and BNC-CTAB in solution (mg). If the adsorption process is long enough, Q_t and C_t will be constant and can be referred to as Q_e and C_e, corresponding to equilibrium adsorption capacity (mg·g^{-1}) and genistein concentration at equilibrium (mg·L^{-1}) [53]. Homogenous monolayer, heterogeneous monolayer, and multilayer (Appendix B) adsorption models were selected to analyze experimental data as described in Table 6.

4.4.3. Determination of Thermodynamic Parameters

The thermodynamic parameters reflect the feasibility and spontaneity of the adsorption process, and can be evaluated through Gibbs free energy, enthalpy, and entropy, which can be calculated according to the method proposed by Pérez et al. (2011) [54]:

$$\Delta G° = -RT \ln(K_c) \tag{4}$$

$$\Delta G^\circ = \Delta H^\circ - T\Delta S^\circ \tag{5}$$

where R (8.314 J·mol^{-1}·K^{-1}) is the universal gas constant, T (K) is the absolute temperature, ΔG° (kJ·mol^{-1}) is the change in Gibbs free energy, ΔH° (kJ·mol^{-1}) is the enthalpy change, ΔS° (J·mol^{-1}·K^{-1}) is the entropy change, and K_c is the equilibrium constant calculated using Equation (6) [53,54]:

$$K_c = \frac{C_{ad}}{C_e} \tag{6}$$

C_{ad} is the concentration of the adsorbate contained on the surface of the adsorbent in equilibrium (mg·L^{-1}). For the Langmuir isotherm, the equilibrium constant K_c is equal to K_L (L·mol^{-1}) [55]. Arranging Equations (4) and (5), the Van't Hoff equation is obtained [40,54]:

$$\ln(K_c) = \frac{-\Delta H^\circ}{RT} + \frac{\Delta S^\circ}{R} \tag{7}$$

plotting $\ln(K_c)$ vs. T^{-1}, where the linear intercept is $\frac{\Delta S^\circ}{R}$ and the slope is $\frac{\Delta H^\circ}{R}$.

Table 6. Adsorption models of homogeneous and heterogenous monolayers [38,53].

Model	Parameters	Description
Langmuir $Q_e = \frac{Q_{mL}K_L C_e}{1 + K_L C_e}$	Q_{mL}, maximum adsorption capacity (mg·g^{-1}) K_L, Langmuir constant (L·mg^{-1}).	Assumes monolayer adsorption and homogeneous surface. Additionally, adsorption is localized.
Freundlich $Q_e = K_f C_e^{1/n_f}$	K_f, Freundlich constant (mg^{1-n}·L^n·g^{-1}) n_f, adsorption intensity	Describes a heterogeneous surface of adsorption with the interaction between adsorbed molecules.
Sips $Q_e = \frac{Q_{ms}a_s C_e^{B_s}}{1 + a_s C_e^{B_s}}$	Q_{ms}, maximum adsorption capacity (mg·g^{-1}) a_s, Sips constant (L·mg^{-1}) B_s, model exponent	Describes heterogeneous systems, localized adsorption without adsorbate–adsorbate interactions. B_s is known as the heterogeneity factor.
Toth $Q_e = \frac{Q_{mT}C_e}{(A_T + C_e^z)^{\frac{1}{z}}}$	Q_{mT}, maximum adsorption capacity (mg·g^{-1}) A_T, Toth constant (L·mg^{-1}) z, model exponent	Describes heterogeneous systems. Parameter z is related to system heterogeneity.

4.4.4. Adsorption kinetics

Adsorption kinetics were determined by mixing BNC and BNC-CTAB at 0.5 wt.% with genistein at 42.28 µg·mL^{-1} in a 45 mL conic tube. Tubes were placed in a water bath at 23 °C for 2 h, and 2 mL aliquots were taken at 5, 10, 20, 40, 60, 80, 100, 110, and 120 min. Samples were vacuum-filtered, and 600 µL of this solution was diluted in 10 mL of ethanol to determine genistein concentration in the solution via UV–Vis spectroscopy. The measurement was performed at a wavelength of 260 nm on a UV–Vis Evolution 600 spectrophotometer.

The experimental data were modeled using the following models:

- Pseudo-first-order—This model assumes that there is an adsorption site in the adsorbent for each adsorbate molecule. The kinetics are described by the following equation [53],

$$Q_t = Q_e\left(1 - e^{-K_1 t}\right) \tag{8}$$

where K_1 is the pseudo-first-order constant (min^{-1}) and t is the time (min);

- Pseudo-second-order—This model assumes that the adsorbate is adsorbed onto two active sites [56]. It is described by the following equations [53,56],

$$\frac{t}{Q_t} = \frac{1}{K_2 Q_e^2} + \frac{t}{Q_e} \tag{9}$$

$$Q_t = \frac{K_2 Q_e^2 t}{1 + K_2 Q_e t} \tag{10}$$

where K_2 is the pseudo-second-order constant ($g \cdot mg^{-1} \cdot min^{-1}$). Equation (10) corresponds to the standard form of the model. Initial adsorption rate h ($mg \cdot g^{-1} \cdot min^{-1}$) can be determined as shown in the following equation [57],

$$h = K_2 Q_e^2 \tag{11}$$

- Elovich—This model assumes that the adsorbent active sites are heterogeneous; thus, it shows different activation energies [53,55,56]. It is described by the following equation [53,55,56],

$$Q_t = \frac{1}{\beta} \ln(\alpha\beta) + \frac{1}{\beta} \ln(t) \tag{12}$$

where α ($mg \cdot g^{-1} \cdot min^{-1}$) and β ($g \cdot mg^{-1}$) are the Elovich constants corresponding to the initial rate of adsorption, surface coverage, and activation energy;

- Intra-particle diffusion—This model assumes the transfer of the adsorbate through the internal structure of the adsorbent; therefore, the adsorbent acquires a homogenous structure [55,56]. It is described by the following equation [55,56],

$$Q_t = K_3 \sqrt{t} \tag{13}$$

where K_3 is the diffusion constant ($mg \cdot g^{-1} \cdot min^{-1/2}$).

4.5. Development of Thin-Film Drug Delivery Systems

Thin films were prepared by adding 2.11 mL of GEN (42.28 mg·L^{-1}) into 7.89 mL of adsorbent solution (0.63 wt.%) at 23 °C. After 2 h, the samples were vacuum-filtered, and films were obtained. The drug delivery films were dried oven gravimetrically at 60 °C.

Thin Films' Characterization

Atomic force microscopy (AFM) was used to characterize the surface morphology and texture of the films. Square thin films of 1 cm^2 were analyzed using a Nanosurf FlexAFM mounted on an isostatic table. Images were recorded under static force of 10 nN, PID (100-800-100) in a 6.25 μm^2 area using a triangle cantilever (HYDRA-ALL-G D) from AppNano. The recorded images were analyzed to determine the area roughness using C3000i software tools, and they were exported using the Gwyddion software. Five samples of BNC and BNC-CTAB were tested. Furthermore, films were analyzed by FTIR and TGA, as described in Sections 4.3.5 and 4.3.6.

4.6. In Vitro Gastrointestinal Fluids Release Study

For genistein release profiles, the membrane dialysis method [58,59] was used in simulated fluids of the stomach, small intestine, and colon with pH values of 1.2, 6.0, and 7.4, respectively [14]. The dry thin films of BNC-GEN and BNC-CTAB-GEN were placed inside the dialysis membranes (MWCO: 12–14 kD). The sealed membranes were put into 78 mL of simulated fluid at 37 °C with a stirring speed of 80 rpm. During the first 120 min, the membranes remained in stomach fluid, from 120 to 1920 min in the small intestine fluid, and from 1920 to 4320 min in the colon fluid. The aliquots were taken, and immediately, the volume was replaced with the corresponding fluid. Genistein concentration was determined by UV–Vis spectroscopy at a wavelength of 260 nm by diluting 300 μL of aliquots in 5 mL of ethanol.

The release profile (%) of the adsorbate is described by the following equation [23]:

$$D = \frac{C_d}{C_L} * 100 \tag{14}$$

where D is the percentage of desorption, C_d is the concentration of adsorbate in the release medium ($mg \cdot L^{-1}$) and C_L is the loaded concentration of the active compound in the adsorbent ($mg \cdot L^{-1}$). The amount of solute released, Q_d, is described by:

$$Q_d = \frac{C_d V_d}{W} \qquad (15)$$

The experimental data were modeled to pseudo-first- and pseudo-second-order kinetics according to Equations (8)–(10).

5. Conclusions

In this study, thin films of genistein loaded in BNC and BNC-CTAB were successfully prepared after the evaluation of surface modification, thermodynamic parameters, and adsorption. Several techniques, such as TEM, FTIR, TGA, and AFM, were used to analyze the modification of BNC and the correct adsorption of GEN on adsorbents. The results show an increase of two-fold in contact angle measurements as a result of BNC modification. The FTIR and TGA spectra reveal genistein incorporation into BNC and BNC-CTAB via the appearance of new peaks. The genistein successfully encapsulated in BNC and BNC-CTAB films was characterized by hydrogen bond interactions and electrostatic and hydrophobic interactions, respectively. Adsorption studies have demonstrated that the process was spontaneous and endothermic at all temperatures; however, the best adsorption capacity was reached at 23 °C. Furthermore, kinetics models indicate a better adjustment of BNC-CTAB, due to its hydrophobic character, which imparted genistein with more stability in aqueous systems. To conclude, in vitro studies showed a genistein release rate of 85% after 72 h and the low desorption of the compound into stomach fluid, thus indicating that BNC and BNC-CTAB acted as protective agents of genistein in acid conditions, and BNC and BNC-CTAB can be considered as genistein nanocarriers for CRC, as it can deliver concentrations with therapeutic effects, according to in vitro studies. Further work should be focused on animal evaluation and the pharmacokinetics of the system.

Author Contributions: Conceptualization, M.O. and C.C.; methodology M.C., E.M. and M.O.; validation, M.C. and E.M.; formal analysis, M.C., E.M. and M.O.; investigation, M.C. and E.M.; writing—original draft preparation, M.C.; writing—review and editing, M.C., M.O. and C.C.; visualization, M.C. and M.O.; supervision, M.O. and C.C.; project administration, M.O. and C.C.; funding acquisition, C.C. All authors have read and agreed to the published version of the manuscript.

Funding: The work has been funded by MINCIENCIAS, MINEDUCACIÓN, MINCIT and ICETEX, through the Program Ecosistema Científico Cod. FP44842-211-2018, project number 58674.

Institutional Review Board Statement: Not applicable.

Data Availability Statement: The data presented in this study are available upon request to the corresponding author.

Conflicts of Interest: The authors declare no conflict of interest.

Sample Availability: Samples are not available from the authors.

Appendix A

Experimental data have been fitted to multilayer isotherm models (BET, FHH, and GAB models). Nevertheless, adjustments were poor. Table A1 shows the obtained results.

Table A1. Multilayer adsorption isotherm models for genistein adsorption in BNC and BNC-CTAB [38].

	Isotherm Models	BNC-GEN			BNC-CTAB-GEN		
		0 °C	23 °C	40 °C	0 °C	23 °C	40 °C
BET	Q_m ($mg \cdot g^{-1}$)	3.08	3.66	3.11	3.501	3.8613	2.848
	C_{BET}	1.00	1.00	1.00	0.999	0.9998	1.000
	C_S ($mg \cdot L^{-1}$)	0.02	0.00	0.03	0.026	0.0171	0.001
	R^2	0.96	0.96	0.95	0.792	0.8455	0.659

Table A1. *Cont.*

Isotherm Models		BNC-GEN			BNC-CTAB-GEN		
		0 °C	23 °C	40 °C	0 °C	23 °C	40 °C
FHH	A_{FHH}	0.568	1.00	0.90	1.00	0.25	0.89
	B_{FHH}	7.780	1.00	1.90	2.31	1.00	2.00
	C_S (mg·L^{-1})	0.299	25.00	0.50	29.57	0.70	25.24
	R^2	−6.702	−2.17	−1.93	−5.06	−5.23	−1.92
GAB	Q_{mGAB} (mg·g^{-1})	0.45	0.34	0.00	0.28	1.00	1.00
	C_{GAB}	11,060.30	6228.31	1.37	43.98	12.15	1.00×10^8
	K_G	328.404	0.01	0.99	0.18	1.24	1.16×10^8
	C_S (mg·L^{-1})	3053.60	0.10	0.89	2.00	0.634	0.03
	R^2	−4.68	6.83	−5.63	−3.28	−5.19	−4.48

Appendix B

The multilayer adsorption models selected to analyze experimental data were Bet, FHH, and GAB models. Table A2 describes the models.

Table A2. Multilayer adsorption isotherm models [38].

Model	Parameters	Description
BET $Q_e = \dfrac{Q_{mBET}C_{BET}C_e}{(C_e - C_s)\left[1 + (C_{BET} - 1)\frac{C_e}{C_s}\right]}$	Q_{mBET}, maximum adsorption capacity corresponding to the monolayer saturation (mg·g^{-1}). C_s, saturation concentration of the monolayer (mg·L^{-1}). C_{BET}, model constant.	Assumes a uniform surface and homogeneous adsorption energy at all sites. The second, third, and other layers have adsorption energies equal to the enthalpy of fusion.
FHH $Q_e = \left[\frac{-1}{A_{FFH}}\ln\left(\frac{C_e}{C_s}\right)\right]^{-\frac{1}{B_{FFH}}}$	A_{FFH}, model constant. B_{FFH}, model constant.	Assumes the variation in adsorption potential based on the distance between adsorbent and molecule. Additionally, accepts that surface heterogeneity affects all the layers.
GAB $Q_e = \dfrac{Q_{mG}C_{GAB}K_GC_e}{(C_s - K_GC_e)\left[1 + (C_{GAB} - 1)K_G\frac{C_e}{C_s}\right]}$	Q_{mG}, maximum adsorption capacity corresponding to the monolayer (mg·g^{-1}). C_{GABs}, model constant. K_G, model constant.	This model includes the K_G parameter to differentiate the chemical potential between the molecules of the second and subsequent layers and molecules in the liquid state.

References

1. Organisatión Mondiale de la Santé (OMS). Enfermedades no Transmisibles. 2018. Available online: https://www.who.int/es/news-room/fact-sheets/detail/noncommunicable-diseases (accessed on 28 April 2020).
2. International Agency for Research on Cancer. Cancer Today. Global Cancer Observatory. Available online: https://gco.iarc.fr/today/online-analysis-dual-bars-2?v=2020&mode=cancer&mode_population=regions&population=900&populations=900&key=crude_rate&sex=0&cancer=39&type=0&statistic=5&prevalence=0&population_group=0&ages_group%5B%5D=0&ages_group%5B%5D=17&nb_ite (accessed on 13 January 2021).
3. GLOBOCAN—World Health Organization (WHO). Estimated number of deaths in 2020, both sexes, all ages. *Int. Agency Res. Cancer* **2020**, *144*, 100.
4. Pointet, A.-L.; Taieb, J. Manual AEC Cirugía Colorrectal. Capítulo 20: Cáncer de colon. In *Colloids and Surfaces A: Physicochemical and Engineering Aspects*; Elsevier B.V.: Amsterdam, The Netherlands, 2017; Volume 21, pp. 213–228.
5. Cáncer de Colon—Diagnóstico y Tratamiento. Mayo Clinic. 2019. Available online: https://www.mayoclinic.org/es-es/diseases-conditions/colon-cancer/diagnosis-treatment/drc-20353674 (accessed on 8 June 2020).
6. Pool, H.; Campos-Vega, R.; Herrera-Hernández, M.G.; García-Solis, P.; García-Gasca, T.; Sánchez, I.C.; Luna-Bracenas, G.; Vergara-Castaneda, H. Development of genistein-PEGylated silica hybrid nanomaterials with enhanced antioxidant and antiproliferative properties on HT29 human colon cancer cells. *Am. J. Transl. Res.* **2018**, *10*, 2306–2323. [PubMed]
7. Zhu, J.; Ren, J.; Tang, L. Genistein inhibits invasion and migration of colon cancer cells by recovering WIF1 expression. *Mol. Med. Rep.* **2018**, *17*, 7265–7273. [CrossRef] [PubMed]
8. Abotaleb, M.; Samuel, S.M.; Varghese, E.; Varghese, S.; Kubatka, P.; Liskova, A.; Büsselberg, D. Flavonoids in cancer and apoptosis. *Cancers* **2019**, *11*, 28. [CrossRef]
9. Dixon, R.A.; Ferreira, D. Genistein. *Phytochemistry* **2002**, *60*, 205–211. [CrossRef]

10. Mukund, V.; Mukund, D.; Sharma, V.; Mannarapu, M.; Alam, A. Genistein: Its role in metabolic diseases and cancer. *Crit. Rev. Oncol. Hematol.* **2017**, *119*, 13–22. [CrossRef]

11. Jerez, M.H. Antocianos: Metabolismo y Actividad Biológica. Ph.D. Thesis, Universidad Complutense de Madrid, Madrid, Spain. 2013. Available online: https://eprints.ucm.es/20093/1/T34345.pdf (accessed on 12 June 2020).

12. Wang, Y.; Wang, Y.; Liu, Y.; Liu, Q.; Jang, J.; Han, J. Preparation, characterization, and antioxidant activities of cellulose nanocrystals/genistein nanocomposites. *BioResources* **2019**, *14*, 336–348. [CrossRef]

13. Jaiswal, N.; Akhtar, J.; Singh, S.P.; Ahsan, F. An Overview on Genistein and its Various Formulations. *Drug Res.* **2019**, *69*, 305–313. [CrossRef]

14. Rahmani, F.; Karimi, E.; Oskoueian, E. Synthesis and characterisation of chitosan-encapsulated genistein: Its anti-proliferative and anti-angiogenic activities. *J. Microencapsul.* **2020**, *37*, 305–313. [CrossRef]

15. De Barud, H.G.; da Silva, R.R.; da Silva Barud, H.; Tercjak, A.; Gutierrez, J.; Lustri, W.R.; de Oliveira, O.B.J.; Ribeiro, S.J.L. A multipurpose natural and renewable polymer in medical applications: Bacterial cellulose. *Carbohydr. Polym.* **2016**, *153*, 406–420. [CrossRef]

16. Osorio, M.; Cañas, A.; Puerta, J.; Díaz, L.; Naranjo, T.; Ortiz, I.; Castro, C. Ex Vivo and In Vivo Biocompatibility Assessment (Blood and Tissue) of Three-Dimensional Bacterial Nanocellulose Biomaterials for Soft Tissue Implants. *Sci. Rep.* **2019**, *9*, 10553. [CrossRef] [PubMed]

17. Al Balawi, A.N. *Antioxidative Response Mechanisms of Nanocelluloses and Nanohydrogels Matrices: A Review*; INC: Geneva, Switzerland, 2020.

18. Perez, M.A. La Nanocelulosa Desafía al Grafeno como Nuevo Material Prodigio. 2013. Available online: https://blogthinkbig.com/nanocelulosa-grafeno-material-prodigio (accessed on 28 April 2020).

19. Osorio, M.; Velásquez-Cock, J.; Restrepo, L.M.; Zuluaga, R.R.; Gañán, P.; Rojas, O.J.; Ortiz-Trujillo, I.; Castro, C. Bioactive 3D-Shaped Wound Dressings Synthesized from Bacterial Cellulose: Effect on Cell Adhesion of Polyvinyl Alcohol Integrated In Situ. *Int. J. Polym. Sci.* **2017**, *2017*, 3728485. [CrossRef]

20. Zuluaga Gallego, R.O.; Serpa Guerra, A.M.; Velez Acosta, L.M.; Gómez Hoyos, C.; Castro Herazo, C.I.; Casas Botero, A.E.; Osorio Delgado, A.M.; Velazquez Cock, J.; Torres Taborda, M.M.; Gañán Rojo, P.F.; et al. La nanocelulosa: Una estructura producida por la naturaleza. In *Nanotecnología, Fundamentos y Aplicaciones*; Álvarez-Láinez, M.L., Martínez-Tejada, H.V., Jaramillo Isaza, F., Eds.; Editorial Universidad de Antioquia: Medellín, Colombia, 2019; pp. 99–111.

21. Cacicedo, M.L.; León, I.E.; Gonzalez, J.S.; Porto, L.M.; Alvarez, V.A.; Castro, G.R. Modified bacterial cellulose scaffolds for localized doxorubicin release in human colorectal HT-29 cells. *Colloids Surf. B Biointerfaces* **2016**, *140*, 421–429. [CrossRef]

22. Subtaweesin, C.; Woraharn, W.; Taokaew, S.; Chiaoprakobkij, N.; Sereemaspun, A.; Phisalaphong, M. Characteristics of Curcumin-Loaded Bacterial Cellulose Films and Anticancer Properties against Malignant Melanoma Skin Cancer Cells. *Appl. Sci.* **2018**, *8*, 1188. [CrossRef]

23. Putro, J.N.; Ismadji, S.; Gunarto, C.; Yuliana, M.; Santoso, S.P.; Soetaredjo, F.E.; Ju, Y.H. The effect of surfactants modification on nanocrystalline cellulose for paclitaxel loading and release study. *J. Mol. Liq.* **2019**, *282*, 407–414. [CrossRef]

24. Akagi, S.; Ando, H.; Fujita, K.; Shimizu, T.; Ishima, Y.; Tajima, K.; Matsushima, T.; Kusano, T.; Ishida, T. Therapeutic efficacy of a paclitaxel-loaded nanofibrillated bacterial cellulose (PTX/NFBC) formulation in a peritoneally disseminated gastric cancer xenograft model. *Int. J. Biol. Macromol.* **2021**, *174*, 494–501. [CrossRef] [PubMed]

25. Qu, J.; Yuan, Z.; Wang, C.; Wang, A.; Liu, X.; Wei, B.; Wen, Y. Enhancing the redispersibility of TEMPO-mediated oxidized cellulose nanofibrils in N,N-dimethylformamide by modification with cetyltrimethylammonium bromide. *Cellulose* **2019**, *26*, 7769–7780. [CrossRef]

26. Qin, J.; Chen, J.X.; Zhu, Z.; Teng, J.A. Genistein inhibits human colorectal cancer growth and suppresses MiR-95, Akt and SGK1. *Cell. Physiol. Biochem.* **2015**, *35*, 2069–2077. [CrossRef]

27. Chen, X.; Gu, J.; Wu, Y.; Liang, P.; Shen, M.; Xi, J.; Qin, J. Clinical characteristics of colorectal cancer patients and anti-neoplasm activity of genistein. *Biomed. Pharmacother.* **2020**, *124*, 109835. [CrossRef]

28. Shafiee, G.; Saidijam, M.; Tavilani, H.; Ghasemkhani, N.; Khodadadi, I. Genistein Induces Apoptosis and Inhibits Proliferation of HT29 Colon Cancer Cells. *Int. J. Mol. Cell. Med.* **2016**, *5*, 178–191. Available online: https://www.ncbi.nlm.nih.gov/pmc/articles/PMC5125370/ (accessed on 10 June 2022).

29. Gómez, R.L.; Moreno, Q.G.; Herrera, R.A.; Castrillón, L.W.; Yepes, A.F.; Cardona, G.W. New hybrid scaffolds based on ASA/genistein: Synthesis, cytotoxic effect, molecular docking, drug-likeness, and in silico ADME/Tox modeling. *J. Appl. Pharm. Sci.* **2022**, *12*, 15–30. [CrossRef]

30. Tong, Y.; Wang, M.; Huang, H.; Zhang, J.; Huang, Y.; Chen, Y.; Pan, H. Inhibitory effects of genistein in combination with gefitinib on the hepatocellular carcinoma Hep3B cell line. *Exp. Ther. Med.* **2019**, *18*, 3793–3800. [CrossRef] [PubMed]

31. Susanikova, I.; Puchl'ová, M.; Lachová, V.; Švajdlenka, E.; Mučaji, P.; Smetana, K.; Gál, P. Genistein and selected phytoestrogen-containing extracts differently modulate antioxidant properties and cell differentiation: An in vitro study in NIH-3T3, HaCaT and MCF-7 cells. *Folia Biol.* **2019**, *65*, 24–35.

32. Ly, M. Hydrophobic Modifications of Cellulose Nanocrystals for Anticorrosion and Polymer Coating Applications. Master's Thesis, University of Waterloo, Waterloo, ON, Canada, 2020.

33. Rauniyar, B.S.; Bhattarai, A. Study of conductivity, contact angle and surface free energy of anionic (SDS, AOT) and cationic (CTAB) surfactants in water and isopropanol mixture. *J. Mol. Liq.* **2021**, *323*, 114604. [CrossRef]

34. Çiftçi, H.; Ersoy, B.; Evcin, A. Purification of Turkish Bentonites and Investigation of the Contact Angle, Surface Free Energy and Zeta Potential Profiles of Organo-Bentonites as a Function of Ctab Concentration. *Clays Clay Miner.* **2020**, *68*, 250–261. [CrossRef]
35. Abdel-Halim, E.S. Chemical modification of cellulose extracted from sugarcane bagasse: Preparation of hydroxyethyl cellulose. *Arab. J. Chem.* **2014**, *7*, 362–371. [CrossRef]
36. Castro, C.; Zuluaga, R.; Putaux, J.L.; Caro, G.; Mondragon, I.; Gañán, P. Structural characterization of bacterial cellulose produced by Gluconacetobacter swingii sfrom Colombian agroindustrial wastes. *Carbohydr. Polym.* **2011**, *84*, 96–102. [CrossRef]
37. Zainuddin, N.; Ahmad, I.; Kargarzadeh, H.; Ramli, S. Hydrophobic kenaf nanocrystalline cellulose for the binding of curcumin. *Carbohydr. Polym.* **2017**, *163*, 261–269. [CrossRef]
38. Saadi, R.; Saadi, Z.; Fazaeli, R.; Fard, N.E. Monolayer and multilayer adsorption isotherm models for sorption from aqueous media. *Korean J. Chem. Eng.* **2015**, *32*, 787–799. [CrossRef]
39. Sheela, T.; Nayaka, Y.A.; Viswanatha, R.; Basavanna, S.; Venkatesha, T.G. Kinetics and thermodynamics studies on the adsorption of Zn(II), Cd(II) and Hg(II) from aqueous solution using zinc oxide nanoparticles. *Powder Technol.* **2012**, *217*, 163–170. [CrossRef]
40. Lyubchik, S.; Lyubchik, A.; Lygina, O.; Lyubchik, S.; Fonseca, I. Comparision of the thermodynamic parameters estimation for the adsorption process of the metals from liquid phase on activated carbons. In *Thermodynamics, Interaction Studies, Solids, Liquids and Gases*; IntechOpen: London, UK, 2020; pp. 95–122.
41. Scheufele, F.B.; Módenes, A.N.; Borba, C.E.; Ribeiro, C.; Espinoza-Quiñones, F.R.; Bergamasco, R.; Pereira, N.C. Monolayer-multilayer adsorption phenomenological model: Kinetics, equilibrium and thermodynamics. *Chem. Eng. J.* **2016**, *284*, 1328–1341. [CrossRef]
42. Adebayo, G.B.; Jamiu, W.; Okoro, H.K.; Okeola, F.O.; Adesina, A.K.; Feyisetan, O.A. Kinetics, thermodynamics and isothermal modelling of liquid phase adsorption of methylene blue onto moringa pod husk activated carbon. *South African J. Chem.* **2019**, *72*, 263–273. [CrossRef]
43. Sahoo, T.R.; Prelot, B. *Adsorption Processes for the Removal of Contaminants from Wastewater: The Perspective Role of nanomaterials and Nanotechnology*; Elsevier Inc.: Amsterdam, The Netherlands, 2020.
44. Adepu, S.; Khandelwal, M. Ex-situ modification of bacterial cellulose for immediate and sustained drug release with insights into release mechanism. *Carbohydr. Polym.* **2020**, *249*, 116816. [CrossRef] [PubMed]
45. Wang, H.; Li, Q.; Chen, H. Genistein affects histone modifications on Dickkopf-related protein 1 (DKK1) gene in SW480 human colon cancer cell line. *PLoS ONE* **2012**, *7*, e40955. [CrossRef] [PubMed]
46. Sun, J.; Li, M.; Lin, T.; Wang, D.; Chen, J.; Zhang, Y.; Mu, Q.; Su, H.; Wu, N.; Liu, A.; et al. Cell cycle arrest is an important mechanism of action of compound Kushen injection in the prevention of colorectal cancer. *Sci. Rep.* **2022**, *12*, 4384. [CrossRef] [PubMed]
47. Castro, C.; Zuluaga, R.; Álvarez, C.; Putaux, J.-L.; Caro, G.; Rojas, O.J.; Mondragon, I.; Gañán, P. Bacterial cellulose produced by a new acid-resistant strain of Gluconacetobacter genus. *Carbohydr. Polym.* **2012**, *89*, 1033–1037. [CrossRef] [PubMed]
48. Syverud, K.; Xhanari, K.; Chinga-Carrasco, G.; Yu, Y.; Stenius, P. Films made of cellulose nanofibrils: Surface modification by adsorption of a cationic surfactant and characterization by computer-assisted electron microscopy. *J. Nanoparticle Res.* **2011**, *13*, 773–782. [CrossRef]
49. Liu, W.; Zhang, S.; Zu, Y.-G.; Fu, Y.-J.; Ma, W.; Zhang, D.-Y.; Kong, Y.; Li, X.-J. Preliminary enrichment and separation of genistein and apigenin from extracts of pigeon pea roots by macroporous resins. *Bioresour. Technol.* **2010**, *101*, 4667–4675. [CrossRef]
50. Agudelo, C.D.; Arango, S.; Cortes-Mancera, F.; Rojano, B.; Maldonado, M.E. Antiproliferative and pro-apoptotic effects of Andean berry juice (Vaccinium meridionale Swartz) on human colon adenocarcinoma SW480 cells. *J. Med. Plants Res.* **2017**, *11*, 393–402. [CrossRef]
51. Bertrand, D.; Bertrand, S.; Ballivet, M. Pharmacological properties of the homomeric α7 receptor. *Neurosci. Lett.* **1992**, *146*, 87–90. [CrossRef]
52. Cañas-Gutiérrez, A.; Martinez-Correa, E.; Suárez-Avendaño, D.; Arboleda-Toro, D.; Castro-Herazo, C. Influence of bacterial nanocellulose surface modification on calcium phosphates precipitation for bone tissue engineering. *Cellulose* **2020**, *27*, 10747–10763. [CrossRef]
53. Figueroa, D.; Moreno, A.; Hormaza, A. Equilibrio, termodinámica y models cinéticos en la adsorción de Rojo 40 sobre tuza de maíz. *Rev. Ing. Univ. Medellín* **2015**, *14*, 105–120. [CrossRef]
54. Pérez, N.; González, J.; Delgado, L.A. Estudio termodinámico del proceso de adsorción de iones de ni Y V por parte de ligninas precipitadas del licor negro kraft. *Rev. Latinoam. Metal. Mater.* **2011**, *31*, 168–181.
55. Zhang, W.; An, Y.; Li, S.; Liu, Z.; Chen, Z.; Ren, Y.; Wang, S.; Zhang, X.; Wang, X. Enhanced heavy metal removal from an aqueous environment using an eco-friendly and sustainable adsorbent. *Sci. Rep.* **2020**, *10*, 16453. [CrossRef] [PubMed]
56. Tejada, C.; Herrera, A.; Ruiz, E. Kinetic and isotherms of biosorption of Hg(II) using citric acid treated residual materials. *Ing. Compet.* **2016**, *18*, 113. [CrossRef]
57. Kavitha, D.; Namasivayam, C. Experimental and kinetic studies on methylene blue adsorption by coir pith carbon. *Bioresour. Technol.* **2007**, *98*, 14–21. [CrossRef]

58. Xiao, Y.; Ho, C.-T.; Chen, Y.; Wang, Y.; Wei, Z.; Dong, M.; Huang, Q. Synthesis, Characterization, and Evaluation of Genistein-Loaded Zein/Carboxymethyl Chitosan Nanoparticles with Improved Water Dispersibility, Enhanced Antioxidant Activity, and Controlled Release Property. *Foods* **2020**, *9*, 1604. [CrossRef]

59. Shehata, E.M.M.; Elnaggar, Y.S.R.; Galal, S.; Abdallah, O.Y. Self-emulsifying phospholipid pre-concentrates (SEPPs) for improved oral delivery of the anti-cancer genistein: Development, appraisal and ex-vivo intestinal permeation. *Int. J. Pharm.* **2016**, *511*, 745–756. [CrossRef]

 *molecules*

Article

Chemopreventive Effect on Human Colon Adenocarcinoma Cells of Styrylquinolines: Synthesis, Cytotoxicity, Proapoptotic Effect and Molecular Docking Analysis

Vanesa Bedoya-Betancur [1,*], Elizabeth Correa [1], Juan Pablo Rendón [1], Andrés F. Yepes-Pérez [2], Wilson Cardona-Galeano [2] and Tonny W. Naranjo [1,3,*]

[1] Medical and Experimental Mycology Group, CIB-UPB-UdeA-UDES, Medellin 050034, Colombia
[2] Chemistry of Colombian Plants Group, Faculty of Exact and Natural Sciences, Institute of Chemistry, University of Antioquia (UdeA), Medellin 050010, Colombia
[3] School of Health Sciences, Pontifical Bolivarian University, Medellin 050034, Colombia
* Correspondence: vbedoya24@gmail.com (V.B.-B.); tonny.naranjo@upb.edu.co (T.W.N.)

Abstract: Seven styrylquinolines were synthesized in this study. Two of these styrylquinolines are new and were elucidated by spectroscopic analysis. The chemopreventive potential of these compounds was evaluated against SW480 human colon adenocarcinoma cells, its metastatic derivative SW620, and normal cells (HaCaT). According to the results, compounds **3a** and **3d** showed antiproliferative activity in SW480 and SW620 cells, but their effect seemed to be caused by different mechanisms of action. Compound **3a** induced apoptosis independent of ROS production, as evidenced by increased levels of caspase 3, and had an immunomodulatory effect, positively regulating the production of different immunological markers in malignant cell lines. In contrast, compound **3d** generated a pro-oxidant response and inhibited the growth of cancer cells, probably by another type of cell death other than apoptosis. Molecular docking studies indicated that the most active compound, **3a,** could efficiently bind to the proapoptotic human caspases-3 protein, a result that could provide valuable information on the biochemical mechanism for the in vitro cytotoxic response of this compound in SW620 colon carcinoma cell lines. The obtained results suggest that these compounds have chemopreventive potential against CRC, but more studies should be carried out to elucidate the molecular mechanisms of action of each of them in depth.

Keywords: styrylquinolines; colorectal cancer; antiproliferation; cell death; apoptosis; inflammation; reactive oxygen species; molecular docking

Citation: Bedoya-Betancur, V.; Correa, E.; Rendón, J.P.; Yepes-Pérez, A.F.; Cardona-Galeano, W.; Naranjo, T.W. Chemopreventive Effect on Human Colon Adenocarcinoma Cells of Styrylquinolines: Synthesis, Cytotoxicity, Proapoptotic Effect and Molecular Docking Analysis. *Molecules* **2022**, *27*, 7108. https://doi.org/10.3390/molecules27207108

Academic Editor: Anne-Marie Caminade

Received: 22 September 2022
Accepted: 19 October 2022
Published: 21 October 2022

Publisher's Note: MDPI stays neutral with regard to jurisdictional claims in published maps and institutional affiliations.

1. Introduction

According to the World Health Organization (WHO), cancer is one of the leading causes of death worldwide [1]. Colorectal cancer (CRC) was classified as the third most common type of cancer in the world and the second most frequent cause of death in 2020, presenting more than 1.9 million new cases and more than 900 thousand deaths [2].

Currently, different treatments are used for CRC, such as surgery, radiotherapy, chemotherapy, immunotherapy, and targeted therapy. However, the treatment of choice mainly depends on the stage of cancer and the general health of the patient. Among the most commonly used treatments in CRC, chemotherapy stands out as adjuvant therapy that is characterized by the administration of different drugs for the elimination of cancer cells, with 5-fluorouracil (5-FU) being one of the most frequently used in combination with other drugs (such as oxaliplatin, irinotecan, and leucovorin). However, these therapeutic options have been associated with multiple adverse effects, such as alopecia, gastrointestinal disorders, nausea, and vomiting, among other signs and symptoms that affect the quality of life of patients [3–5]. For this reason, the amount of research focused on cancer chemoprevention—defined as the use of natural, synthetic, or biological compounds to reduce the risk or delay the development of cancer—has increased in recent years.

One of the most studied natural compounds for cancer chemoprevention is resveratrol, a polyphenol belonging to the stilbene family that is found in grapes, peanuts, blackberries, and other foods of plant origin [6]. Antitumor, antioxidant, anti-inflammatory, cardioprotective and neuroprotective activities have been described for resveratrol, and they have been evidenced in both in vitro and in vivo assays [7–9]. Additionally, some studies have evaluated this compound in combination with 5-FU, demonstrating the ability of this stilbene to improve the effectiveness of 5-FU in CRC therapy [10]. Its antitumor effect has mainly been associated with its ability to modulate oxidative stress, inflammation, platelet aggregation, and the induction of tumor cell apoptosis [11–13]. On the other hand, 8-hydroxyquinoline is a heterocyclic organic compound known to be a chelating agent with antimicrobial and anticancer activity [14–17]. Specifically, its anticancer activity has been associated with its possible ability to inhibit the proliferation and migration of cancer cells through the induction of apoptosis and the generation of ROS, among other mechanisms [18–20].

In recent years, the synthesis of hybrid compounds has gained importance for their use as therapeutic agents in different diseases, including cancer. These hybrid compounds have been defined as chemical entities comprising the partial or total structure of two or more molecules with different biological activities. This has allowed for the generation of new compounds that have shown greater safety, effectiveness at low doses, and tolerability to treatment due to improvements in pharmacokinetic properties and reductions in adverse effects related to the toxicity produced by the administration of multiple drugs [21,22]. An example of these hybrid compounds are the styrylquinolines, each comprising a quinoline nucleus attached to a styryl group (Figure 1), which have aroused significant interest in recent years given their potential as antiparasitic and antitumor agents [23–25]. However, there is not enough information about the mechanisms of action carried out by these hybrid compounds to exert their antitumor effect, specifically on colon cancer cells. On the other hand, there is no great evidence about their selectivity [25–27]. Thus, in order to find new molecules with antitumor potential, greater effectiveness and fewer side effects than currently available treatments for CRC, we evaluate the selectivity and antiproliferative capacity of hybrids based on 8-hydroxyquinoline (8-HQ) and resveratrol in vitro in human colon adenocarcinoma cells (SW480) and their metastatic derivative (SW620). Additionally, we determine the effect of these hybrids on different biological processes, such as apoptosis, the production of reactive oxygen species, and the regulation of different markers associated with inflammation. In this way, we will obtain preliminary results about the biological activity of these compounds that will be the basis for further research in more complex models, such as animal models.

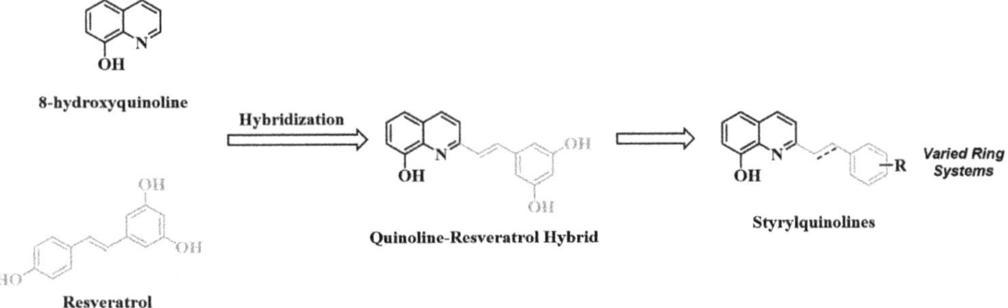

Figure 1. Design of styrylquinolines.

2. Materials and Methods

2.1. Chemistry

2.1.1. General Remarks

Microwave reactions were carried out in a CEM Discover (CEM, Matthews, NC, USA) microwave reactor in sealed vessels (monowave, maximum power of 300 W, temperature fixed with an IR sensor). ^{1}H and ^{13}C nuclear magnetic resonance (NMR) spectra were recorded on a Varian instrument (Palo alto, CA, USA) operating at 600 and 150 MHz. The signals of the deuterated solvent (CDCl$_3$) were used as reference (CDCl$_3$: δ = 7.27 ppm for ^{1}H NMR and δ = 77.00 ppm for ^{13}C NMR). Silica gel 60 (0.063–0.200 mesh, Merck, Whitehouse Station, NJ, USA) was used for column chromatography, and precoated silica gel plates (60 F254 0.2 mm, Merck, Whitehouse Station, NJ, USA) were used for thin-layer chromatography (TLC).

2.1.2. General Procedure for the Synthesis of Styrylquinolines

Styrylquinolines **3a–e**

8-Hydroxyquinaldine (1 eq) and benzaldehyde (2 eq) were dissolved in acetic anhydride (20 mL) in a 50 mL flatbottomed flask equipped with a magnetic stirring bar. The mixture was stirred and heated to reflux under microwave irradiation for a period of 3 h. The crude reaction mixture was evaporated under reduced pressure, the residue was dissolved in methanol, and then 3 eq of KOH was added. This solution was stirred for 1 h. Finally, this solution was added to cold water in an ice bath. The resulting yellow solid was filtered, washed with water, and dried. The obtained solid was purified by column chromatography over silica gel eluting with mixtures of hexane and ethyl acetate of different ratios to obtain styrylquinolines with yields between 60% and 75%. The monitoring of the reaction progress and product purification was carried out by TLC.

Obtention of Compounds **4a** and **4b**

A solution of **3a** or **3d** (1 eq) in methanol was added under hydrogen to a suspension of Pd-C 10% (0.05%) in dry methanol (10 mL). The reaction was monitored by NMR until the consumption of the starting material. Filtration afforded compound **4a** at 90% and **4b** at 94%.

2-(4-hydroxyphenethyl)quinolin-8-ol (**4a**): ^{1}H NMR (600 MHz, chloroform-d) δ 8.03 (d, *J* = 8.4 Hz, 1H), 7.39 (t$_{apparent}$, *J* = 7.8 Hz, 1H), 7.29 (d, *J* = 8.0 Hz, 1H), 7.22 (d, *J* = 8.4 Hz, 1H), 7.15 (d, *J* = 7.8 Hz, 1H), 6.83 (d, *J* = 8.0 Hz, 1H), 6.75–6.67 (m, 2H), 3.81 (s, 3H), 3.26 (t, *J* = 7.8 Hz, 2H), 3.10 (t, *J* = 7.8 Hz, 2H). ^{13}C NMR (151 MHz, Chloroform-d) δ 159.63, 151.76, 146.35, 143.84, 137.68, 136.19, 133.35, 126.87, 126.83, 122.49, 120.98, 117.60, 114.28, 111.10, 109.78, 55.85, 40.48, 35.09. EIMS: *m/z* 266.1292 [M + H]$^{+}$, Calcd. for C$_{17}$H$_{16}$NO$_2$: 266.1287.

2-(4-hydroxy-3-methoxyphenethyl)quinolin-8-ol (**4b**): ^{1}H NMR (600 MHz, Chloroform-d) δ 8.02 (d, *J* = 8.4 Hz, 1H), 7.39 (t, *J* = 7.9 Hz, 1H), 7.28 (d, *J* = 8.4 Hz, 1H), 7.23 (d, *J* = 8.4 Hz, 1H), 7.15 (d, *J* = 7.8 Hz, 1H), 7.08 (d, *J* = 7.9 Hz, 2H), 6.75 (d, *J* = 7.8 Hz, 2H), 3.24 (t, *J* = 7.8 Hz, 2H), 3.10 (t, *J* = 7.8 Hz, 2H). ^{13}C NMR (151 MHz, Chloroform-d) δ 159.63, 153.77, 151.74, 137.68, 136.19, 133.63, 129.57 (2C), 126.87, 126.81, 122.45, 117.61, 115.25 (2C), 109.77, 40.38, 34.49. EIMS: *m/z* 296.1395 [M + H]$^{+}$, Calcd. for C$_{18}$H$_{18}$NO$_3$: 296.1393.

2.2. Biological Activity Assays

2.2.1. Cell Lines and Culture Conditions

For the biological assays, the human colon adenocarcinoma cell line SW480 and its metastatic derivative SW620 were used. Additionally, the nonmalignant cell line HaCaT was used to find the selectivity index of the compounds. All cell lines were obtained by the Colombian Plant Chemistry Group of the Faculty of Exact and Natural Sciences of the University of Antioquia (Medellín, Colombia) from the European Collection of Authenticated Cell Cultures (ECACC, Salisbury, UK). Cells were cultured in a DMEM medium with 4500 mg/L of glucose and L-glutamine (Sigma-Aldrich, Burlington, MA, USA), supplemented with 10% heat-inactivated (56 °C) horse serum (Gibco, Waltham,

MA, USA), 1% penicillin/streptomycin (Sigma-Aldrich, Burlington, MA, USA), and 1% non-essential amino acids (Sigma-Aldrich, Burlington, MA, USA). For all experiments, horse serum was reduced to 3%, and the medium was supplemented with 5 mg/mL of transferrin, 5 ng/mL of selenium, and 10 mg/mL of insulin (ITS; Sigma-Aldrich, Burlington, MA, USA) [28]. All cell lines were incubated at 37 °C in a 5% CO_2 atmosphere. Additionally, cell cultures were constantly monitored with PCR (Sigma-Aldrich, Burlington, MA, USA) for *Mycoplasma* spp. [29] to control contamination with this agent.

2.2.2. Cytotoxic Activity

To evaluate the effect of styrylquinolines on the viability of the SW480, SW620, and HaCaT cell lines, sulforhodamine B (SRB) staining was used. SRB is a colorimetric assay that indirectly estimates the number of living cells based on the ability of SRB to bind to protein components of adherent cells [30]. Briefly, the malignant cell lines were seeded at a final density of 2.0×10^4 cells/well, and the nonmalignant cell line was seeded at a final density of 1.5×10^4 cells/well in 96-well tissue culture plates. All cell lines were incubated at 37 °C in a 5% CO_2 atmosphere for 24 h to enable cell adherence, and they were then treated for 24 and 48 h with increasing concentrations (0.01–160 µM) of the styrylquinolines or their respective precursors (8-hydroxyquinoline and resveratrol), as well as 1% DMSO (negative control) and 5-FU (standard drug). After treatment, cells were fixed with trichloroacetic acid (50% *v/v*; PanReac AppliChem, Barcelona, Spain) for a period of one hour at 4 °C. After this, the cells were incubated for 30 min at room temperature with 0.4% (*w/v*) SRB (Sigma-Aldrich, Burlington, MA, USA). To remove unbound SRB, the cells were washed with 1% acetic acid, and the plates were allowed to dry at room temperature. Protein-bound SRB was solubilized with 10 mM Tris-base (Amresco, Cleveland, OH, USA) for 30 min at room temperature under constant agitation, and absorbance was measured at 490 nm in a microplate reader (Bio-Rad iMark™, Hercules, CA, USA). Finally, the concentration of the compound that inhibits 50% of cell growth (IC50) and the selectivity index (SI) were determined.

2.2.3. Antiproliferative Activity

The antiproliferative activity of styrylquinolines with higher selective cytotoxicity toward malignant cells was also tested with SRB staining [30]. Briefly, cells were seeded to a final density of 2.5×10^3 cells/well in 96-well tissue culture plates and incubated under the same conditions described for cytotoxic activity. After 24 h, the cells were treated with increasing concentrations (5–80 µM) of the selected hybrids or with 1% DMSO (negative control) for 0, 2, 4, 6, and 8 days. Culture media were replaced every 48 h to guarantee the basic nutrients required for cell growth and viability, maintaining the concentrations of each of the selected hybrids. After each incubation time, cells were fixed and stained, and the absorbance was measured as described above.

2.2.4. Reactive Oxygen Species (ROS) Levels

In order to evaluate the effect of the chosen styrylquinolines on the production of ROS in malignant cells, the 2′,7′-dichlorofluorescein diacetate (2′,7′-DCFDA) probe (Calbiochem, San Diego, CA, USA) was used as described by Kim et al. [31]. Briefly, malignant cells were seeded at a final density of 2.5×10^5 cells/well in 6-well tissue culture plates for 24 h at 37 °C in a 5% CO_2 atmosphere. Afterward, cells were treated for 24 and 48 h with 1% DMSO (negative control) or with the IC50 of the styrylquinolines obtained at 24 and 48 h of each cell line. After treatment, 2′,7′-DCFDA was added to a final concentration of 10 µM and incubated at 37 °C for 30 min. Finally, representative images of each well were taken using a fluorescence microscope (Axio Vert. A1; ZEISS, Jena, Germany) and Zen blue 3.4 software. The cell lysate was obtained from each well, and relative fluorescence units (RFU) were measured at excitation/emission wavelength (Ex/Em) = 485/525 nm using the Varioskan Lux microplate reader (Thermo Fisher Scientific, Waltham, MA, USA). The total protein concentration was quantified with the BCA method using the Pierce™ BCA kit (Thermo Fisher Scientific, Waltham, MA, USA) to normalize the UFR.

2.2.5. Assessment of Apoptosis

To assess whether the chosen styrylquinolines generated apoptosis in malignant cells, the APO-DIRECT™ kit (Chemicon^R International, Temecula, CA, USA) was used following the manufacturer's instructions. Briefly, the malignant cells were seeded at a final density of 1.1×10^6 cells in a T75 culture flask and incubated for 24 h at 37 °C in a 5% CO_2 atmosphere. Then, cells were treated for 24 and 48 h with 1% DMSO (negative control) or the IC50 obtained at 24 and 48 h for each compound. After incubation time, cells were fixed and stained with FITC-dUTP and propidium iodide (PI), respectively, according to the manufacturer's protocol. Analysis was conducted via flow cytometry (LSR Fortessa; BD Biosciences, San Jose, CA, USA) and FlowJo 7.6 software. All PI-FITC-positive cells were considered to be apoptotic cells.

2.2.6. Determination of Inflammatory Cytokines and Apoptotic Proteins

To assess whether styrylquinolines had any effect on the expression of immunological markers associated with the inflammatory process, malignant cells were seeded at a final density of 2.5×10^5 cells/well in 6-well tissue culture plates, and cell adherence was allowed. After this, the cells were treated for 24 and 48 h with either 1% DMSO (negative control) or the IC50 obtained at 24 and 48 h of treatment with the compound that presented the best results in the aforementioned biological assays. After the incubation time, the supernatant was collected, and the levels of the following analytes were measured using the ProcartaPlex Human Th1/Th2/Th9/Th17/Th22/Treg 18-plex panel (Invitrogen) according to the manufacturer's protocol: GM-CSF, IFN-γ, TNF-α, IL-10, IL-12p70, IL-13, IL-17A, IL-18, IL-1β, IL-2, IL-21, IL-22, IL-23, IL-27, IL-4, IL-5, IL-6, and IL-9. Furthermore, to evaluate the participation of these hybrids in the production of some markers associated with the apoptosis process (Bcl-2, active Caspase-3 and cleaved PARP) in colorectal cancer cells, the human apoptosis panel (Invitrogen, Waltham, Massachusetts, United States) was used following the manufacturer's instructions. To normalize the concentration of each marker, total proteins were quantified using the PierceTM BCA kit (Thermo Fisher Scientific, Waltham, MA, USA).

In both cases, the reading was performed in the MAGPIX marker multiplex analyzer (Luminex XMAP, Austin, TX, USA). The concentration of each molecule was extrapolated from the calibration curve (individual for each marker) obtained from the standards provided by the kit.

2.2.7. Statistical Analysis

All experiments were performed at least three times. The normality of the variables was evaluated using the Kolmogorov–Smirnov test. Data are expressed as the mean $\pm$ SE (standard error). IC50 values were evaluated by non-linear regression. Statistical differences between the negative control group (cells treated with 1% DMSO) and the treated cells at the different evaluation times were analyzed by two-way ANOVA followed by Dunnett's test. Values with $p \leq 0.05$ were considered significant. Data were analyzed using GraphPad Prism version 8 software for Windows (Graph Pad Software 8, San Diego, CA, USA).

2.3. Computational Methods

The 2D chemical structures of the most active styrylquinolines were drawn using ChemDraw 17.0 software (Cambridge Soft, Cambridge, MA, USA) and then saved as MDL MoL files. Chem3D 17.0 software (Cambridge Soft, Cambridge, MA, USA) was used to generate 3D structures of ligands, and optimization was performed using the MM2 Force-Field in Chem3D Ultra 8.0 Software CS, ChemOffice Chem3D Ultra 8.0, and Cambridge Soft. AutoDockTools (ADT) was used to parameterize ligands: non-polar hydrogens were merged, rotatable bonds were assigned, full hydrogens were added, and

Kollman united partial atom charges were added to the individual protein atoms. The 3D protein structure of the caspase-3 (PDB ID: 5i9b) was downloaded from the Protein Data Bank website (accessed on 18 June 2022). Co-crystallized ligands, ions, and water molecules were removed from the protein structure by using DS Visualizer 2.5 program. For docking analysis, grid map dimensions ($32 \times 32 \times 32$ Å) were set surrounding the active site at x, y, and z coordinates of $x = 1.5$, $y = -8.1$, and $z = -13.4$ at an exhaustiveness of 20 for each protein–compound pair and a grid spacing of 1 Å. The AutoDock Vina v.1.2.0 software package by The Scripps Research Institute [32] was used with a flexible-ligand/rigid-receptor protocol and binding affinity/free energy estimated in kcal/mol. Finally, to inspect docking solutions, DS Visualizer 2.5 and PyMOL Molecular Graphics System Version 2.0 Schrodinger, LLC (2015) were used.

3. Results

3.1. Chemistry

Styrylquinolines **3a–e** were obtained via microwave-assisted, Perkin-type condensation between 8-hydroxyquinaldine (1) and benzaldehydes with different substituents (2) [33]. The reaction yields ranged between 60 and 75%. These compounds have already been reported [27,34–37]. However, our synthetic strategy involves microwave-assisted reactions that allow us to create compounds with shorter reaction times than those created using conventional heating methods. Here, the products were obtained in good-to-excellent yields and without appreciable by-product formation. Then, compounds **3a** and **3d** were reduced using catalytic hydrogenation and yielded **4a** and **4b**, respectively, at yields greater than 90% [38] (Scheme 1).

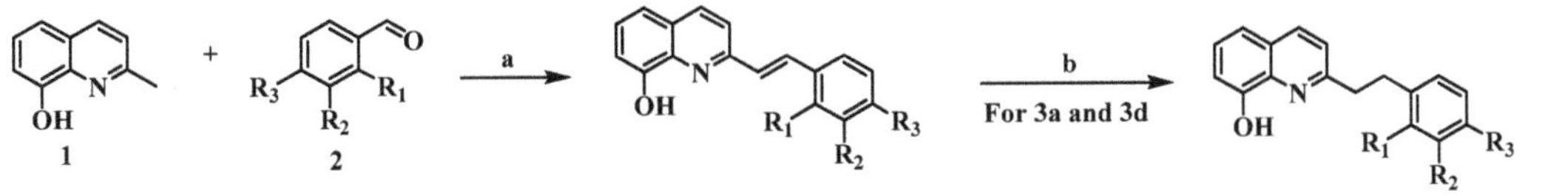

3a, R_3 = OH, R_1 = R_2 = H,
3b, R_1 = R_2 = OH, R_3 = H
3c, R_1 = R_3 = OH, R_2 = H
3d, R_2 = R_3 = OH, R_1 = H
3e, R_2 = OMe, R_3 = OH, R_1 = H

4a, R_3 = OH, R_1 = R_2 = H,
4b, R_2 = OMe, R_3 = OH, R_1 = H

Reagents and conditions: a) Ac_2O, MW, then KOH, MeOH, 60-75%; b) H_2, Pd-C, MeOH, > 90%

Scheme 1. Synthetic pathway to the obtention of styrylquinolines.

3.2. Effect of Styrylquinolines on Cell Viability of Malignant and Nonmalignant Cells

To determine the effect of hybrids based on resveratrol and 8-hydroxyquinoline on cell viability, different concentrations of styrylquinolines were evaluated in malignant cell lines (SW480 and SW620) and a nonmalignant cell line (HaCaT). Data are reported in terms of cytotoxicity, finding inhibitory concentration 50 (IC50), as shown in Table 1.

IC_{50} values were obtained from dose-response curves for each compound. The selectivity index (SI) was calculated as the ratio of IC_{50} values in nonmalignant HaCaT cells to the IC_{50} of SW480 cells or SW620 cells. Data are presented as the mean $\pm$ SE of at least three independent experiments. Compound 3e was not evaluated due to solubility problems.

After 24 h of treatment, it was observed that hybrid **3a** presented more selective cytotoxicity towards both malignant cell lines compared to the other styrylquinolines, as well as the precursors and the reference drug (5-FU). This was evidenced by the high IC50 values (231.9 $\pm$ 29.2 μM) in the non-malignant cell line (HaCaT) and the high SI values ($SI_{SW480-24h}$ = 3.4; $SI_{SW620-24h}$ = 3.8). In addition, hybrid **3c** also exhibited high selectivity on SW480 cells (SI = 3.0), and compound **3d** was selective towards the SW620 cell line (SI = 3.0). Although compound **3b** exhibited the lowest IC50 values among all tested hybrids (IC50 of 57.2 $\pm$ 6.3 μM, 72.6 $\pm$ 11.0 μM, and 49.1 $\pm$ 4.6 μM in HaCaT, SW480, and SW620 cells, respectively), its selectivity was significantly low because this hybrid also showed high cytotoxicity in the nonmalignant cell line. Similar results were obtained with hybrids **4a** and **4b**.

Table 1. Cytotoxic effect (IC50) of styrylquinolines on SW480, SW620 and HaCaT cell lines at 24 and 48 h post-treatment.

Compound	24 h					48 h				
	HaCaT IC$_{50}$ (μM)	SW480 IC$_{50}$ (μM)	SI	SW620 IC$_{50}$ (μM)	SI	HaCaT IC$_{50}$ (μM)	SW480 IC$_{50}$ (μM)	SI	SW620 IC$_{50}$ (μM)	SI
3a	231.9 ± 29.2	67.7 ± 2.8	3.4	60.5 ± 2.9	3.8	47.9 ± 3.3	26.5 ± 4.0	1.8	16.6 ± 2.2	2.9
3b	57.2 ± 6.3	72.6 ± 11.0	0.8	49.1 ± 4.6	1.2	33.5 ± 4.9	39.0 ± 3.6	0.9	9.7 ± 2.0	3.4
3c	110.5 ± 12.7	37.1 ± 3.8	3.0	62.0 ± 4.0	1.8	38.4 ± 6.2	38.4 ± 3.8	1.0	20.7 ± 3.1	1.9
3d	163.9 ± 14.1	159.7 ± 16.6	1.0	54.6 ± 7.5	3.0	78.9 ± 2.3	42.5 ± 5.2	1.9	6.4 ± 1.1	12.3
4a	65.1 ± 8.4	117.4 ± 5.3	0.6	51.0 ± 5.6	1.3	69.7 ± 3.0	63.2 ± 7.1	1.1	28.9 ± 1.2	2.4
4b	109.3 ± 8.1	93.4 ± 16.6	1.2	58.8 ± 5.5	1.9	53.6 ± 2.2	43.2 ± 2.0	1.2	18.6 ± 2.0	2.9
Resveratrol	179.9 ± 7.2	217.1 ± 12.8	0.8	168.9 ± 3.9	1.1	98.2 ± 7.0	135.7 ± 9.1	0.7	125.2 ± 4.6	0.8
8-HQ	87.1 ± 9.4	89.6 ± 11.8	1.0	55.3 ± 3.6	1.6	61.6 ± 4.8	38.8 ± 3.2	1.6	36.3 ± 4.8	1.7
5-FU	>2000	>2000	2.0	>2000	0.9	38.78 ± 7.2	748.3 ± 157.6	0.4	295.5 ± 39.7	0.1

IC50 values were obtained from dose-response curves for each compound. The selectivity index (SI) was calculated by the ratio of IC50 values in nonmalignant cells (HaCaT) to the IC50 of malignant cells (SW480 or SW620). Data are presented as the mean ± SE of at least three independent experiments. Compound 3e was not evaluated due to solubility problems. 8-HQ: 8-hydroxyquinoline; 5-FU: 5-Fluorouracil.

After 48 h of treatment, the SW620 cell line showed greater susceptibility to the evaluated hybrid compounds since it showed greater cytotoxicity in the metastatic line (IC50 of between 6.4 and 28.9 μM) compared to the SW480 cell line (IC50 of between 26.5 and 63.2 μM). Furthermore, hybrids **3a** and **3d** had the highest SI values (SI = 1.8 and 1.9, respectively) in the SW480 cell line. However, higher selectivities were observed in the SW620 cell line (SI ≥ 1.9), mainly for compound **3d** (SI = 12.3).

On a structure-activity relationship basis, it was observed in SW480 cells that there was a decrease in activity in the presence of dihydroxylated compounds compared to the presence of monohydroxylated compounds (**3a** vs. **3b–d**). This relationship was not clear in SW620 cells. On the other hand, the presence of a double bond in a side chain is important for activity, i.e., a decrease in the effect was observed when the reduction was carried out (**3a** vs. **4a**). Similar results were obtained in other studies with cinnamic acid alkyl ester derivatives [39].

In accordance with the cytotoxicity and high SI, compounds **3a** and **3d** were chosen to continue with the other biological assays. The IC50 values obtained for each cell line at the different treatment times (24 and 48 h) were taken into account.

3.3. Antiproliferative Effect of Styrylquinolines

In order to evaluate the antiproliferative activity of compounds **3a** and **3d** on colorectal cancer cell lines, Sulforhodamine B (SRB) staining was used. As shown in Figure 2, hybrids **3a** and **3d** showed concentration- and time-dependent antiproliferative activity on malignant cells, as evidenced by the statistically significant decrease ($p < 0.05$) in the cell viability percentage compared to the negative control. Compound **3a** decreased the cell viability percentage to 0% using concentrations from 10 μM in both SW480 and SW620 cell lines. In contrast, compound **3d** required higher concentrations (fourfold more than those used with compound **3a**) to reach the same effect. This compound produces a high percentage of reduction in cell viability just from 40 μM concentration in both malignant cell lines.

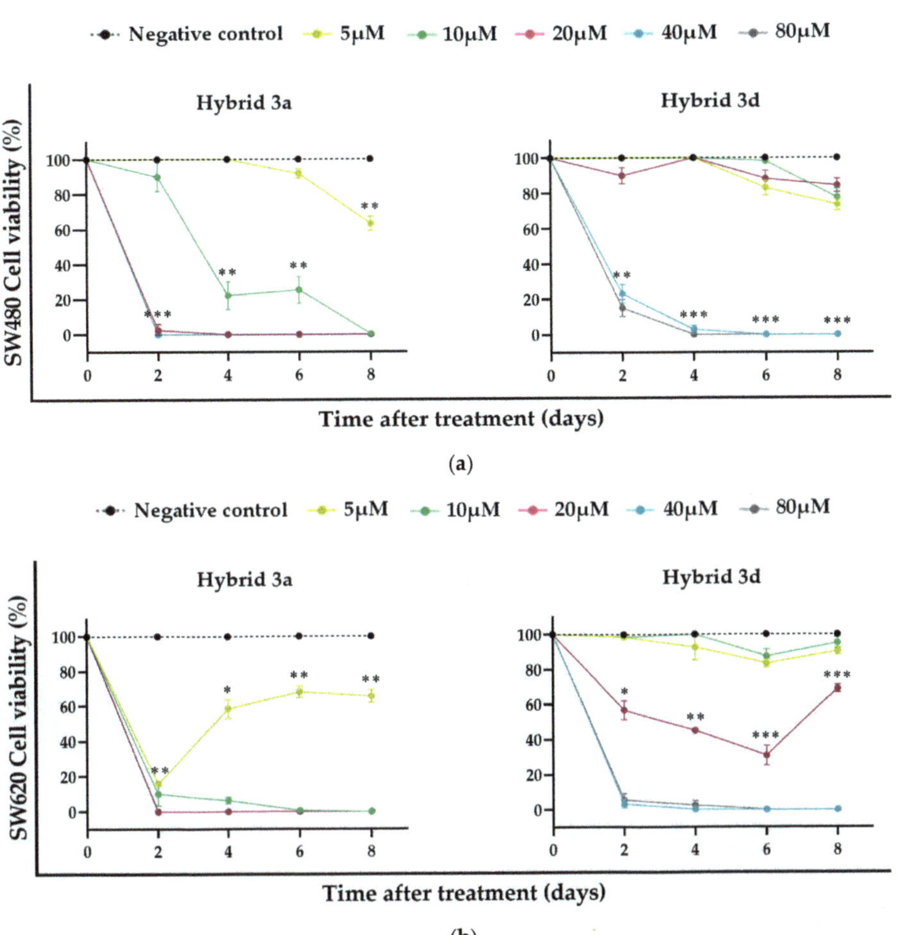

Figure 2. Antiproliferative effect of styrylquinolines **3a** and **3d** in the (**a**) SW480 and (**b**) SW620 cell lines. Data are presented as the mean $\pm$ SE of at least three replicates (* $p < 0.05$; ** $p < 0.01$; *** $p < 0.001$ vs. negative control). The negative control was assumed as 100% cell viability.

The SW620 cells exhibited greater susceptibility to treatments compared to SW480 cells since the latter required higher concentrations (20 and 40 μM for **3a** and **3d**) to significantly decrease the cell viability percentage from day 2 post-treatment than did the metastatic cell line (5 and 20 μM for **3a** and **3d**, respectively) at the same time.

3.4. ROS Production Induced by Styrylquinolines

To assess the intracellular ROS production induced by hybrids **3a** and **3d** in colorectal cancer cell lines, a 2′,7′-DCFDA probe was used. According to the results, compound **3d** was the only one capable of inducing a statistically significant increase in ROS production compared to the control ($p < 0.05$; Figure 3a). The significant increase in ROS levels in the SW620 and SW480 cell lines occurred at 24 h (1.8 ± 0.1 RFU) and 48 h (3.3 ± 0.2 RFU) after treatment, respectively. The increase in ROS production by compound **3d** in the SW620 cell line was lower than in the SW480 cell line, which was evidenced by the low fluorescence observed in Figure 3c. Importantly, the cells showed morphological changes regarding size and shape after being treated (Figure 3b,c).

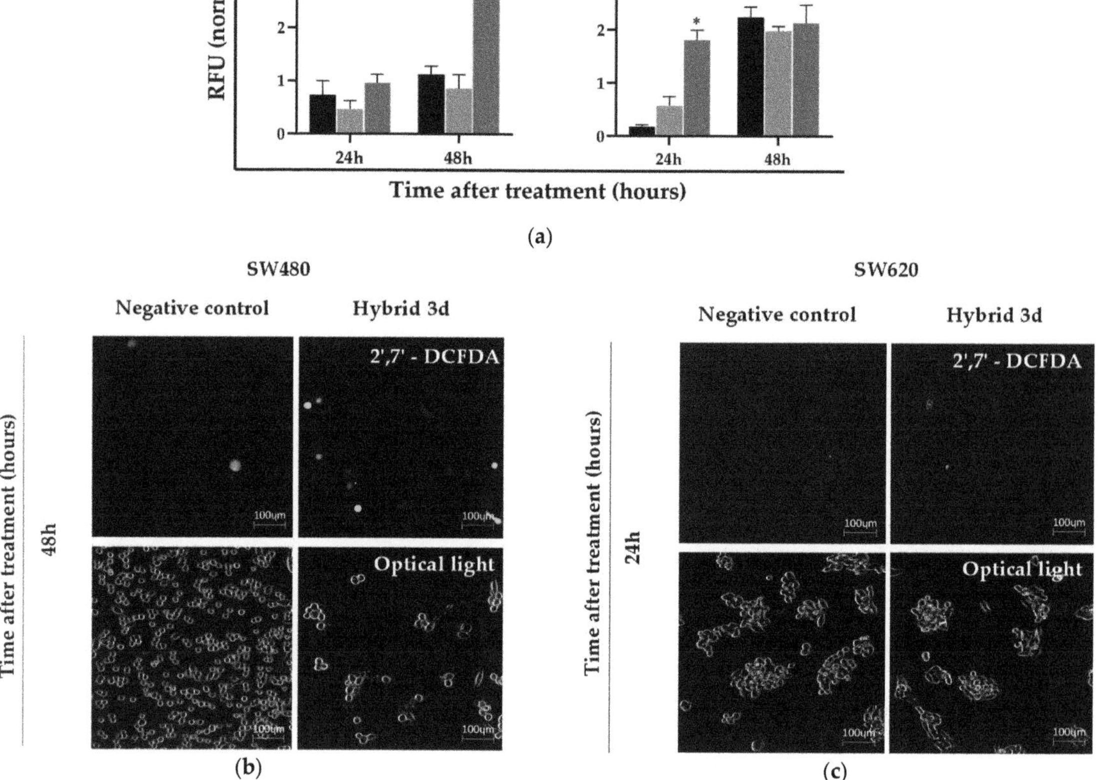

Figure 3. Intracellular ROS production: (**a**) quantification of ROS in SW480 and SW620 cell lines after 24 and 48 h of treatment with hybrids **3a** and **3d**; (**b**) representative images of the SW480 cell line at 48 h post-treatment with hybrid **3d** and (**c**) representative images of the SW620 cell line at 24 h post-treatment with hybrid **3d**. Magnification: 20×. Data are presented as the mean ± SE of at least three replicates (* $p < 0.05$ vs. negative control). RFU: Relative Fluorescence Units.

3.5. Apoptosis Induction by Styrylquinolines

In order to investigate whether the **3a** and **3d** hybrids caused apoptosis in CRC cell lines, a flow cytometry assay was performed using FITC-dUTP to label fragmented DNA. Hybrids **3a** and **3d** generated 6.8 and 5.9%, respectively, of cell death by apoptosis in the SW620 cell line after 24 h of treatment. Compound **3a** also generated an increase in apoptotic cells (21.5%) at 48 h post-treatment (Figure 4a). None of the hybrids generated apoptosis in the SW480 cell line (data not shown).

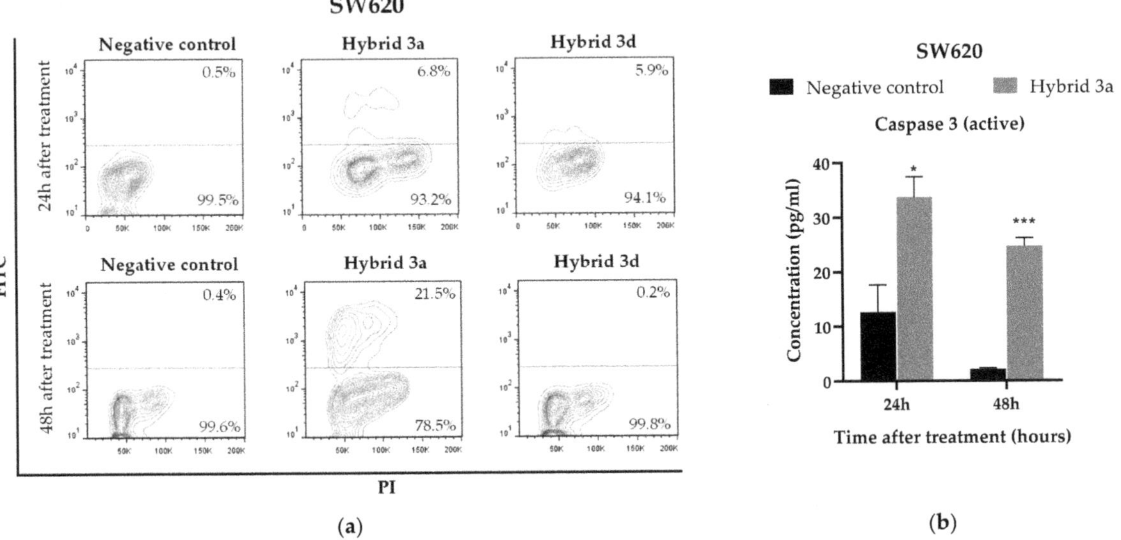

Figure 4. (**a**) Population of apoptotic cells of SW620 cell line treated with hybrids **3a** and **3d** for 24 and 48 h. (**b**) The concentration of markers associated with the apoptosis process in the SW620 cell line at 24 and 48 h after treatment with hybrid **3a**. Data are presented as the mean ± SE of at least three replicates (* $p < 0.05$; *** $p < 0.001$ vs. negative control). PI: Propidium Iodide.

Following the aforementioned results, the MAGPIX platform was used to evaluate whether compound **3a** regulated some markers (Bcl-2, active caspase 3 and cleaved PARP) involved in the cell death process in the SW620 cell line. Hybrid **3a** significantly increased the levels of active caspase 3 at both 24 and 48 h post-treatment in the metastatic cells (Figure 4b). The other evaluated markers (Bcl-2 and cleaved PARP) did not show significant differences when compared to the negative control.

3.6. Effect of Styrylquinolines on the Production of Immunological Markers Associated with the Inflammatory Response

To evaluate the immunomodulatory response of styrylquinolines in colorectal cancer cell lines, the hybrid with the best results obtained in the aforementioned biological assays was selected. For this purpose, the levels of some representative immunological markers of Th1 (GM-CSF, IFN-γ, TNF-α, IL-12p70, IL-1β, and IL-2), Th2 (IL-4, IL-5, IL-10, and IL-13), Th17/Treg (IL-17A, IL-6, IL-18, IL-21 and IL-23), Th22 (IL-22 and IL-27) and Th9 (IL-9) response were determined. As shown in Figure 5a, compound **3a** induced significant increases in the levels of GM-CSF, IFN gamma, IL-12p70, IL-4, IL-6, IL-10, and IL-17A in SW480 cells at 24 h post-treatment. Evidently, 48 h after treatment, a greater immunomodulatory effect was observed to maintain significant increases in GM-CSF, IFN gamma, IL-4, IL-6, IL-10 and IL-17A levels. Additionally, this hybrid induced significant increases in the levels of IL-1 beta, IL-13, IL-18, IL-2, TNF alpha, IL-22, IL-27, and IL-9 in this same post-treatment time.

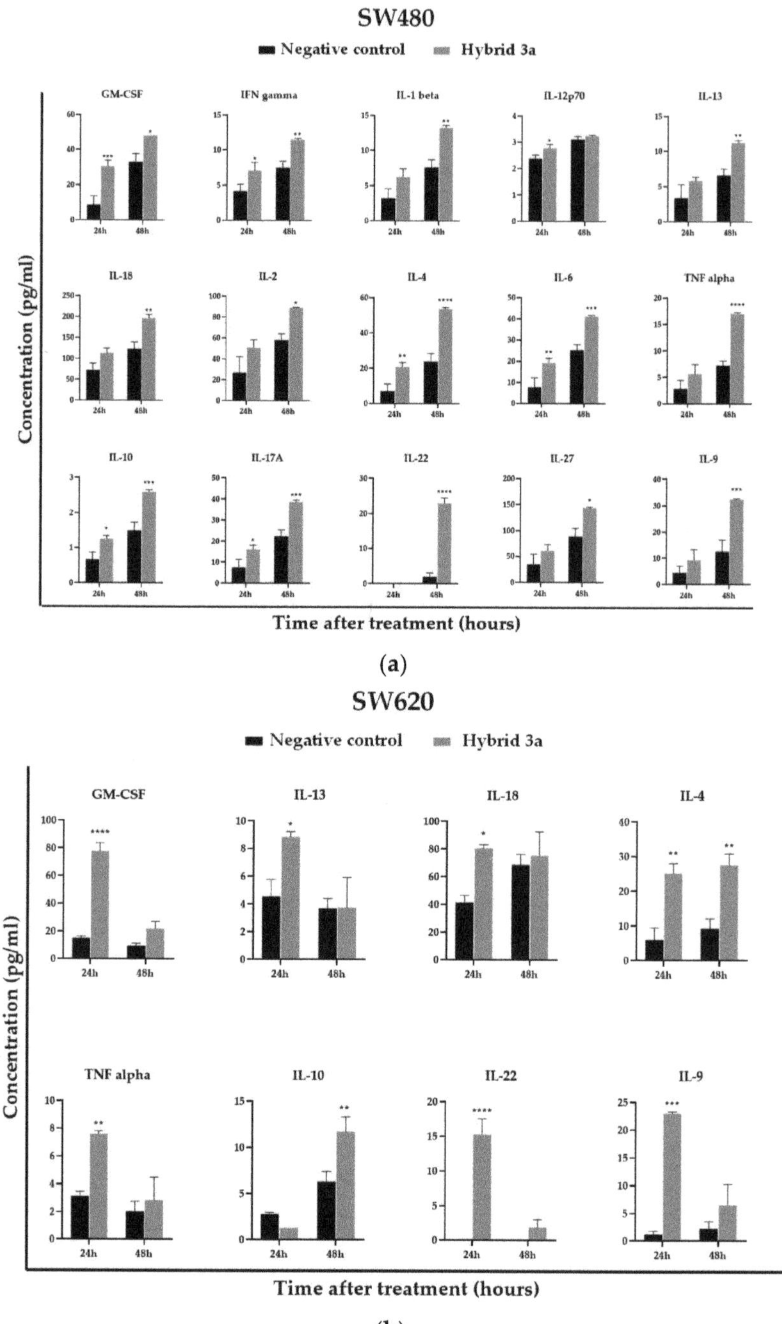

Figure 5. Levels of immunological markers associated with the inflammatory process in the (**a**) SW480 and (**b**) SW620 cell lines at 24 and 48 h post-treatment with hybrid **3a**. Data are presented as the mean $\pm$ SE of at least three replicates (* $p < 0.05$; ** $p < 0.01$; *** $p < 0.001$; **** $p < 0.0001$ vs. negative control).

Furthermore, hybrid **3a** induced a significant increase in the levels of GM-CSF, IL-13, IL-18, IL-4, TNF alpha, IL-22, and IL-9 at 24 h post-treatment in the SW620 cell line. At 48 h of treatment, the levels of IL-4 and IL-10 were significantly increased by this hybrid in the same cell line (see Figure 5b). The other evaluated cytokines did not show significant differences compared to the control group.

3.7. Molecular Docking Studies

According to biological assays, hybrid **3a** (4-hydroxy-styryl-substituted) caused a remarkable apoptotic effect in colorectal cancer cells. The in vitro cytotoxic response produced for **3a** in SW620 colon carcinoma cell lines appeared to be strongly associated with the modulation of caspase-3. Therefore, we hypothesized that **3a** targets caspase-3, thus altering its activity or function. In this scenario, we performed computational studies with the aim of exploring a possible binding mechanism of caspase-3 for compound **3a** using the docking program AutoDock Vina v.1.2.0. To accomplish this goal, compound **3a** was docked inside the catalytic domain of the X-ray crystallographic structures of caspase-3 (PDB code: 5i9b) protein, and their protein-ligand binding affinities (in kcal/mol) together with binding modes were estimated.

In our docking scheme, we first proceeded with self-docking simulations in order to validate our AutoDock Vina docking protocol. For this purpose, we carried out a comparison of the binding modes of the re-dock Ac-DEVD-CMK inhibitor (in yellow) and their crystallographic binding mode (in red) deposited in the PDB archive (PDB code: 5i9b) [40]. The results indicated that our docking procedure was able to reproduce the binding mode of the co-crystallized inhibitor Ac-DEVD-CMK (in red) with a strong root mean square deviation (RMSD) of 1.075 Å, showing a close homology (Figure 6a). This finding indicated a high level of feasibility in our protein-ligand docking protocol. After the docking procedure was validated, compound **3a** was docked into the caspase-3 catalytic domain. We found that hybrid **3a** (in blue) not only efficiently bound to caspase-3 with a closer binding affinity (−7.6 kcal/mol) than the current inhibitor Ac-DEVD-CMK (−8.2 kcal/mol) but it also fits well inside the catalytic cavity of caspase-3, as can be seen in Figure 6b. These facts support our experimental evidence suggesting that compound **3a** could prevent cell growth and proliferation in colorectal cancer cells by modulating caspase-3 function. Considering that the active site of caspase-3 comprises eighteen "hotspot" amino acid residues (Arg64, Leu119, Ser120, His121, Gln161, Ala162, Cys163, Ser198, Tyr204, Ser205, Trp206, Asn208, Ser209, Trp214, Ser249, Phe250, Ser251, and Phe252) [40], our modeling work also suggested that **3a** could bind to caspase-3 through several non-covalent interactions with those critical amino acid residues vital for caspase-3 function (Figure 6c) [40]. A close view of the 2D ligand-protein interaction plot after the docking procedure showed that **3a** interacted with the Arg207 residue via one hydrogen bond at a distance of 3.66 Å. Similarly, the styryl portion was found to create one hydrogen bond and one π–cation contact with the Glu123 residue at distances of 3.54 and 3.64 Å, respectively. We also noted that both the quinoline ring and the styryl moiety were able to bind to the caspase-3 via two π–alkyl contacts with Cys163. Figure 6c also shows numerous hydrophobic contacts that could play important roles in stabilizing the **3a**/caspase-3 complex following the binding event. These results suggest that two hydrogen bonding interactions, one π−cation and two π–alkyl contacts with those critical "hotspot" residues, could have important roles in the effective modulation of caspase-3 in **3a**-induced cytotoxicity.

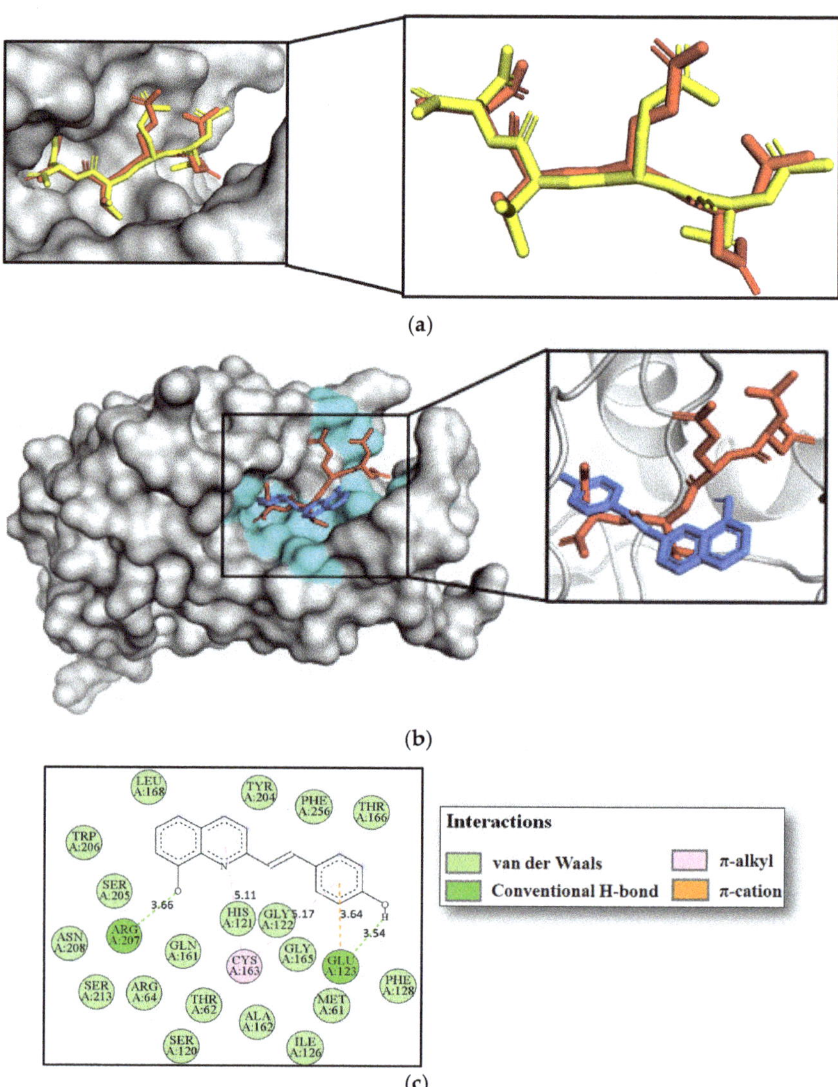

Figure 6. Docking studies of the **3a**-caspase-3 complex. (**a**) Self-docking study. 3D superposition of the best-docked pose of Ac-DEVD-CMK (in yellow) and its crystallographic binding pose (in red). (**b**) Alignment of the best-docked conformation of **3a** (in blue) and crystallographic binding mode of inhibitor Ac-DEVD-CMK (in red) within the caspase-3 catalytic domain. (**c**) 2D interaction diagram between **3a** and caspase-3. Hotspot amino acid residues are colored in cyan.

4. Discussion

Previous studies have evaluated the antimicrobial activity of different styrylquinolines and antitumor activity in different cancer cell lines. However, more information is needed about the anticancer effect of these hybrids on CRC. In our study, the styrylquinoline compounds showed better activity in all cell lines than resveratrol and 5-Fu. Specifically, styrylquinolines **3a** and **3b** were more active than 8-HQ. These results show the importance of hybridization in the design of new drugs. Most of the evaluated hybrids showed a cytotoxic effect at low concentrations on the malignant cells compared with the reference drug, which only showed cytotoxic activity 48 h after treatment. These results are consistent with those of previous studies showing that other styrylquinolines also have cytotoxic activity at low concentrations (even < 10 µM) in different types of cancer cells [25,26]. It has been seen that 5-FU requires high concentrations to achieve a cytotoxic effect on different malignant cell lines, including SW480 and SW620 [30]. For this reason, many studies have focused on administering 5-FU in combination with other compounds to increase chemosensitivity in cancer cells and improve the effectiveness of this drug [41]. It should be noted that the chemosensitivity of cancer cells may be due to their heterogeneity, which may explain the variability in IC50 found in the literature for 5-FU in different cell lines and the different responses to the drugs used in cancer chemotherapy [42]. Even so, hybrids **3a** and **3d** were chosen to continue with the other biological assays because they showed greater selectivity towards malignant lines.

Regarding the antiproliferative activity in malignant colon cell lines, it was found that hybrids **3a** and **3d** inhibited cell proliferation in direct proportion to the time of treatment and to the concentration of each hybrid in cancer cells. Compound **3a** presented a significant inhibition of cell proliferation at lower concentrations than those required by compound **3d**, which may have been due to the difference in the number of functional groups and their positions in the structure of each hybrid, which provided them different properties and reactivity. Furthermore, greater cytotoxicity of the treatments was observed in the SW620 cell line, possibly because it has been described that SW480 and SW620 cells present differences in the karyotype and expression profile of microRNAs, which have been associated with CRC progression through regulation of some signaling pathways and with chemoresistance to some drugs [43–46]. In the same sense, it has been seen that certain post-translational modifications of some cellular proteins have been gaining importance in the study of many diseases, including different types of cancer [47]. Many studies have investigated the role of different proteins susceptible to glycosylation and have described their association with resistance to some drugs used in cancer chemotherapy [48,49]. In this way, the differences in the response of the cell lines used in our study to the treatment with the hybrids evaluated could be explained.

In order to study the mechanisms by which these hybrids exert an antiproliferative effect on colon cancer cells, the levels of ROS production, apoptosis, and some immunological markers associated with inflammation were evaluated. ROS production occurs as a consequence of the normal physiological aerobic metabolism of a cell [50]. The production of ROS occurs as a consequence of the normal physiological aerobic metabolism of a cell, but when there is an imbalance between the production of ROS and the antioxidant mechanisms used by the cell to remove ROS, an accumulation of these free radicals occurs within the cell and causes damage to lipids, proteins and DNA; this process has been associated with the development of various diseases, including cancer [51]. In parallel, the overproduction of ROS in cancer cells induced by different compounds has been associated with cell death, which is why some current therapeutic strategies are focused on evaluating this phenomenon as an antitumor mechanism [52,53]. The results obtained in this study showed that compound **3d** had a pro-oxidant effect on SW620 and SW480 cells but did not induce apoptosis in either cell line. This effect of styrylquinolines may be due to the chelating properties of 8-hydroxyquinoline, one of the precursors of these hybrid compounds since this chelation process can lead to the formation of ROS and cause oxidative damage to cells [27,54]. It is important to highlight that ROS overproduction has also been associated with other types of cell death that are not characterized by nuclear fragmentation, such as ferroptosis and necroptosis [55–57]. A study by Lee SH. and Lee YJ. showed that resveratrol, in combination with docetaxel (a drug used to treat different types of cancer), concurrently induced apoptosis and necroptosis in prostate carcinoma cells [58]. In another study conducted by Lee J. et al., it was shown that resveratrol increased ferroptosis in head and neck cancer cells through the induction of the activation of the protein sirtuin 1, which has been associated with increased susceptibility to this type of cell death [59]. Thus, it is possible that the mechanism used by compound **3d** to induce cell death in colorectal cancer cells was something other than apoptosis.

It is known that apoptosis is a type of programmed cell death that can occur in two ways depending on the stimulus that triggers it. The intrinsic or mitochondrial pathway is activated by multiple factors that generate cellular stress, e.g., DNA damage, the increased production of free oxygen radicals, and endoplasmic reticulum stress. On the other hand, the extrinsic pathway is activated by the binding of different ligands to death receptors expressed in the cell, such as Fas and TRAIL receptors (TNF-related apoptosis-inducing ligand) [60]. Our results showed that compound **3a** induced apoptosis in SW620 cells and generated a significant increase in caspase 3 levels in this same cell line. Moreover, molecular docking analysis showed that compound **3a** effectively bound with caspase 3 protein, obtaining a comparable binding affinity (-7.6 kcal/mol) to that of the Ac-DEVD-CMK inhibitor. Therefore, combined experimental and computational findings indicated that the modulation of this protein might be a possible molecular mechanism to understand the cytotoxic response of **3a** in the SW620 colon carcinoma cell line. However, determining the pathway by which apoptosis is carried out requires further evaluation because caspase 3 is one of the effector proteins shared by both pathways of apoptosis (intrinsic and extrinsic) [61].

Compound **3a** presented the best results throughout this study, which is why it was chosen to evaluate its effect on the regulation of different immunological markers in CRC cells. There is significant evidence of an association between increases in the levels of certain cytokines related to the inflammatory process and the progression of cancer. However, cytokines have pleiotropic properties that enable them to play a dual role in cancer. They can participate in pro- or anti-tumor responses depending on the stage and microenvironment of the tumor [62]. Thus, in vitro and in vivo studies and some clinical trials have shown an association between increases in the levels of some cytokines and antitumor responses. Some studies have demonstrated that IL-2 can induce T cell proliferation and differentiation, as well as cause its activation [63]. A study by Ding et al. demonstrated the ability of IL-27 to enhance T cell anti-tumor immunity enhancing cell survival and memory T cell differentiation [64]. Another cytokine that has been associated with an antitumor response is IL-9, which favors the activation of cytotoxic T lymphocytes by recruiting dendritic cells to tumor tissues for the presentation of these antigens and the subsequent elimination of malignant cells [65,66]. Although the role of IL-22 in cancer has not been well-elucidated and it is controversial, it has been described that it has a positive prognosis in this type of cancer, relying on its capacity to induce a cross-talk between tumor cells and immune cells associated with a favorable clinical outcome [67]. IL-4 seems to have an important role in Th9 cell priming and differentiation, which have been associated with a powerful antitumor capacity [68]. IL-13 promotes the migration of dendritic cells and the activation of cytotoxic T lymphocytes [69]. In this same sense, Chen et al. demonstrated that IL-17 has the ability to modulate neutrophil-mediated antitumor immunity in cells [70]. IL-6 seems to promote antitumor immunity mediated by a Th17 response [71,72], and IFN-γ promotes tumor antigen presentation [73]. GM-CSF production seems to have a synergistic effect with Toll-Like Receptor—2 (TLR2) to inhibit tumor growth and modulate tumor-infiltrating Antigen Presenting Cells (APCs) [74], and IL-18 has been associated with better survival rates in cancer patients [75]. Finally, our results showed that compound **3a** upregulated different immunological markers such as GM-CSF, IFN gamma, IL-4, IL-6, IL-10, IL-17A, IL-1 beta, IL-13, IL-18, IL-2, TNF alpha, IL-22, IL-27, and IL-9, indicating that its antitumor effect is probably caused by its high immunomodulatory capacity.

In conclusion, the findings obtained in this study suggest that hybrids **3a** and **3d** have chemopreventive potential against CRC. Both compounds inhibited the proliferation of SW480 human colon adenocarcinoma cells and their metastatic derivative SW620. Compound **3a** was shown to be more effective at lower concentrations than those required by compound **3d**. The mechanisms by which these compounds exert their antiproliferative effect on malignant cell lines appear to be different. Hybrid **3d** promoted SW480 and SW620 cell death, probably through another mechanism different from apoptosis that may be related to ROS production. Hybrid **3a** induced apoptosis in SW620 cells, as evidenced by nuclear fragmentation and increased levels of active caspase 3 in these cells. Additionally, hybrid **3a** exhibited a high immunomodulatory effect, upregulating most of the immunological markers evaluated in this study. However, further experimental and computational studies are needed to clearly delineate the cytotoxic mechanism associated with styrilquinoline **3a**, preferably in other more complex models, in order to assess its effect on the immune response in a tumor microenvironment.

Author Contributions: Conceptualization, T.W.N.; Formal analysis, V.B.-B.; Funding acquisition, T.W.N.; Investigation, V.B.-B., E.C. and J.P.R.; Methodology, V.B.-B., A.F.Y.-P. and T.W.N.; Project administration, T.W.N.; Resources, W.C.-G.; Supervision, T.W.N.; Validation, V.B.-B.; Writing—original draft, V.B.-B., A.F.Y.-P. and W.C.-G.; Writing—review & editing, E.C., J.P.R., A.F.Y.-P., W.C.-G. and T.W.N. All authors have read and agreed to the published version of the manuscript.

Funding: This work has been funded by MINCIENCIAS, MINEDUCACIÓN, MINCIT and ICETEX through the Program Ecosistema Científico Cod. FP44842-211-2018 (Project number 58478 and 58537).

Institutional Review Board Statement: Not applicable.

Informed Consent Statement: Not applicable.

Data Availability Statement: Not applicable.

Acknowledgments: The authors thank Pontifical Bolivarian University, MINCIENCIAS, MINEDUCACIÓN, MINCIT, ICETEX, University of Antioquia, Corporation for Biological Research and the sustainability grant from the Research Committee (CODI) of the University of Antioquia for their support.

Conflicts of Interest: The authors declare no conflict of interest.

Sample Availability: Samples of the compounds are available from the authors.

References

1. Organización Mundial de la Salud Cáncer. 2021. pp. 1–7. Available online: https://www.who.int/es/news-room/fact-sheets/detail/cancer (accessed on 20 September 2022).
2. The Global Cancer Observatory (GLOBOCAN) Colorectal Cancer. 2021. p. 1. Available online: https://gco.iarc.fr/today/data/factsheets/cancers/10_8_9-Colorectum-fact-sheet.pdf (accessed on 20 September 2022).
3. American Cancer Society Treating Colorectal Cancer. 2020. pp. 1–52. Available online: https://www.cancer.org/cancer/colon-rectal-cancer/treating/by-stage-colon.html (accessed on 20 September 2022).
4. Herrmann, J. Adverse cardiac effects of cancer therapies: Cardiotoxicity and arrhythmia. *Nat. Rev. Cardiol.* **2020**, *17*, 474–502. Available online: http://www.ncbi.nlm.nih.gov/pubmed/32231332 (accessed on 20 September 2022). [CrossRef] [PubMed]
5. Herrmann, J. Vascular toxic effects of cancer therapies. *Nat. Rev. Cardiol.* **2020**, *17*, 503–522. [CrossRef] [PubMed]
6. Alesci, A.; Nicosia, N.; Fumia, A.; Giorgianni, F.; Santini, A.; Cicero, N. Resveratrol and Immune Cells: A Link to Improve Human Health. *Molecules* **2022**, *27*, 424. [CrossRef]
7. Kazemirad, H.; Kazerani, H.R. Cardioprotective effects of resveratrol following myocardial ischemia and reperfusion. *Mol. Biol. Rep.* **2020**, *47*, 5843–5850. [CrossRef]
8. Miguel, C.A.; Noya-Riobó, M.V.; Mazzone, G.L.; Villar, M.J.; Coronel, M.F. Antioxidant, anti-inflammatory and neuroprotective actions of resveratrol after experimental nervous system insults. Special focus on the molecular mechanisms involved. *Neurochem. Int.* **2021**, *150*, 105188. [CrossRef]
9. Yang, M.-D.; Sun, Y.; Zhou, W.-J.; Xie, X.-Z.; Zhou, Q.-M.; Lu, Y.-Y.; Su, S.-B. Resveratrol Enhances Inhibition Effects of Cisplatin on Cell Migration and Invasion and Tumor Growth in Breast Cancer MDA-MB-231 Cell Models In Vivo and In Vitro. *Molecules* **2021**, *26*, 2204. [CrossRef]
10. Moutabian, H.; Majdaeen, M.; Ghahramani-Asl, R.; Yadollahi, M.; Gharepapagh, E.; Ataei, G.; Falahatpour, Z.; Bagheri, H.; Farhood, B. A systematic review of the therapeutic effects of resveratrol in combination with 5-fluorouracil during colorectal cancer treatment: With a special focus on the oxidant, apoptotic, and anti-inflammatory activities. *Cancer Cell Int.* **2022**, *22*, 142. [CrossRef]
11. He, L.; Fan, F.; Hou, X.; Gao, C.; Meng, L.; Meng, S.; Huang, S.; Wu, H. Resveratrol suppresses pulmonary tumor metastasis by inhibiting platelet-mediated angiogenic responses. *J. Surg. Res.* **2017**, *217*, 113–122. [CrossRef]

12. Fu, Y.; Ye, Y.; Zhu, G.; Xu, Y.; Sun, J.; Wu, H.; Feng, F.; Wen, Z.; Jiang, S.; Li, Y.; et al. Resveratrol induces human colorectal cancer cell apoptosis by activating the mitochondrial pathway via increasing reactive oxygen species. *Mol. Med. Rep.* **2020**, *23*, 170. [CrossRef]

13. Kumar, S.; Chang, Y.-C.; Lai, K.-H.; Hwang, T.-L. Resveratrol, a Molecule with Anti-Inflammatory and Anti-Cancer Activities: Natural Product to Chemical Synthesis. *Curr. Med. Chem.* **2021**, *28*, 3773–3786. [CrossRef]

14. National Center for Biotechnology Information PubChem Compound Summary for CID 1923, 8-Hydroxyquinoline. 2021. p. 54. Available online: https://pubchem.ncbi.nlm.nih.gov/compound/8-Hydroxyquinoline (accessed on 20 September 2022).

15. Balthazar, J.D.; Soosaimanickam, M.P.; Emmanuel, C.; Krishnaraj, T.; Sheikh, A.; Alghafis, S.F.; Ibrahim, H.-I.M. 8-Hydroxyquinoline a natural chelating agent from Streptomyces spp. inhibits A549 lung cancer cell lines via BCL2/STAT3 regulating pathways. *World J. Microbiol. Biotechnol.* **2022**, *38*, 182. [CrossRef] [PubMed]

16. Joaquim, A.R.; Boff, R.T.; Adam, F.C.; Lima-Morales, D.; Cesare, M.A.; Kaminski, T.F.; Teixeira, M.L.; Fuentefria, A.M.; Andrade, S.F.; Martins, A.F. Antibacterial and synergistic activity of a new 8-hydroxyquinoline derivative against methicillin-resistant Staphylococcus aureus. *Future Microbiol.* **2022**, *17*, 425–436. [CrossRef]

17. Reginatto, P.; Joaquim, A.R.; Rocha, D.A.; Berlitz, S.J.; Külkamp-Guerreiro, I.C.; De Andrade, S.F.; Fuentefria, A.M. 8-hydroxyquinoline and quinazoline derivatives as potential new alternatives to combat Candida spp. Biofilm. *Lett. Appl. Microbiol.* **2022**, *74*, 395–404. [CrossRef]

18. Chen, C.; Yang, X.; Fang, H.; Hou, X. Design, synthesis and preliminary bioactivity evaluations of 8-hydroxyquinoline derivatives as matrix metalloproteinase (MMP) inhibitors. *Eur. J. Med. Chem.* **2019**, *181*, 111563. [CrossRef]

19. Wang, L.; Deng, K.; Gong, L.; Zhou, L.; Sayed, S.; Li, H.; Sun, Q.; Su, Z.; Wang, Z.; Liu, S.; et al. Chlorquinaldol targets the β-catenin and T-cell factor 4 complex and exerts anti-colorectal cancer activity. *Pharmacol. Res.* **2020**, *159*, 104955. [CrossRef]

20. Chan, L.-P.; Tseng, Y.-P.; Ding, H.-Y.; Pan, S.-M.; Chiang, F.-Y.; Wang, L.-F.; Chou, T.-H.; Lien, P.-J.; Liu, C.; Kuo, P.-L.; et al. Tris(8-Hydroxyquinoline)iron induces apoptotic cell death via oxidative stress and by activating death receptor signaling pathway in human head and neck carcinoma cells. *Phytomedicine* **2019**, *63*, 153005. [CrossRef]

21. Stefanachi, A.; Leonetti, F.; Pisani, L.; Catto, M.; Carotti, A. Coumarin: A Natural, Privileged and Versatile Scaffold for Bioactive Compounds. *Molecules* **2018**, *23*, 250. [CrossRef]

22. Nepovimova, E.; Kuca, K. Multi-Target-Directed Ligands in Alzheimer's Disease Therapy. In *Neurodegenerative Diseases—Molecular Mechanisms and Current Therapeutic Approaches*; IntechOpen: London, UK, 2021. Available online: https://www.intechopen.com/books/neurodegenerative-diseases-molecular-mechanisms-and-current-therapeutic-approaches/multi-target-directed-ligands-in-alzheimer-s-disease-therapy (accessed on 20 September 2022).

23. Szemik-Hojniak, A.; Deperasińska, I.; Nizhnik, Y.P. Photophysical behavior of a potential drug candidate, trans-[2-(4-methoxystyryl)]quinoline-1-oxide tuned by environment effects. *Spectrochim. Acta. A. Mol. Biomol. Spectrosc.* **2017**, *187*, 198–206. [CrossRef]

24. Espinosa, R.; Robledo, S.; Guzmán, C.; Arbeláez, N.; Yepes, L.; Santafé, G.; Sáez, A. Synthesis and evaluation of the in vitro and in vivo antitrypanosomal activity of 2-styrylquinolines. *Heliyon* **2021**, *7*, e07024. [CrossRef]

25. Mrozek-Wilczkiewicz, A.; Spaczynska, E.; Malarz, K.; Cieslik, W.; Rams-Baron, M.; Kryštof, V.; Musiol, R. Design, Synthesis and In Vitro Activity of Anticancer Styrylquinolines. The p53 Independent Mechanism of Action. *PLoS ONE* **2015**, *10*, e0142678. [CrossRef]

26. Mirzaei, S.; Eisvand, F.; Hadizadeh, F.; Mosaffa, F.; Ghasemi, A.; Ghodsi, R. Design, synthesis and biological evaluation of novel 5,6,7-trimethoxy-N-aryl-2-styrylquinolin-4-amines as potential anticancer agents and tubulin polymerization inhibitors. *Bioorg. Chem.* **2020**, *98*, 103711. [CrossRef] [PubMed]

27. Mrozek-Wilczkiewicz, A.; Kuczak, M.; Malarz, K.; Cieślik, W.; Spaczyńska, E.; Musiol, R. The synthesis and anticancer activity of 2-styrylquinoline derivatives. A p53 independent mechanism of action. *Eur. J. Med. Chem.* **2019**, *177*, 338–349. [CrossRef]

28. Herrera-R, A.; Castrillón, W.; Otero, E.; Ruiz, E.; Carda, M.; Agut, R.; Naranjo, T.; Moreno, G.; Maldonado, M.E.; Cardona-G, W. Synthesis and antiproliferative activity of 3- and 7-styrylcoumarins. *Med. Chem. Res.* **2018**, *27*, 1893–1905. [CrossRef]

29. Puerta, J.D.; Usme-Ciro, J.A.; Gallego-Gómez, J.C. Implementación de un control interno en la detección molecular de las principales especies de micoplasmas contaminantes de cultivos celulares. *Salud Uninorte* **2013**, *29*, 160–173.

30. Hernández, C.; Moreno, G.; Herrera-R, A.; Cardona-G, W. New Hybrids Based on Curcumin and Resveratrol: Synthesis, Cytotoxicity and Antiproliferative Activity against Colorectal Cancer Cells. *Molecules* **2021**, *26*, 2661. [CrossRef]

31. Kim, H.; Xue, X. Detection of Total Reactive Oxygen Species in Adherent Cells by 2′,7′-Dichlorodihydrofluorescein Diacetate Staining. *J. Vis. Exp.* **2020**, e60682. [CrossRef] [PubMed]

32. Eberhardt, J.; Santos-Martins, D.; Tillack, A.F.; Forli, S. AutoDock Vina 1.2.0: New Docking Methods, Expanded Force Field, and Python Bindings. *J. Chem. Inf. Model.* **2021**, *61*, 3891–3898. [CrossRef]

33. Podeszwa, B.; Niedbala, H.; Polanski, J.; Musiol, R.; Tabak, D.; Finster, J.; Serafin, K.; Milczarek, M.; Wietrzyk, J.; Boryczka, S.; et al. Investigating the antiproliferative activity of quinoline-5,8-diones and styrylquinolinecarboxylic acids on tumor cell lines. *Bioorg. Med. Chem. Lett.* **2007**, *17*, 6138–6141. [CrossRef]

34. Cieslik, W.; Musiol, R.; Nycz, J.E.; Jampilek, J.; Vejsova, M.; Wolff, M.; Machura, B.; Polanski, J. Contribution to investigation of antimicrobial activity of styrylquinolines. *Bioorg. Med. Chem.* **2012**, *20*, 6960–6968. [CrossRef] [PubMed]

35. Chang, F.-S.; Chen, W.; Wang, C.; Tzeng, C.-C.; Chen, Y.-L. Synthesis and antiproliferative evaluations of certain 2-phenylvinylquinoline (2-styrylquinoline) and 2-furanylvinylquinoline derivatives. *Bioorg. Med. Chem.* **2010**, *18*, 124–133. [CrossRef] [PubMed]

36. PHILLIPS, J.P.; BREESE, R.; BARRALL, E.M. Styryl Derivatives of 8-Quinolinol. *J. Org. Chem.* **1959**, *24*, 1104–1106. [CrossRef]

37. Ouali, M.; Laboulais, C.; Leh, H.; Gill, D.; Desmaële, D.; Mekouar, K.; Zouhiri, F.; D'Angelo, J.; Auclair, C.; Mouscadet, J.F.; et al. Modeling of the inhibition of retroviral integrases by styrylquinoline derivatives. *J. Med. Chem.* **2000**, *43*, 1949–1957. [CrossRef] [PubMed]

38. Otero, E.; García, E.; Palacios, G.; Yepes, L.M.; Carda, M.; Agut, R.; Vélez, I.D.; Cardona, W.I.; Robledo, S.M. Triclosan-caffeic acid hybrids: Synthesis, leishmanicidal, trypanocidal and cytotoxic activities. *Eur. J. Med. Chem.* **2017**, *141*, 73–83. [CrossRef]

39. Otero, E.; Robledo, S.M.; Díaz, S.; Carda, M.; Muñoz, D.; Paños, J.; Vélez, I.D.; Cardona, W. Synthesis and leishmanicidal activity of cinnamic acid esters: Structure–activity relationship. *Med. Chem. Res.* **2014**, *23*, 1378–1386. [CrossRef]

40. Maciag, J.J.; Mackenzie, S.H.; Tucker, M.B.; Schipper, J.L.; Swartz, P.; Clark, A.C. Tunable allosteric library of caspase-3 identifies coupling between conserved water molecules and conformational selection. *Proc. Natl. Acad. Sci. USA* **2016**, *113*, E6080–E6088. [CrossRef] [PubMed]

41. Yang, Y.; Han, J.; Ma, Y.; Zhang, J.; Zhang, Z.; Wang, G. Demethylzeylasteral inhibits cell proliferation and enhances cell chemosensitivity to 5-fluorouracil in Colorectal Cancer cells. *J. Cancer* **2020**, *11*, 6059–6069. [CrossRef] [PubMed]

42. Arul, M.; Roslani, A.C.; Cheah, S.H. Heterogeneity in cancer cells: Variation in drug response in different primary and secondary colorectal cancer cell lines in vitro. *Vitr. Cell. Dev. Biol.-Anim.* **2017**, *53*, 435–447. [CrossRef]

43. Yan, W.; Yang, W.; Liu, Z.; Wu, G. Characterization of microRNA expression in primary human colon adenocarcinoma cells (SW480) and their lymph node metastatic derivatives (SW620). *Onco. Targets. Ther.* **2018**, *11*, 4701–4709. [CrossRef]

44. Ge, T.; Zhang, Y. Tanshinone IIA reverses oxaliplatin resistance in colorectal cancer through microRNA-30b-5p/AVEN axis. *Open Med.* **2022**, *17*, 1228–1240. [CrossRef]

45. Zou, Y.; Liu, L.; Meng, J.; Dai, M. Circular RNA circ_0068464 combined with microRNA-383 regulates Wnt/β-catenin pathway to promote the progression of colorectal cancer. *Bioengineered* **2022**, *13*, 5113–5125. [CrossRef] [PubMed]

46. Huang, X.; Xu, X.; Ke, H.; Pan, X.; Ai, J.; Xie, R.; Lan, G.; Hu, Y.; Wu, Y. microRNA-16-5p suppresses cell proliferation and angiogenesis in colorectal cancer by negatively regulating forkhead box K1 to block the PI3K/Akt/mTOR pathway. *Eur. J. Histochem.* **2022**, *66*. [CrossRef] [PubMed]

47. Gao, G.; Li, C.; Fan, W.; Zhang, M.; Li, X.; Chen, W.; Li, W.; Liang, R.; Li, Z.; Zhu, X. Brilliant glycans and glycosylation: Seq and ye shall find. *Int. J. Biol. Macromol.* **2021**, *189*, 279–291. [CrossRef] [PubMed]

48. Shen, L.; Dong, X.; Wang, Y.; Qiu, L.; Peng, F.; Luo, Z. β3GnT8 regulates oxaliplatin resistance by altering integrinα5β1 glycosylation in colon cancer cells. *Oncol. Rep.* **2018**, *39*, 2006–2014. [CrossRef] [PubMed]

49. Gao, T.; Wen, T.; Ge, Y.; Liu, J.; Yang, L.; Jiang, Y.; Dong, X.; Liu, H.; Yao, J.; An, G. Disruption of Core 1-mediated O-glycosylation oppositely regulates CD44 expression in human colon cancer cells and tumor-derived exosomes. *Biochem. Biophys. Res. Commun.* **2020**, *521*, 514–520. [CrossRef]

50. Zhang, B.; Pan, C.; Feng, C.; Yan, C.; Yu, Y.; Chen, Z.; Guo, C.; Wang, X. Role of mitochondrial reactive oxygen species in homeostasis regulation. *Redox Rep.* **2022**, *27*, 45–52. [CrossRef] [PubMed]

51. Ginckels, P.; Holvoet, P. Oxidative Stress and Inflammation in Cardiovascular Diseases and Cancer: Role of Non-coding RNAs. *Yale J. Biol. Med.* **2022**, *95*, 129–152.

52. Mihanfar, A.; Yousefi, B.; Ghazizadeh Darband, S.; Sadighparvar, S.; Kaviani, M.; Majidinia, M. Melatonin increases 5-flurouracil-mediated apoptosis of colorectal cancer cells through enhancing oxidative stress and downregulating survivin and XIAP. *Bioimpacts* **2021**, *11*, 253–261. [CrossRef]

53. Zhang, P.; Yuan, X.; Yu, T.; Huang, H.; Yang, C.; Zhang, L.; Yang, S.; Luo, X.; Luo, J. Lycorine inhibits cell proliferation, migration and invasion, and primarily exerts in vitro cytostatic effects in human colorectal cancer via activating the ROS/p38 and AKT signaling pathways. *Oncol. Rep.* **2021**, *45*, 19. [CrossRef]

54. Malarz, K.; Mrozek-Wilczkiewicz, A.; Serda, M.; Rejmund, M.; Polanski, J.; Musiol, R. The role of oxidative stress in activity of anticancer thiosemicarbazones. *Oncotarget* **2018**, *9*, 17689–17710. [CrossRef]

55. Chen, Y.; Fan, Z.; Hu, S.; Lu, C.; Xiang, Y.; Liao, S. Ferroptosis: A New Strategy for Cancer Therapy. *Front. Oncol.* **2022**, *12*, 830561. [CrossRef]

56. Zhang, L.; Jia, R.; Li, H.; Yu, H.; Ren, K.; Jia, S.; Li, Y.; Wang, Q. Insight into the Double-Edged Role of Ferroptosis in Disease. *Biomolecules* **2021**, *11*, 1790. [CrossRef] [PubMed]

57. Florean, C.; Song, S.; Dicato, M.; Diederich, M. Redox biology of regulated cell death in cancer: A focus on necroptosis and ferroptosis. *Free Radic. Biol. Med.* **2019**, *134*, 177–189. [CrossRef] [PubMed]

58. Lee, S.-H.; Lee, Y.-J. Synergistic anticancer activity of resveratrol in combination with docetaxel in prostate carcinoma cells. *Nutr. Res. Pract.* **2021**, *15*, 12–25. [CrossRef] [PubMed]

59. Lee, J.; You, J.H.; Kim, M.-S.; Roh, J.-L. Epigenetic reprogramming of epithelial-mesenchymal transition promotes ferroptosis of head and neck cancer. *Redox Biol.* **2020**, *37*, 101697. [CrossRef]

60. Roberts, J.Z.; Crawford, N.; Longley, D.B. The role of Ubiquitination in Apoptosis and Necroptosis. *Cell Death Differ.* **2022**, *29*, 272–284. [CrossRef] [PubMed]

61. Yadav, P.; Yadav, R.; Jain, S.; Vaidya, A. Caspase-3: A primary target for natural and synthetic compounds for cancer therapy. *Chem. Biol. Drug Des.* **2021**, *98*, 144–165. [CrossRef] [PubMed]

62. Zhenqwen, A.; Flores-Borja, F.; Irshad, S.; Deng, J.; Ng, T. Pleiotropic Role and Bidirectional Immunomodulation of Innate Lymphoid Cells in Cancer. *Front. Immunol.* **2020**, *10*, 3111. [CrossRef]

63. Choudhry, H.; Helmi, N.; Abdulaal, W.H.; Zeyadi, M.; Zamzami, M.A.; Wu, W.; Mahmoud, M.M.; Warsi, M.K.; Rasool, M.; Jamal, M.S. Prospects of IL-2 in Cancer Immunotherapy. *Biomed Res. Int.* **2018**, *2018*, 9056173. [CrossRef]

64. Ding, M.; Fei, Y.; Zhu, J.; Ma, J.; Zhu, G.; Zhen, N.; Zhu, J.; Mao, S.; Sun, F.; Wang, F.; et al. IL-27 improves adoptive CD8 + T cells' antitumor activity via enhancing cell survival and memory T cell differentiation. *Cancer Sci.* **2022**, *113*, 2258–2271. [CrossRef]

65. Wang, J.; Sun, M.; Zhao, H.; Huang, Y.; Li, D.; Mao, D.; Zhang, Z.; Zhu, X.; Dong, X.; Zhao, X. IL-9 Exerts Antitumor Effects in Colon Cancer and Transforms the Tumor Microenvironment In Vivo. *Technol. Cancer Res. Treat.* **2019**, *18*, 1533033819857737. [CrossRef]

66. Xiao, L.; Ma, X.; Ye, L.; Su, P.; Xiong, W.; Bi, E.; Wang, Q.; Xian, M.; Yang, M.; Qian, J.; et al. IL-9/STAT3/fatty acid oxidation–mediated lipid peroxidation contributes to Tc9 cell longevity and enhanced antitumor activity. *J. Clin. Investig.* **2022**, *132*. [CrossRef]

67. Droeser, R.A.; Lezzi, G. IL-22-mediates Cross-talk between Tumor Cells and Immune Cells Associated with Favorable Prognosis in Human Colorectal Cancer. *J. Cell. Immunol.* **2021**, *3*, 118. [CrossRef] [PubMed]

68. Xue, G.; Jin, G.; Fang, J.; Lu, Y. IL-4 together with IL-1β induces antitumor Th9 cell differentiation in the absence of TGF-β signaling. *Nat. Commun.* **2019**, *10*, 1376. [CrossRef]

69. Loyon, R.; Jary, M.; Salomé, B.; Gomez-Cadena, A.; Galaine, J.; Kroemer, M.; Romero, P.; Trabanelli, S.; Adotévi, O.; Borg, C.; et al. Peripheral Innate Lymphoid Cells Are Increased in First Line Metastatic Colorectal Carcinoma Patients: A Negative Correlation With Th1 Immune Responses. *Front. Immunol.* **2019**, *10*, 2121. [CrossRef] [PubMed]

70. Chen, C.-L.; Wang, Y.; Huang, C.-Y.; Zhou, Z.-Q.; Zhao, J.-J.; Zhang, X.-F.; Pan, Q.-Z.; Wu, J.-X.; Weng, D.-S.; Tang, Y.; et al. IL-17 induces antitumor immunity by promoting beneficial neutrophil recruitment and activation in esophageal squamous cell carcinoma. *Oncoimmunology* **2018**, *7*, e1373234. [CrossRef]

71. Knochelmann, H.M.; Dwyer, C.J.; Smith, A.S.; Bowers, J.S.; Wyatt, M.M.; Nelson, M.H.; Rangel Rivera, G.O.; Horton, J.D.; Krieg, C.; Armeson, K.; et al. IL6 Fuels Durable Memory for Th17 Cell–Mediated Responses to Tumors. *Cancer Res.* **2020**, *80*, 3920–3932. [CrossRef] [PubMed]

72. Chonov, D.C.; Ignatova, M.M.K.; Ananiev, J.R.; Gulubova, M.V. IL-6 Activities in the Tumour Microenvironment. Part 1. *Open Access Maced. J. Med. Sci.* **2019**, *7*, 2391–2398. [CrossRef]

73. Ren, J.; Li, N.; Pei, S.; Lian, Y.; Li, L.; Peng, Y.; Liu, Q.; Guo, J.; Wang, X.; Han, Y.; et al. Histone methyltransferase WHSC1 loss dampens MHC-I antigen presentation pathway to impair IFN-γ-stimulated anti-tumor immunity. *J. Clin. Investig.* **2022**, *132*, e153167. [CrossRef]

74. Yan, W.-L.; Wu, C.-C.; Shen, K.-Y.; Liu, S.-J. Activation of GM-CSF and TLR2 signaling synergistically enhances antigen-specific antitumor immunity and modulates the tumor microenvironment. *J. Immunother. Cancer* **2021**, *9*, e002758. [CrossRef]

75. Feng, X.; Zhang, Z.; Sun, P.; Song, G.; Wang, L.; Sun, Z.; Yuan, N.; Wang, Q.; Lun, L. Interleukin-18 Is a Prognostic Marker and Plays a Tumor Suppressive Role in Colon Cancer. *Dis. Markers* **2020**, *2020*, 6439614. [CrossRef]

Article

Evaluation of the Effects of Genistein In Vitro as a Chemopreventive Agent for Colorectal Cancer—Strategy to Improve Its Efficiency When Administered Orally

Juan Pablo Rendón [1],*, Ana Isabel Cañas [2], Elizabeth Correa [1], Vanesa Bedoya-Betancur [1], Marlon Osorio [2], Cristina Castro [2] and Tonny W. Naranjo [1,3],*

[1] Medical and Experimental Mycology Group, CIB-UPB-UdeA-UDES, Corporación para Investigaciones Biológicas, Carrera 72 A # 78B-141, Medellin 050034, Colombia
[2] School of Engineering, Universidad Pontificia Bolivariana, Circular 1 # 70-01, Medellin 050031, Colombia
[3] School of Health Sciences, Universidad Pontificia Bolivariana, Calle 78 B # 72 A-109, Medellin 050034, Colombia
*   Correspondence: juan.rendonm@udea.edu.co (J.P.R.); tonny.naranjo@upb.edu.co (T.W.N.)

Abstract: Colorectal Cancer (CRC) ranks third in terms of incidence and second in terms of mortality and prevalence worldwide. In relation to chemotherapy treatment, the most used drug is 5-fluorouracil (5-FU); however, the use of this drug generates various toxic effects at the systemic level. For this reason, new therapeutic strategies are currently being sought that can be used as neoadjuvant or adjuvant treatments. Recent research has shown that natural compounds, such as genistein, have chemotherapeutic and anticancer effects, but the mechanisms of action of genistein and its molecular targets in human colon cells have not been fully elucidated. The results reported in relation to non-malignant cell lines are also unclear, which does not allow evidence of the selectivity that this compound may have. Therefore, in this work, genistein was evaluated in vitro in both cancer cell lines SW480 and SW620 and in the non-malignant cell line HaCaT. The results obtained show that genistein has selectivity for the SW480 and SW620 cell lines. In addition, it inhibits cell viability and has an antiproliferative effect in a dose-dependent manner. Increased production of reactive oxygen species (ROS) was also found, suggesting an association with the cell death process through various mechanisms. Finally, the encapsulation strategy that was proposed made it possible to demonstrate that bacterial nanocellulose (BNC) is capable of protecting genistein from the acidic conditions of gastric fluid and also allows the release of the compound in the colonic fluid. This would allow genistein to act locally in the mucosa of the colon where the first stages of CRC occur.

Keywords: genistein; encapsulation; colon cancer; chemoprevention

Citation: Rendón, J.P.; Cañas, A.I.; Correa, E.; Bedoya-Betancur, V.; Osorio, M.; Castro, C.; Naranjo, T.W. Evaluation of the Effects of Genistein In Vitro as a Chemopreventive Agent for Colorectal Cancer—Strategy to Improve Its Efficiency When Administered Orally. *Molecules* **2022**, *27*, 7042. https://doi.org/10.3390/molecules27207042

Academic Editor: Alejandro Baeza

Received: 21 September 2022
Accepted: 14 October 2022
Published: 19 October 2022

Publisher's Note: MDPI stays neutral with regard to jurisdictional claims in published maps and institutional affiliations.

1. Introduction

According to the Global Cancer Observatory (GCO), in 2020, colorectal cancer (CRC) ranked as the type of cancer with the fourth highest incidence worldwide (approximately 19.5%). In women, an incidence rate of 23.4% is estimated and in men a rate of 16.2%. This type of cancer represents the third most common cause of death and is the second most prevalent for men and women of all ages worldwide [1]. Currently, examinations through colonoscopy have successfully detected early CRC, which makes it possible to search for different therapies to treat this pathology; however, 25% of patients are diagnosed with metastatic disease. Nonetheless, thanks to advances in medicine, the survival of these patients can be improved to more than two years with the combination of chemotherapy and biological agents [2].

In the clinical field, most cases of CRC are diagnosed as a terminal chronic condition or in a metastatic state due to the non-obvious development of the disease and symptoms during its initial phase. Additionally, the inevitable drug tolerance and side effects associated with chemotherapy make it more difficult to treat CRC effectively. Consequently, any candidate component with potential application for the treatment of metastatic CRC should be explored in depth [3]. In general, the choice for the treatment of CRC depends on several factors such as the clinical and health conditions of the patient, the size of the tumor, its location, and the presence of metastases. However, surgery remains the most common treatment option, especially for localized lesions [4]. Chemoradiation is sometimes required for locally advanced rectal cancer after surgical removal. Immunotherapy is also an option for metastatic CRCs that are microsatellite unstable [5]. When surgery is not necessary, treatment may include radiofrequency ablation, cryosurgery, chemotherapy, radiation therapy, or targeted therapy. For these cases, chemotherapy treatment is the most common and consists of the use of drugs that hinder tumor growth through the destruction of cancer cells, but the toxicity of chemotherapy increases with the age of the patients [6]. The drug most used in the chemotherapy of various types of cancer is 5-fluorouracil (5-FU), and it has been used with great success in CRC [7].

5-FU is an analog of uracil with a fluorine atom at the C-5 position instead of hydrogen, allowing it to rapidly enter the cell using the same facilitated transport mechanism as uracil. The mechanism of 5-FU cytotoxicity has been attributed to the misincorporation of fluorinated nucleotides into RNA and DNA and the inhibition of the nucleotide-synthesizing enzyme thymidylate synthase (TS) [7]. This drug can be given by continuous pump, 48 h infusion, weekly injections, or daily injections [2]. More than 80% of administered 5-FU is primarily catabolized in the liver and has shown toxic effects such as myelosuppression and other adverse gastrointestinal, hematologic, neural, and dermatologic side effects. Therefore, new therapeutic strategies are currently being sought to treat CRC that have fewer toxic effects at the systemic level [8].

Recent research has shown that various natural compounds have chemotherapeutic and anticancer effects; these investigations focus on the relationship of these effects with the specific biological targets associated with cancer on which these compounds act [9]. Among these compounds are flavonoids, which are a class of natural polyphenolic compounds present in vegetables, fruits, and soybeans. These compounds have been studied extensively in recent years in an attempt to understand the specific proteins on which they act to exert their anticancer functions. These functions and mechanisms have been evaluated in both in vitro and in vivo studies showing that flavonoids suppress carcinogenesis in various models of cancer cells, acting on multiple pathways involved in cell metabolism, apoptosis, adhesion, migration, and angiogenesis, as well as the immune response [10].

Genistein is a biologically active flavonoid found in high amounts in soybeans. Many studies have described a relationship between a soy-rich diet and cancer prevention, further demonstrating the pharmacological effects of genistein, including antiestrogenic action, antioxidant action, inhibition of angiogenesis, and anticancer activity against breast and ovarian cancer tumor cells. Therefore, it is considered as a promising chemopreventive agent in the treatment of cancer [3,11]. In this sense, chemoprevention is defined as a blockage, delay or reversal of a carcinogenic process through chemical and/or natural agents. Clinically, chemoprevention is classified as primary, secondary, or tertiary. Primary chemoprevention applies to the general population and to those who may be at risk of developing the disease. Secondary chemoprevention applies to patients with premalignant lesions that may progress to invasive disease. Tertiary chemoprevention is aimed at preventing disease recurrence in those who have already undergone potentially curative therapy. At the molecular level, cancer chemoprevention is characterized by the interruption, or at least the delay, of multiple pathways and processes in any of the three stages of carcinogenesis: initiation, promotion, and progression [12].

In CRC, several in vitro studies have shown that genistein exhibits growth-inhibitory activity and promotes apoptosis in a dose-dependent manner. Genistein causes cell cycle arrest in colon cancer cell lines HCT-116 and SW480, mainly participating in cell cycle regulation and apoptosis [13]. In other studies with the HT-29 cell line, it was observed that genistein inhibits EGF-induced proliferation, reverses the Epithelial–Mesenchymal Transition (EMT) and promotes the activation of apoptosis via caspase 3 [14,15]. Recently, genistein was shown to be able to inhibit cell invasion and migration of colon cancer cells by altering the expression of migration-associated factors and genes, including MMP9, MMP2, TIMP1, E-cadherin, β-catenin, c-Myc, and cyclin D1 [16]. Finally, other authors demonstrated that the treatment of HCT-116 cells with genistein causes inhibition of cell proliferation and induces apoptosis [17].

The following information demonstrates the ability of genistein to inhibit the proliferation and migration of tumor cells due to the inhibition of the activity of several molecular targets, making it a promising natural compound for the chemoprevention and treatment of CRC. However, the effect of this compound on non-malignant cell lines that show the selectivity of this compound has not yet been reported. By definition, an ideal compound should have a relatively high toxic concentration but a very low active concentration. Under this premise, the compound would affect cancer cells but should not affect non-malignant ones [18–20]. To date, there have been no reports of the evaluation of genistein on two cell lines derived from the same patient but with different stages of evolution of colorectal adenocarcinoma, and its role in other mechanisms of action such as necroptosis or its ability to regulate the expression of immunological markers in cells. Therefore, in this work, the effect of genistein on cancer cell lines and a non-malignant cell line was evaluated by means of in vitro tests. This would allow us not only to determine cell viability, but also to obtain the calculation of the selectivity index. Likewise, the effect that this natural compound exerts on antiproliferative and apoptotic activity, the production of reactive oxygen species (ROS) and the expression of immunological markers in these cell lines was evaluated as an approach strategy towards the possible mechanism of action involved.

Additionally, in this research, the encapsulation of genistein is proposed as a strategy to improve pharmacokinetic activity, achieving a better local effect at the target site of this compound when administered orally. This is because, despite the beneficial properties of genistein, the use of this compound in vivo is limited due to its low water solubility, rapid biotransformation to inactive metabolites, poor accumulation in target tissues and cells, and low concentration in the blood after oral administration [21,22].

2. Results

2.1. Encapsulation of Genistein

Figure 1 shows the empty BNC capsules (Figure 1a,b) and the BNC capsules loaded with genistein (Figure 1c,d), prepared by the spray-drying method. Empty BNC capsules have a particle size distribution between 1 and 5 µm, while BNC/GEN capsules have a particle size distribution between 1 and 6 µm, but in both cases, the particles are most often around 3 µm. In Figure 1b,d, it can be seen that the surfaces of the capsules are composed of a network of collapsed nanofibers in response to the evaporation of water and the establishment of irreversible hydrogen bonds. The ratio of genistein and BNC in the capsules was 5.52 mg GEN/1 g BNC and a spray drying yield of 52% was obtained.

2.2. In Vitro Release Assays in Gastrointestinal Fluids

To evaluate the maximum desorption capacity of genistein when it is encapsulated in BNC, a release of the compound was performed in simulated physiological fluids of the stomach, small intestine, and colon. In the release profiles found in Figure 2, it was observed that 8.7% of the compound is released in gastric fluid at 2 h, while 44.6% is released in small intestine fluid at 24 h. Finally, a genistein release of 92.5% in the colonic fluid was observed at 48 h.

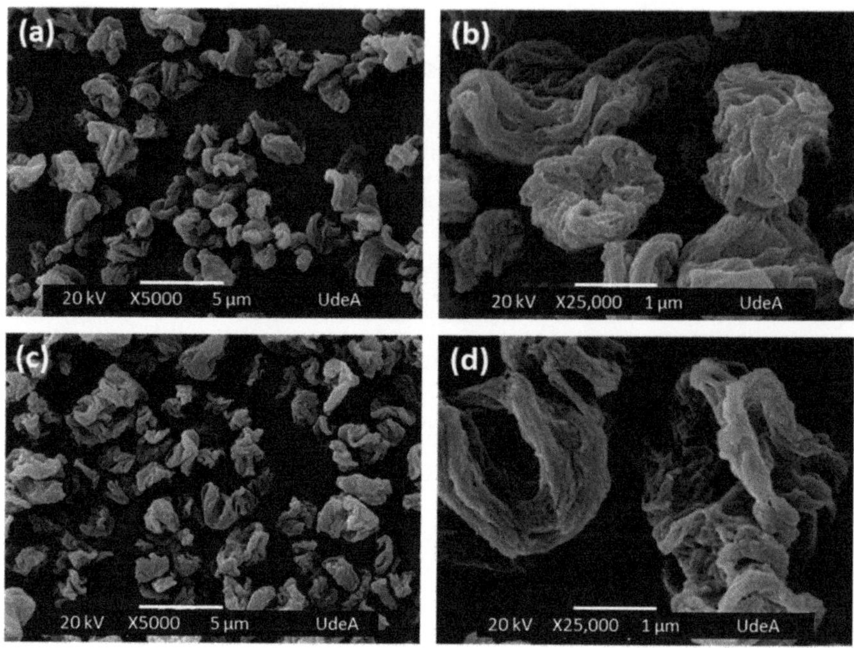

Figure 1. SEM images of spray-dried capsules. (**a**,**b**) BNC capsules and (**c**,**d**) BNC/GEN capsules.

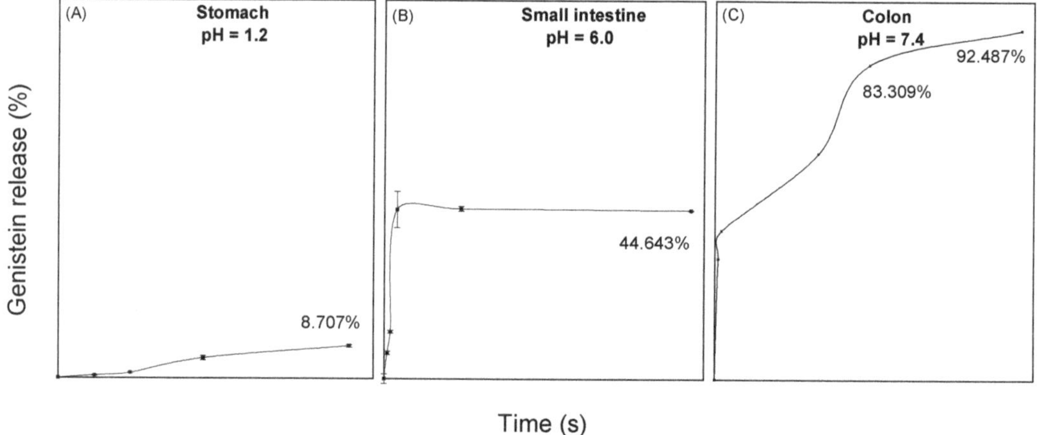

Figure 2. Release curves of genistein encapsulated in BNC at different pH and at 37 °C. (**A**) Simulated stomach fluid at pH 1.2; (**B**) simulated small intestine fluid at pH 6, and (**C**) simulated colon fluid at pH 7.4.

For release into the stomach and intestine fluid, the experimental data had a better fit to a pseudo-first-order kinetic model with $R^2 = 0.968$ and $R^2 = 0.888$, respectively. In physiological stomach fluid, a gradual release of genistein is observed with a release rate of $0.001\ \text{min}^{-1}$, and for the small intestine, a release rate of $0.021\ \text{min}^{-1}$ was obtained, which stabilizes after 60 min. This is due to the protection generated by bacterial nanocellulose to the compound. On the other hand, the experimental data of the release in colonic fluid fit with $R^2 = 0.927$ to a pseudo-second-order model. The release of genistein in this medium was much greater and in a controlled and prolonged manner.

2.3. Effect of Free Genistein on Cell Viability

Tables 1 and 2 show the results obtained in the cell viability assay. The inhibitory concentration 50 (IC_{50}) was determined by dose–response curves and the selectivity index by the formula IS = IC_{50} Control cells (HaCat)/IC_{50} Tumor cells (SW480 and SW620). A selectivity index greater than 1 indicates that the treatment is more cytotoxic to tumor cells than to control cells. The results show that genistein has selectivity in the SW480 and SW620 cell lines at 24 h and 48 h of treatment against the non-malignant HaCat cell line. Figure 3 shows that cell viability was inhibited in a dose-dependent manner in the three cell lines evaluated compared to the growth control.

Table 1. Cytotoxic effect of genistein on HaCaT and SW480 cells.

Compound	24 h			48 h		
	IC50 (μM) HaCat	IC50 (μM) SW480	SI	IC50 (μM) HaCat	IC50 (μM) SW480	SI
Genistein	>740	280.93 ± 28.56	3.17	225.01 ± 7.06	128.80 ± 15.87	1.75
5-FU	>962.10	>962.10	2	18.5 ± 3.44	355.23 ± 75.85	0.04

The Inhibitory Concentration 50 (IC_{50}) was found using dose–response curves. The selectivity index (SI) was calculated using the formula SI = IC_{50} Non-malignant cells (HaCat)/IC_{50} Tumor cells (SW480).

Table 2. Cytotoxic effect of genistein in HaCaT and SW620 cells.

Compound	24 h			48 h		
	IC50 (μM) HaCat	IC50 (μM) SW620	SI	IC50 (μM) HaCat	IC50 (μM) SW620	SI
Genistein	>740	667.17 ± 33.52	1.34	225.01 ± 7.06	152.60 ± 19.13	1.47
5-FU	>962.10	>962.10	0.9	18.5 ± 3.44	139.57 ± 19.09	0.1

The Inhibitory Concentration 50 (IC_{50}) was found using dose–response curves. The selectivity index (SI) was calculated using the formula SI = IC_{50} Non-malignant cells (HaCat)/IC_{50} Tumor cells (SW620).

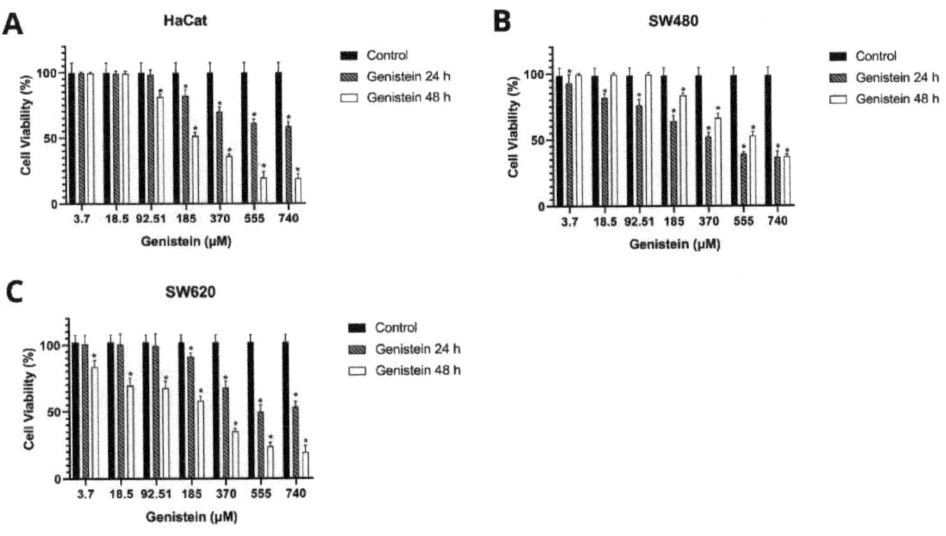

Figure 3. Effect of genistein on cell viability after 24 h and 48 h of treatment compared to the control (**A**) non-malignant HaCaT cell line, (**B**) SW480 colorectal adenocarcinoma cell line, and (**C**) SW620 colorectal adenocarcinoma cell line. All tests were performed in quintuplicate. *p*-value < 0.05 *.

Based on the IC50 value of free genistein, cytotoxicity assays of empty BNC capsules and BNC/GEN were performed. Figure 4 shows that the empty NCB capsules do not have a cytotoxic effect on any of the cell lines evaluated, since they showed a percentage of cell viability greater than 96% both at 24 and 48 h of incubation. Comparing the evaluated concentration of the IC50 of free and encapsulated genistein, it can be seen that there is a greater effect on the decrease in cell viability of BNC/GEN both at 24 and 48 h. This is associated with the protection offered by the BNC to the compound from the pH of the medium, which allows a controlled and prolonged release of it over time.

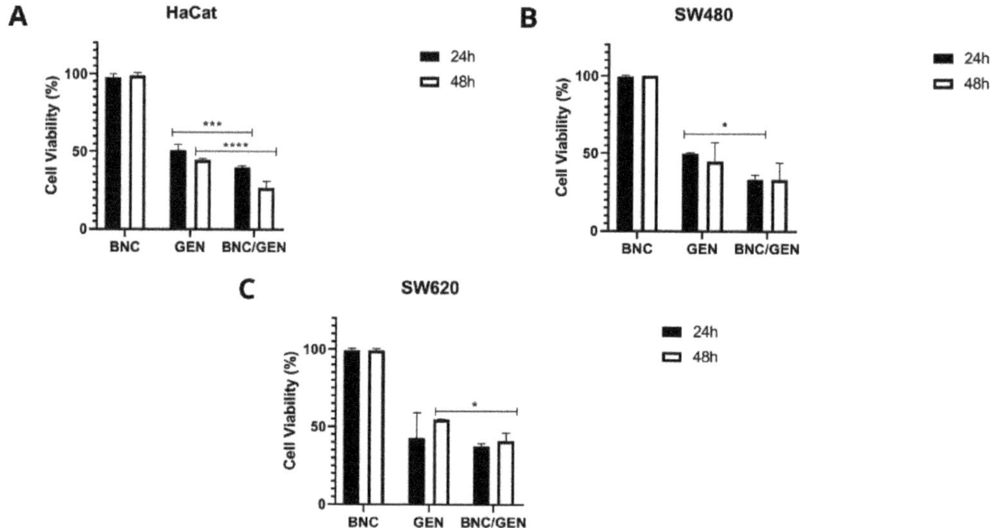

Figure 4. Cell viability of empty BNC capsules and BNC/GEN capsules, taking into account the IC_{50} value of free genistein, after 24 and 48 h (**A**) non-malignant HaCaT cell line, (**B**) SW480 colorectal adenocarcinoma cell line, and (**C**) SW620 colorectal adenocarcinoma cell line. p-value < 0.05 *; $p < 0.001$ ***; and $p < 0.0001$ ****.

2.4. Antiproliferative Effect of Genistein

In Figure 5, it can be seen that genistein showed an antiproliferative effect throughout the evaluated times in a dose-dependent manner for both cell lines SW480 and SW620. In SW480 cells, genistein induced an inhibitory effect on viability from day 2 of treatment with high concentrations of the compound (37, 185, and 370 µM), while for days 4 and 6, this effect was observed at low concentrations (from 18.5 µM). Similar results were observed in the metastatic cell line (SW620), in which genistein was found to have a significant effect on cell viability on days 2 and 4 at all concentrations tested (3.7–379 µM). While for day 6, concentrations higher than 18.5 µM were required and for day 8, concentrations higher than 37 µM, concluding that the antiproliferative effect of genistein is conditioned by the dose and treatment time in these cell models. These results again demonstrate that there is a greater effect of genistein when the dose and exposure time to the compound are increased.

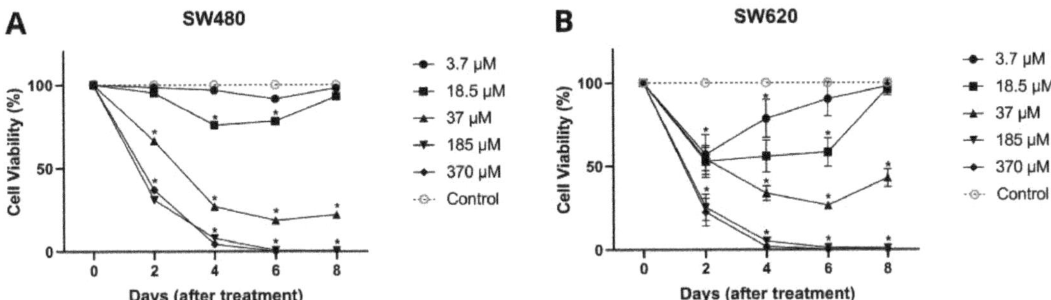

Figure 5. Antiproliferative effect of genistein. (A) Colorectal adenocarcinoma cell line SW480. (B) Colorectal adenocarcinoma cell line SW620. All tests were performed in quintuplicate. *p*-value < 0.0001 *.

2.5. Production of Reactive Oxygen Species (ROS)

To evaluate ROS production in SW480 and SW620 cell lines, the IC_{50} value of genistein was used. Figure 6 shows that there is a greater production of ROS in the cells treated with genistein. In the SW480 cell line, a greater production of oxidative stress is observed at 24 h of treatment and in the SW620 cell line, this effect is observed both at 24 h and 48 h, showing in this last evaluation time higher levels of ROS, indicating that genistein can generate oxidative stress in this colorectal cancer cell line, possibly leading to the process of cell death.

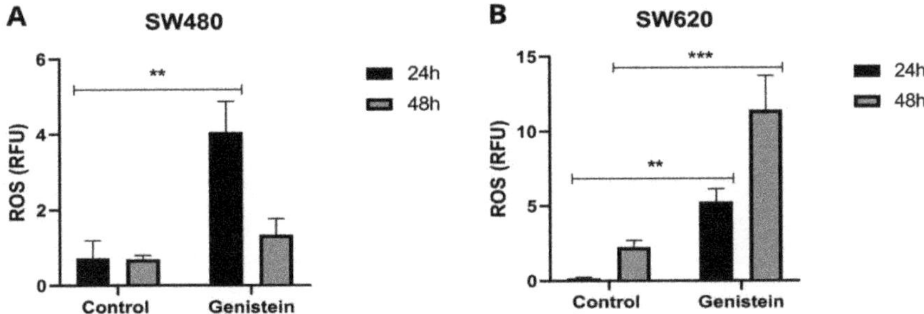

Figure 6. Effect of genistein on the production of reactive oxygen species (ROS) after treatment with genistein in two periods of time (24 h and 48 h). (A) SW480 cell line and (B) SW620 cell line. All tests were performed in triplicate. RFU: Relative Fluorescence Units. *p*-value < 0.01 **; and *p* < 0.001 ***.

2.6. Apoptotic Capacity of Genistein

The apoptotic capacity of genistein on cell lines SW480 and SW620 can be seen in Figures 7 and 8. The results showed an increase in the population where DNA fragmentation was evidenced, with respect to the control in the treatments with genistein, both at 24 h and 48 h in both cell lines. By means of this assay, the APO-DIRECT^{TM} kit (Chemicon Cat. N°APT110) with PI/F-dUTP makes it possible to differentiate, at the top of each graph, the population where a cell death process takes place. For the SW480 cell line, higher percentages of cell death were observed compared to the control (77.8% and 21.9% at 24 h and 48 h, respectively). For the SW620 cell line, higher percentages of cell death were also observed compared to the control (44.2% and 30.3% at 24 h and 48 h, respectively).

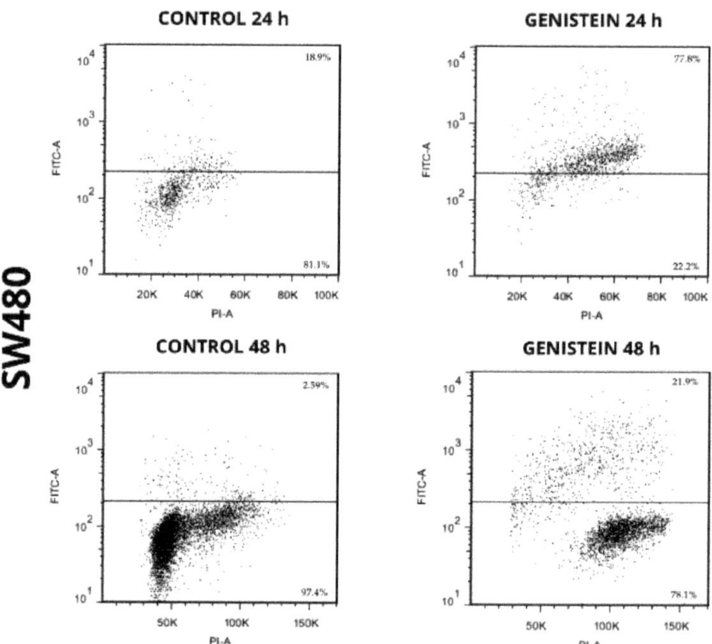

Figure 7. Genistein induces cell death in SW480 cell line. Dot plot of PI/FITC in SW480 cell line after treatment with genistein in two periods of time (24 h and 48 h).

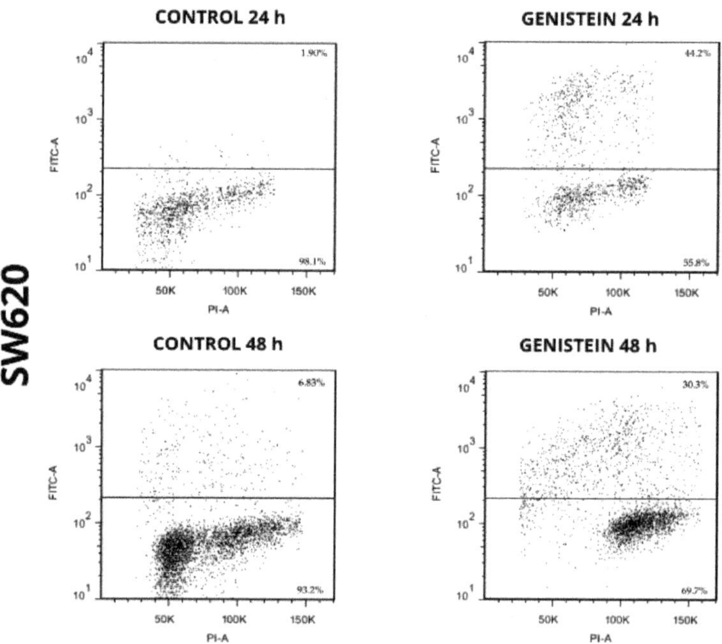

Figure 8. Genistein induces cell death in SW620 cell line. Dot plot of PI/FITC in SW620 cell line after treatment with genistein in two periods of time (24 h and 48 h).

To determine the intracellular proteins that participate in the apoptosis process of the SW480 and SW620 cell lines, Caspase 3, p53, Cytochrome c, BCL2, and cleaved PARP proteins were analyzed. In the SW620 cell line, after 24 h of treatment with genistein, a significant increase in Caspase 3 and cleaved PARP proteins was observed, and after 48 h of treatment, a significant increase in Caspase 3, p53, Cytochrome c, and cleaved PARP proteins was observed. The antiapoptotic protein BCL2 was not detected at 24 h and did not show significant differences compared to the control at 48 h of treatment (Figure 9). In the SW480 cell line, no levels of these proteins were detected at any of the two evaluation times with genistein (Results not shown).

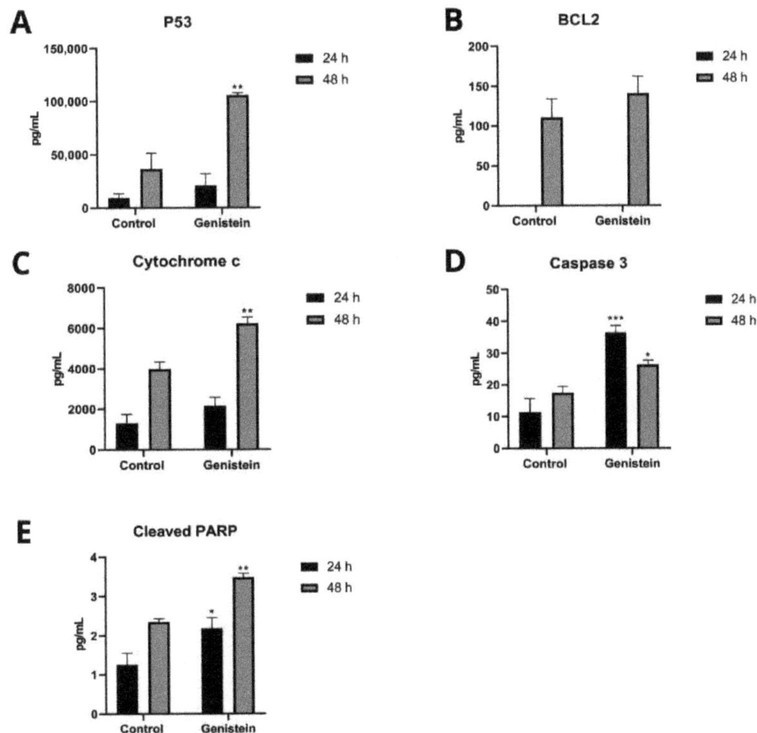

Figure 9. Effect of genistein on intracellular proteins that participate in the apoptosis process in SW620 cell line. Levels of each protein after treatment with genistein in two periods of time (24 h and 48 h) compared to the control. (**A**) p53. (**B**) BCL2. (**C**) Cytochrome c. (**D**) Caspase 3. (**E**) Cleaved PARP. $p < 0.05$ *; $p < 0.01$ **; and $p < 0.001$ ***.

2.7. Evaluation of Cytokine Expression

Cell supernatants were analyzed after 24 and 48 h incubation, with and without genistein. In the SW480 cell line, there were no significant changes in the cytokines tested (results not shown).

On the other hand, in the SW620 cell line, several cytokines showed a significant increase both at 24 h and 48 h after genistein treatment (Figure 10). At 24 h, a significant increase over the control was observed after treatment in the cytokines IL-1B, IL-2, IL-6, IL-13, IL-17A, IL-27, and GM-CSF. For 48 h of treatment, a significant increase in the cytokines IL-1B, IL-2, IL-4, IL-5, IL-10, IL-17A, IL-18, IL-27, and GM-CSF was observed. The other cytokines did not show significant differences from the control in this cell line (results not shown).

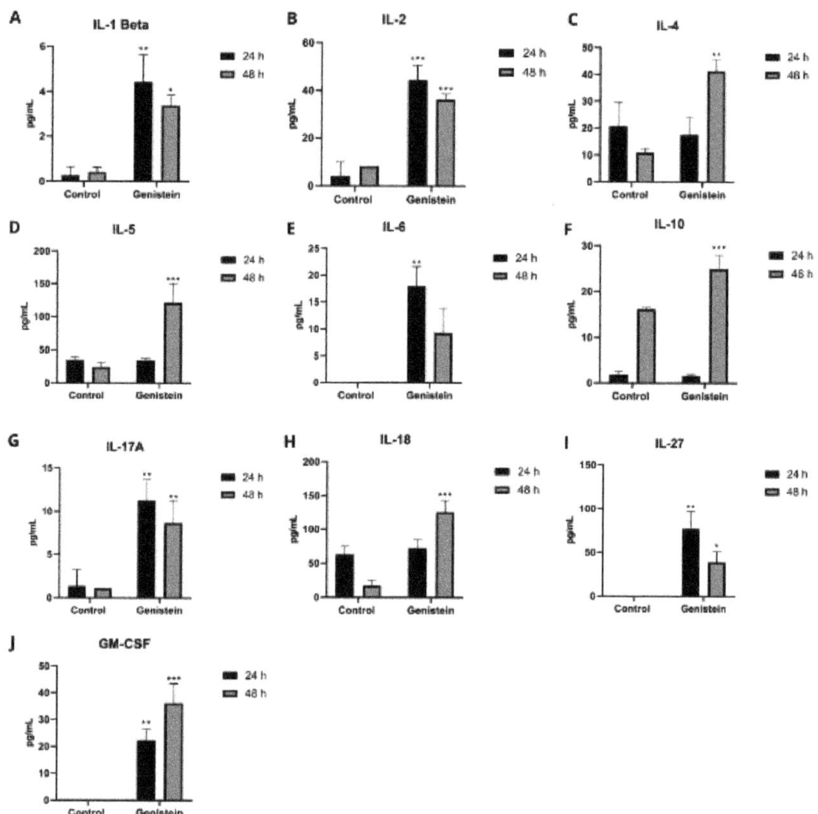

Figure 10. Effect of genistein on important cytokines in the inflammatory and carcinogenic process in SW620 cell line. Levels of each cytokine at 24 and 48 h after treatment with genistein compared to the control. (**A**) IL-1Beta. (**B**) IL-2. (**C**) IL-4. (**D**) IL-5. (**E**) IL-6. (**F**) IL-10. (**G**) IL-17A. (**H**) IL-18. (**I**) IL-27. (**J**) GM-CSF. $p < 0.05$ *; $p < 0.01$ **; and $p < 0.001$ ***.

3. Materials and Methods

This work was carried out in the Laboratory of Medical and Experimental Mycology CIB-UdeA-UPB-UDES located in the Corporación para Investigaciones Biológicas, Medellín, Colombia.

A commercial genistein CAS No. 446-72-0 (Shanghai Yingrui Biopharma Co., Shanghai, China), with 98% purity, was used for the analyses.

3.1. Preparation of Bacterial Nanocellulose

Bacterial nanocellulose (BNC) was used as encapsulating agent, which was obtained through the fermentation of the bacterium *Komagataeibacter medellinensis* NBRC 3288, which was isolated in the Central Retail of Medellín and was identified in the Universidad Pontificia Bolivariana [23]. Nanocellulose was prepared using a modified Hestrin–Schramm (HS) culture medium with glucose at 2% (*w/v*), peptone at 0.5% (*w/v*), yeast at 0.5% (*w/v*), disodium phosphate at 0.267% (*w/v*), and citric acid to adjust the pH up to 3.5. After fermentation for 7 days, the BNC membranes were removed and purified in a 5% (*w/v*) KOH solution to remove biomass and debris. Finally, the NCB was processed in a monobloc blender and passed to a MKCA6-3 ultrafine friction mill (Masuko Sangyo® Co., Ltd.,

Kawaguchi, Japan), with a total of 27 passes for the individualization of the nanofibers. The BNC was sterilized for later use.

3.2. Encapsulation of Genistein

The encapsulation of genistein in BNC was carried out using the spray-drying technique in a (BÜCHI Labortechnik AG, Flawil, Switzerland). First, a solution of 0.1% BNC (1 mg/mL) and 5.52 mg of genistein were prepared. Then, the solution was dried with a feed flow rate of 5 mL/min, an air flow of 35 m^3/h, an air pressure of 6 bars, and an air inlet temperature of 150 °C.

3.3. Morphology and Size of Bacterial Nanocellulose/Genistein (BNC/GEN) Capsules

The morphology and size of the capsules obtained were analyzed by scanning electron microscopy (SEM) using JEOL JSM 6490 LV equipment (JEOL, Tokyo, Japan) in a high vacuum with a secondary electron detector to obtain high-resolution SEI images at an acceleration voltage of 20 kV. The capsules were placed on a carbon ribbon and coated with a thin layer of gold. The size of the capsules was measured using the free distribution software: Image J 1.49, adjusting each capsule to an ellipse. A total of 100 capsules were measured to find the size distribution.

3.4. In Vitro Release Assays in Gastrointestinal Fluids

For genistein release profiles, the membrane dialysis method with a size of 12–14 kDa was used [24] in simulated physiological fluids of the stomach, small intestine and colon with pH values of 1.2, 6.0 and 7.4, respectively, following the formulations proposed by Marques et al. [25]. These formulations allowed us to simulate the basal conditions of each organ taking into account the pH and osmolarity. The capsules inside the membranes were kept at 37 °C with a stirring speed of 80 rpm for 2 h for the stomach, 24 h for the intestine, and 48 h for the colon. At every time point, an aliquot of the fluid was taken and the total volume was replenished. Genistein concentration was determined by UV-Vis spectroscopy at 260 nm, using absolute ethanol as a solvent. Finally, the percentage and amount of genistein released from the capsules were fitted to pseudo-first- and second-order kinetic models [26].

3.5. Cell Lines

For the biological tests, the human colon adenocarcinoma cell line SW480, its metastatic derivative SW620, and the non-malignant cell line HaCaT were used. These cell lines were obtained from the European Collection of Authenticated Cell Cultures (ECACC, Salisbury, UK) and cultured in Dulbecco's Modified Eagle Medium (DMEM), supplemented with 10% horse serum (Gibco, Waltham, Massachusetts, United States) heat-inactivated (60 °C), Penicillin/Streptomycin 1% (Sigma-Aldrich, Burlington, MA, USA) and Non-Essential Amino Acids 1% (Sigma-Aldrich). For the experiments, the serum concentration was reduced to 3% and the medium was supplemented with 5 mg/mL transferrin, 5 ng/mL selenium, and 10 mg/mL insulin (ITS Liquid Media Supplement 100×; Sigma-Aldrich) [27]. Before use, all cell lines were tested for the detection of *Mycoplasma* spp. by PCR, using specific primers [28].

3.6. Cell Viability Assay

Cell viability after genistein treatments was evaluated by the Sulforhodamine B (SRB) assay. A colorimetric assay consisted of staining the cellular protein content of adherent cells. Cell lines were seeded in 96-well plates at a density of 20,000 cells per well at 37 °C with 5% CO_2. They were then incubated for 24 h to allow their adherence, and subsequently, they were treated with different concentrations of genistein (3.7–740 µM), as well as with the vehicle used (Dimethyl sulfoxide DMSO at 1%) as growth control. After each treatment for 24 and 48 h, cells were fixed with trichloroacetic acid (PanReac AppliChem, Barcelona, Spain) for one hour at 4 °C. Cellular proteins were determined by staining with 0.4% SRB

(Sigma-Aldrich) for 30 min at room temperature. Subsequently, 5 washes with 1% acetic acid were performed. For these latter procedures, the plate was allowed to dry at room temperature. SRB-bound proteins were solubilized with 10 mM Tris-base and the reading was performed by absorbance at 490 nm in a microplate reader (Bio-Rad iMark™, Hercules, CA, USA).

3.7. Antiproliferation Assay

The antiproliferative effect of genistein was also evaluated with SRB. Cell lines were seeded in 96-well plates with a density of 2500 cells per well at 37 °C with 5% CO_2. The cells were incubated for 24 h to allow their adherence and were subsequently treated with the vehicle as growth control (1% DMSO) and with 5 different concentrations of genistein based on the IC50 found in the cell viability assay (3.7 μM, 18.5 μM, 37 μM, 185 μM, and 370 μM). The evaluation times were 0, 2, 4, 6, and 8 days. Every 48 h, the culture medium was changed with the respective treatments [29]. After completing each evaluation time, cells were fixed, stained, washed, and read in the same way as in the cell viability assay. For all assays, quintuplicate evaluations were performed.

3.8. Reactive Oxygen Species (ROS) Measurement Assay

For the determination of reactive oxygen species, a quantitative analysis was carried out in the Varioskan Lux reader (Thermo Fisher Scientific, Waltham, MA, USA) by measuring fluorescence in a 96-well plate using fluorescein 2′, 7′ Dichlorodihydrofluorescein Diacetate (DCFH-DA) (Calbiochem, San Diego, CA, USA) that allows ROS detection. In this procedure, cells were incubated in 6-well plates at a density of 250,000 cells per well for the SW480 cell line and 350,000 cells per well for the SW620 cell line at 37 °C with 5% CO_2. Cells were incubated for 24 h to allow their adherence and were subsequently treated with the vehicle as growth control, with the positive control (ferrous sulfate) and the IC50 of genistein corresponding to each cell line and evaluation time. The treatments were evaluated for 24 h and 48 h. After this time, washings were carried out with DMEM medium, DCFH-DA staining, and cell lysis and reading in the varioskan equipment at an excitation wavelength of 485 nm and an emission wavelength of 530 nm. Fluorescence intensity was normalized from the total protein concentration measured for each treatment. Triplicate evaluations were performed for this assay.

3.9. Apoptosis Assay

The detection of DNA fragmentation related to cell death processes was detected by flow cytometry using the APO-DIRECT™ kit (Chemicon[R] International, Temecula, CA, USA) allowing a staining method that marks DNA breaks. In this procedure, cells were incubated in T75 flasks at a density of 1.1 million cells per flask for the SW480 cell line and 1.6 million cells for the SW620 cell line at 37 °C with 5% CO_2. Cells were incubated for 24 h to allow their adherence and were subsequently treated with genistein (IC50 of each cell line) for 24 h and 48 h. After this time, cells underwent cell fixation, the addition of staining solution, incubation, and subsequent reading in the flow cytometer (LSR Fortessa; BD Biosciences, San Jose, CA, USA), according to the kit instructions.

3.10. Determination of Immunological Markers and Apoptosis

For these evaluations, the Magpix equipment was used (Luminex XMAP, Austin, TX, USA), which allows a qualitative and quantitative analysis of proteins. Cell lines were seeded in 6-well plates at a density of 250,000 cells per well for the SW480 cell line and 350,000 cells per well for the SW620 cell line, incubating at 37 °C with 5% CO_2. Cells were incubated for 24 h to allow their adherence. Subsequently, they were treated with the vehicle as growth control (DMSO at the highest concentration used with the compound in the test) and with the concentrations of genistein corresponding to the IC50 of 24 h and 48 h, evaluating these same times. For the immunological marker measurement assay, the supernatant was collected at each of the evaluated times and taken to the equipment for measurement following the protocol indicated by the manufacturer; the kit used in this assay was the Th1/Th2/Th9/Th17 Cytokine 18-Plex Human ProcartaPlex™ Panel. For the apoptosis assay, a lysis buffer was used to release intracellular proteins and was subsequently taken to the equipment for measurement following the protocol indicated by the manufacturer. The kit used in this assay was Apoptosis 6-Plex Human ProcartaPlex™ Panel (Invitrogen, Waltham, MA, USA). Triplicate evaluations were performed for these assays. The samples were normalized in protein concentration for analysis.

3.11. Statistical Analysis

In the cell viability assays, non-linear regression was applied to find the IC50. The results were expressed as the mean $\pm$ standard deviation (SD). For the antiproliferation, apoptosis, ROS, and expression of immunological markers assays, a two-way ANOVA was applied followed by the Sidak test after verifying the assumption of normality of the data using the Kolmogorov–Smirnov test. Statistical analyses and graphs were performed using the GraphPad Prism Version 8 program. In all cases, a value of $p < 0.05$ was considered significant.

4. Discussion

Many studies show that genistein has great potential as a chemopreventive agent for different types of cancer [10,14,15,30,31]. However, when this compound is administered orally, it has very low stability and bioavailability, due to its insolubility in water [11,32]. Therefore, its encapsulation could improve not only its bioavailability, but also its stability (against oxidizing agents) and its efficacy at the target site [33]. The choice of a good encapsulating agent will allow the compounds to move through the gastrointestinal system, without changes or loss of activity, and to be released at the specific site where the action of this chemopreventive agent is required. BNC is a biopolymer that can be obtained in large quantities through the fermentation of obligate aerobic bacteria of the genus *Komagataeibacter* [34]. In the food industry, BNC is known as nata de coco (Okiyama et al., 1993), and in 1992 it was accepted by the US Food and Drug Administration (FDA) as generally recognized as safe (GRAS) [35]. As an encapsulating agent, BNC allows the incorporation of active ingredients and allows their controlled release [36,37]. BNC has a high aspect ratio in its crystalline nanofibers [38], and when dried by the spray method, the nanofiber network collapses in response to water evaporation and the establishment of irreversible hydrogen bonds [36,37]. As observed in the micrographs, this type of structure contributed significantly to the effective trapping of genistein, hindering the absorption of liquids and the dissolution of the compound in the stomach, making the release more controlled and prolonged. As observed in Figure 1, the morphology of the capsules obtained by spray drying is irregular and are called crumpled paper, with diameters at the micrometric scale [39].

The release profiles obtained in this research indicate that the BNC/GEN capsules present a release of the compound that depends on the pH of the solution. This behavior is characteristic of NCB-encapsulated and oven-dried compounds [40]. The low percentage of the release of genistein in the stomach fluid indicates that BNC protects the compound from the acidic conditions of the gastric fluid [33]. For its part, the increased release of genistein in colonic fluid is attributed to the pH of the solution (pH = 7.4), which alters the hydrogen bonds of the BNC nanofibers, allowing it to open or relax its structure, and altering the hydrophobic interactions between genistein and BNC that allow the release of the compound [33,36]. Finally, the curve of the release profile of BNC capsules loaded with genistein in colonic fluid is consistent with a controlled or prolonged release of drugs since it maintains an almost constant rate from 15 h to 72 h. This allows the concentration of the compound to be maintained in the therapeutic window and to be below the toxic level and above the subtherapeutic level [41]. The results obtained show that BNC acts as a protective agent for genistein in the stomach and allows its release and absorption in the small intestine and colon, where the enterocytes would be responsible for metabolizing this compound [42].

Although genistein has been described as an agent with anticancer properties in several types of cancer, including CRC, there are many mechanisms that need to be described in relation to its biological activity [30]. The measurement of cell viability plays a fundamental role in cell culture assays. The results showed that genistein inhibits cell growth in a dose-dependent manner in the SW480 and SW620 colorectal cancer cell lines, as reported in other studies [16,43]. There are also several studies that have shown that treatment with genistein inhibits the growth of other types of colon adenocarcinoma cells, but in general, all these investigations lack the analysis of the compound in healthy cells that would allow one to define the selectivity of genistein [14,15,43]. In our work, and based on the IC_{50} of each cell line after treatment with genistein, it was possible to determine a selectivity of 3.17 for the SW480 cell line at 24 h of evaluation and 1.75 at 48 h of evaluation. For the SW620 cell line, at 24 h the selectivity was 1.34 and at 48 h it was 1.47 against the non-malignant HaCaT cell line. Additionally, to date, the evaluation of the effect of this type of compound on a tumor cell line and its metastatic derivative, both from the same patient, had not been reported in the same study. Based on these IS and IC_{50}s, an antiproliferative effect was observed over the time points tested in a dose-dependent manner for both the SW480 cell line and the SW620 cell line. This inhibition of cell growth is associated with the cumulative dose of genistein that was given every 48 h for up to eight days.

One of the mechanisms by which a compound can inhibit cell growth and proliferation is by inducing apoptosis in malignant cells. In cancer cells, the mechanism of programmed cell death is reduced, demonstrating an imbalance in the proteins involved in the apoptosis process, both proapoptotic and antiapoptotic [44,45]. In this study, an increase in the population in the process of cell death, evaluated by flow cytometry, was observed both at 24 h and 48 h after treatment with genistein in the SW620 cell line. This result agrees with that observed during the measurement of proapoptotic proteins, where a significant increase in the levels of these proteins evaluated by Magpix was observed after treatment with genistein for 24 h and 48 h in this same cell line. After 24 h of treatment with genistein, an increase in Caspase 3 and cleaved PARP proteins were found. Caspase 3 is an essential part of the apoptosis execution pathway. During carcinogenesis, it is deregulated and can be used as an indicator in the progression of the disease. Studies have been described where low levels of caspase 3 indicate decreased apoptosis during tumorigenesis and could be significant in disease progression [15,46]. Caspase 3 is also known to activate other important proteins in this process, including cleaved PARP, a protein that plays several biological roles such as DNA repair and cell cycle regulation. Following DNA damage, a rapid signaling cascade is generated at injury sites to activate cell cycle checkpoints and/or apoptosis to ensure efficient DNA repair. If not repaired, this damage results in an abnormal cell cycle. PARP cleavage occurs by caspase 3, as well as caspase 7, downstream of this signaling cascade to continue the process of apoptosis through the degradation of nuclear material and subsequent cell death [47,48]. PARP cleavage by caspases has also been described as a marker of apoptosis in several studies, specifically in colon cancer cell lines [49]. After 48 h of treatment with genistein, in addition to the proteins mentioned above, a significant increase was found in the proapoptotic proteins p53 and Cytochrome c, and no significant differences were found in the levels of the antiapoptotic protein BCL2, compared to the control. The p53 protein has various functions; among them, it acts as a tumor suppressor, participates in negative regulation of the cell cycle by inhibiting cell division, and has proapoptotic activity depending on the physiological circumstances and the cell type. In the induction of apoptosis, the intrinsic signaling pathway is induced by the p53 protein. This is mediated by the stimulation of the expression of BAX and FAS, as well as by the repression of BCL2 [50]. The p53 protein is also responsible for mediating apoptosis directly in the mitochondria, and after various interactions, it allows the release of Cytochrome c, which regulates the supply of cellular energy. Under conditions of cellular stress, the release of Cytochrome c from the mitochondria is an important step for apoptosis, leading to apoptosome formation, caspase activation, and cell death. The suppression of antiapoptotic proteins, such as BCL2, or the activation of proapoptotic proteins also belonging to the BCL2 family lead to an alteration in the permeability of the mitochondrial membrane that results in the release of Cytochrome c in the cytosol, binding to the Apoptosis Activating Factor 1 (Apaf-1) and triggering the activation of Caspase 9, which then accelerates apoptosis by activating other caspases, such as caspase 3, as previously described [51,52]. The above-described process allows us to identify, in the SW620 cell line, that possibly the apoptosis process observed with genistein treatment occurs through the permeabilization of the mitochondrial membrane in which reactive oxygen species (ROS) may be involved, which were also increased, suggesting an activation of the intrinsic pathway of apoptosis.

ROS are highly reactive molecules and their generation in cells occurs in a balanced manner. They have various functions in normal physiological regulation, such as cell-cycle progression, proliferation, differentiation, and cell death. Additionally, they play an important role in the activation of various cell signaling pathways, and their high levels can lead to damage to proteins, nucleic acids, lipids, membranes, and organelles, leading cells to a process of apoptosis [53]. Higher than normal levels of ROS are typically found in cancer cell lines, helping to promote cancer development and progression. However, some anticancer therapeutics have been described, inducing apoptosis by further increasing cancer cells without affecting non-malignant cells [54]. In this work, after treatment with genistein, an increase in ROS production was observed in the SW620 cell line both at 24 h and 48 h and in the SW480 cell line only at 24 h. There may be a relationship between the production of ROS and the apoptotic process described in the SW620 cell line at both times of treatment with genistein, since an increase in ROS can induce the activation of the ASK1/JNK signaling pathway, which precisely allows the activation of ASK1 sending signals for the activation of JNK and its activation-inducing apoptosis through the mitochondrial signaling pathway, leading to the release of Cytochrome c and continuing the apoptosis process by activating caspases or, also, by activating pro-apoptotic genes [55].

In the SW480 cell line, no levels of these proteins were detected at any of the evaluation times after treatment with genistein. However, a population in the process of cell death was observed in the flow cytometric assay. This test allows us to identify ruptures presented in the process of DNA fragmentation, a process that not only occurs in apoptosis but also in other mechanisms of cell death, including necroptosis, which can also be evaluated and identified with Propidium Iodide (PI) and presents a signaling pathway independent of the caspase pathway and the proteins involved in the apoptosis process [56]. Therefore, it is possible that the cell death process observed for the SW480 line could occur through necroptosis, a non-apoptotic mechanism of cell death that has similarities to necrosis in relation to its morphological characteristics such as loss of membrane integrity and damage to intracellular organelles, as necrosis is an unregulated and unprogrammed cell death where there is a rupture of the cell membrane, the release of intracellular components, and therefore inflammation in adjacent tissues, while although in necroptosis the release of intracellular components is similar to that presented in necrotic cells, the mechanisms are different and occur in a regulated manner. For this reason, necroptosis is described as a form of regulated necrosis or programmed necrosis [56–58].

In CRC, cytokines play a crucial role in the development of this type of cancer. However, the available data are insufficient to describe the changes in the cytokine profile during the development of CRC, as well as the mechanisms that lead to changes in the cytokine levels in this type of cancer [59]. In the SW620 cell line, a significant increase was observed, compared to the control, after treatment for 24 h with genistein in the cytokines IL-1beta, IL-2, IL-6, IL-13, IL-17A, IL-27, and GM-CSF. For 48 h of treatment, a significant increase in the cytokines IL-1beta, IL-2, IL-4, IL-5, IL-10, IL-17A, IL-18, IL-27, and GM-CSF was observed. Many of these cytokines have been described with possible antitumor effects; IL-2 antitumor activity appears to be mediated by its effects on NK cells and other cytotoxic cells [60]. It has been used in phase 1 clinical studies combined with other compounds in patients with advanced metastatic colorectal cancer [61]. It has even been used in combination with the drug 5-FU or with other cytokines, comparing its results with chemotherapy used for this type of cancer [60]. As well as this interleukin, although its role is not completely clear in CRC, IL-1beta has also been reported to have antitumor properties. This cytokine has long been associated with inflammation and innate immunity, now describing a broader role that extends beyond classically defined inflammation. Additionally, in the defense of mucosal surfaces, cytokines of the IL-1 family are required, where IL-1beta is found [62–64]. IL-5 is critical in the development, activation, and survival of eosinophils, which have been associated with an antitumor response in CRC [65]. Its presence in tumors can influence the activation of the immune system and predict a better prognosis. It has been shown mainly with antitumor properties in CRC [59]. Additionally, both IL-5 and GM-CSF could control tumor growth, since the negative regulation of IL-5 and GM-CSF has been shown to increase tumor burden in a murine model of CRC [66]. IL-18, for its part, promotes the antitumor ability of NK cells in colorectal cancer [67]. This has been found to be decreased in tissues from patients with CRC and its low expression has been significantly correlated with tumor size [68]. In relation to IL-27 and CRC, it has been described as a cytokine with antitumor effects, showing not only antiproliferative and antiangiogenic effects by acting directly on cancer cells, but also having indirect antitumor effects driven by immunostimulatory activity in this and other types of cancer [69].

IL-13 has effector functions, including allergic inflammation, tissue remodeling, and fibrosis. It is a structurally and functionally similar cytokine to IL-4, and the components of its receptors are similar. They regulate the immune response and are involved in several neoplasms. Both cytokines are important in the Th2 response. Several studies have shown discrepancies in the antiproliferative effect of these cytokines in various cell lines and CRC clinical studies [70]. These interleukins can have both pro- and antitumor functions, depending on the tumor microenvironment, and among these functions, a possible inhibitory effect on CRC has been described [59].

IL-6 can show both pro- and antineoplastic activity [71] and is responsible for regulating the proliferation of intestinal epithelial cells in relation to its antitumor activity. It presents different mechanisms, such as the promotion of macrophages and the increase in cytotoxic effects of neutrophils on tumor cells. However, the specific mechanism through which IL-6 plays a role during CRC initiation and progression is not completely clear [72] as there is also evidence that there is no increase in this cytokine compared to the control in studies with patients with CRC [73–76]. IL-10 is also an immunomodulatory cytokine that exhibits both pro- and antitumor characteristics, depending on the tumor microenvironment, showing its behavior in the pathogenesis and progression of CRC [77]. Perhaps one of its main roles is tumor suppression [59]. Associated with this description, IL-10 has been described with antitumor activity in the tumor microenvironment and with neovascularization inhibitory activity in a murine model, fulfilling a role in the suppression of tumor growth [78]. Finally, IL-17A is part of the IL-17 family, where IL-17B, IL-17C, IL-17D, IL-17E (also called IL-25), and IL-17F are also found. For this family of cytokines, a dual role in the development of CRC has been suggested, thus, highlighting the need for new studies related to its effects in this pathology [79]. Several studies have shown that IL-17A did not have statistical significance at different measurement times in patients with CRC, being shown as a cytokine that is not of potential prognosis in patients with CRC since its levels are not increased [73,76,80].

It is important to highlight that the effects exhibited by these cytokines will depend on the tumor microenvironment in which they are found. Perhaps, for this reason, there is no clear immunomodulatory effect or profile due to the fact that they are in vitro tests with single cell lines, an effect that could, perhaps, be presented in a tumor microenvironment to evaluate different processes in relation to the evolution of tumors.

Finally, one of the great problems of chemotherapeutic agents is the resistance that occurs against them. For this and the other reasons previously described, it is of great importance to develop a cancer therapy based on natural products such as genistein. In this sense, the efficiency of drugs used in chemotherapy depends on the mechanisms to exert their action. One of these mechanisms is apoptosis, in which the deregulation of pro- or antiapoptotic genes in tumor cells has been related to increased resistance to chemotherapy [81]. Apoptosis can be induced by the intrinsic or extrinsic pathway. The intrinsic pathway is regulated by the BCL-2 family of proteins, which allows the release of cytochrome c from the mitochondria and interacts with other proapoptotic proteins until reaching cell death. It has been reported that the overexpression of BCL-2 antiapoptotic proteins increases resistance in ovarian cancer cells to cisplatin, paclitaxel, and other chemotherapeutic agents, both in vitro and in vivo [82]. Genistein has been proposed as an agent to combat this resistance since this mechanism has been described in several studies after its administration, and in this work, its effect is proposed through the intrinsic pathway of apoptosis. The tumor microenvironment in drug resistance is also one of the main reasons for relapse during the treatment of various types of cancer. In this microenvironment, the cytokines produced can provide signals for the growth and survival of tumor cells, hence the importance of their measurement in this work [81]. Other resistance mechanisms in cancer occur due to drug inactivation, drug release from cancer cells, repair of cell damage induced by chemotherapy, activation of pathways favorable to survival, or heterogeneity intratumorally that is observed in different types of cancer and that occurs due to several factors at the cellular level that generate genetic variations such as deletions, translocations, or chromosomal rearrangements [83].

In conclusion, genistein has a chemopreventive effect on the colorectal adenocarcinoma cell line SW480 and its metastatic derivative SW620 through cytotoxicity and antiproliferation evaluations. In both cell lines, there is evidence of the expression of immunological markers that plays an important role in the carcinogenic process and metastasis of CRC and that have been described in other studies with antitumor properties in this type of cancer. By exerting a dual role in this process, depending on the tumor microenvironment, they can have various effects on cell lines, in this case, an increase in expression. The production of ROS in both cell lines may be associated with the cell death process through different mechanisms. Thus, due to the lack of response in the evaluated proteins, it is hypothesized that a necroptotic process mediated by ROS production could occur in the SW480 cell line, while in the SW620 cell line, it is suggested that genistein induces the formation of ROS with the consequent activation of intrinsic apoptosis mediated by caspases and p53. Although genistein has great potential as a chemotherapeutic agent, when administered orally, its bioavailability and, therefore, its effect on the target site, in this case the colon, are decreased. Therefore, new strategies such as encapsulation should be used to improve its effectiveness. BNC encapsulation improved the release profile of genistein, protecting it from gastric pH and allowing its release in the colon.

Author Contributions: Conceptualization, T.W.N.; Formal analysis, J.P.R. and A.I.C.; Funding acquisition, T.W.N.; Investigation, J.P.R., A.I.C., E.C. and V.B.-B.; Methodology, J.P.R., A.I.C. and T.W.N.; Project administration, T.W.N.; Resources, C.C.; Supervision, T.W.N.; Validation, J.P.R. and A.I.C.; Writing—original draft, J.P.R. and A.I.C.; Writing—review and editing, E.C., V.B.-B., M.O., C.C. and T.W.N. All authors have read and agreed to the published version of the manuscript.

Funding: This research was funded by the Ministry of Sciences MINCIENCIAS, Ministry of Education MINEDUCATION, Ministry of Commerce, Industry and Tourism MINCIT and Colombian Institute of Educational Credit and Technical Studies Abroad ICETEX, through the Scientific Ecosystem component of the Colombia Científica Program (NanoBioCáncer alliance Cod FP44842-211-201-2018, project numbers 58478 and 58674).

Institutional Review Board Statement: Not applicable.

Informed Consent Statement: Not applicable.

Data Availability Statement: The data presented in this study are available on request from the corresponding author.

Acknowledgments: The authors thank Universidad Pontificia Bolivariana, MINCIENCIAS, MINEDUCACIÓN, MINCIT, ICETEX and Corporación para Investigaciones Biológicas for their support.

Conflicts of Interest: The authors declare no conflict of interest.

References

1. Globocan. Global Cancer Observatory. 2020. Available online: http://gco.iarc.fr/ (accessed on 21 September 2022).
2. Ciombor, K.K.; Wu, C.; Goldberg, R.M. Recent therapeutic advances in the treatment of colorectal cancer. *Annu. Rev. Med.* **2015**, *66*, 83–95. [CrossRef] [PubMed]
3. Chen, X.; Gu, J.; Wu, Y.; Liang, P.; Shen, M.; Xi, J.; Qin, J. Clinical characteristics of colorectal cancer patients and anti-neoplasm activity of genistein. *Biomed. Pharmacother.* **2020**, *124*, 109835. [CrossRef] [PubMed]
4. Marley, A.R.; Nan, H. Epidemiology of colorectal cancer. *Int. J. Mol. Epidemiol. Genet.* **2016**, *7*, 105–114.
5. Keum, N.N.; Giovannucci, E. Global burden of colorectal cancer: Emerging trends, risk factors and prevention strategies. *Nat. Rev. Gastroenterol. Hepatol.* **2016**, *16*, 713–732. [CrossRef] [PubMed]
6. Millan, M.; Merino, S.; Caro, A.; Feliu, F.; Escuder, J.; Francesch, T. Treatment of colorectal cancer in the elderly. *World J. Gastrointest. Oncol.* **2015**, *7*, 204–220. [CrossRef] [PubMed]
7. Longley, D.B.; Harkin, D.P.; Johnston, P.G. 5-Fluorouracil: Mechanisms of action and clinical strategies. *Nat. Rev. Cancer* **2003**, *3*, 330–338. [CrossRef] [PubMed]
8. Arias, J.L. Novel strategies to improve the anticancer action of 5-fluorouracil by using drug delivery systems. *Molecules* **2008**, *13*, 2340–2369. [CrossRef] [PubMed]
9. Rejhová, A.; Opattová, A.; Čumová, A.; Slíva, D.; Vodička, P. Natural compounds and combination therapy in colorectal cancer treatment. *Eur. J. Med. Chem.* **2018**, *144*, 582–594. [CrossRef] [PubMed]
10. Chae, H.-S.; Xu, R.; Won, J.-Y.; Chin, Y.-W.; Yim, H. Molecular targets of genistein and its related flavonoids to exert anticancer effects. *Int. J. Mol. Sci.* **2019**, *20*, 2420. [CrossRef] [PubMed]
11. Spagnuolo, C.; Russo, G.L.; Orhan, I.E.; Habtemariam, S.; Daglia, M.; Sureda, A.; Nabavi, S.F.; Devi, K.P.; Loizzo, M.R.; Tundis, R.; et al. Genistein and cancer: Current status, challenges, and future directions. *Adv. Nutr.* **2015**, *6*, 408–419. [PubMed]
12. Landis-Piwowar, K.R.; Iyer, N.R. Cancer chemoprevention: Current state of the art. *Cancer Growth Metastasis* **2014**, *7*, CGM.S11288-25. [CrossRef] [PubMed]
13. Zhang, Z.; Wang, C.-Z.; Du, G.-J.; Qi, L.-W.; Calway, T.; He, T.-C.; Du, W.; Yuan, C.-S. Genistein induces G2/M cell cycle arrest and apoptosis via ATM/p53-dependent pathway in human colon cancer cells. *Int. J. Oncol.* **2013**, *43*, 289–296. [CrossRef] [PubMed]
14. Qi, W.; Weber, C.R.; Wasland, K.; Savkovic, S.D. Genistein inhibits proliferation of colon cancer cells by attenuating a negative effect of epidermal growth factor on tumor suppressor FOXO3 activity. *BMC Cancer* **2011**, *11*, 219. [CrossRef]
15. Zhou, P.; Wang, C.; Hu, Z.; Chen, W.; Qi, W.; Li, A. Genistein induces apoptosis of colon cancer cells by reversal of epithelial-to-mesenchymal via a Notch1/NF-κB/slug/E-cadherin pathway. *BMC Cancer* **2017**, *17*, 813. [CrossRef] [PubMed]
16. Zhu, J.; Ren, J.; Tang, L. Genistein inhibits invasion and migration of colon cancer cells by recovering WIF1 expression. *Mol. Med. Rep.* **2018**, *17*, 7265–7273. [CrossRef] [PubMed]
17. Qin, J.; Chen, J.X.; Zhu, Z.; Teng, J.A. Genistein inhibits human colorectal cancer growth and suppresses miR-95, Akt and SGK1. *Cell. Physiol. Biochem.* **2015**, *35*, 2069–2077. [CrossRef] [PubMed]
18. Muñoz, A. *Efecto Citotóxico del grupo 2,4-diariltetrahidroquinolinas sustituidas y su respuesta en combinación con fármacos antitumorales sobre células de cáncer de mama humano*; Universidad Central de Venezuela: Caracas, Venezuela, 2009; 84p.
19. López-Lázaro, M. Two preclinical tests to evaluate anticancer activity and to help validate drug candidates for clinical trials. *Oncoscience* **2015**, *2*, 91–98. [CrossRef] [PubMed]
20. Indrayanto, G.; Putra, G.S.; Suhud, F. Validation of in-vitro bioassay methods: Application in herbal drug research. *Profiles Drug Subst. Excip. Relat. Methodol.* **2020**, *46*, 273–307. [PubMed]
21. Papaj, K.; Rusin, A.; Szeja, W.; Grynkiewicz, G. Absorption and metabolism of biologically active genistein derivatives in colon carcinoma cell line (Caco-2). *Acta Pol. Pharm. Drug Res.* **2015**, *71*, 1037–1044.
22. Tang, J.; Xu, N.; Ji, H.; Liu, H.; Wang, Z.; Wu, L. Eudragit nanoparticles containing genistein: Formulation, development, and bioavailability assessment. *Int. J. Nanomed.* **2011**, *6*, 2429–2435.
23. Castro, C.; Cleenwerck, I.; Trček, J.; Zuluaga, R.; De Vos, P.; Caro, G.; Aguirre, R.; Putaux, J.-L.; Gañán, P. *Gluconacetobacter medellinensis* sp. nov., cellulose- and non-cellulose-producing acetic acid bacteria isolated from vinegar. *Int. J. Syst. Evol. Microbiol.* **2013**, *63*, 1119–1125. [CrossRef]
24. Xiao, Y.; Ho, C.-T.; Chen, Y.; Wang, Y.; Wei, Z.; Dong, M.; Huang, Q. Synthesis, characterization, and evaluation of genistein-loaded zein/carboxymethyl chitosan nanoparticles with improved water dispersibility, enhanced antioxidant activity, and controlled release property. *Foods* **2020**, *9*, 1604. [CrossRef] [PubMed]
25. Marques, M.R.C.; Loebenberg, R.; Almukainzi, M. Simulated biological fluids with possible application in dissolution testing. *Dissolution Technol.* **2011**, *18*, 15–28. [CrossRef]
26. Ho, Y.S.; McKay, G. Pseudo-second order model for sorption processes. *Process Biochem.* **1999**, *34*, 451–465. [CrossRef]
27. Herrera-R, A.; Castrillón, W.; Otero, E.; Ruiz, E.; Carda, M.; Agut, R.; Naranjo, T.; Moreno, G.; Maldonado, M.E.; Cardona, G.W. Synthesis and antiproliferative activity of 3- and 7-styrylcoumarins. *Med. Chem. Res.* **2018**, *27*, 1893–1905. [CrossRef]

28. Arias, J.D.P.; Al, J.; Ciro, U.; Gómez, J.C.G. Implementación de un control interno en la detección molecular de las principales especies de micoplasmas contaminantes de cultivos celulares. *Rev. Científica Salud Uninorte* **2013**, *29*, 160–173.

29. Castrillón, W.; Herrera, R.A.; Prieto, L.J.; Conesa-Milián, L.; Carda, M.; Naranjo, T.; Maldonado, M.E.; Cardona, G.W. Synthesis and in-vitro evaluation of s-allyl cysteine ester-caffeic acid amide hybrids as potential anticancer agents. *Iran. J. Pharm. Res.* **2019**, *18*, 1770–1789. [PubMed]

30. Shafiee, G.; Saidijam, M.; Tavilani, H.; Ghasemkhani, N.; Khodadadi, I. Genistein induces apoptosis and inhibits proliferation of HT29 colon cancer cells. *Int. J. Mol. Cell. Med.* **2016**, *5*, 178–191. [PubMed]

31. Pavese, J.M.; Farmer, R.L.; Bergan, R.C. Inhibition of cancer cell invasion and metastasis by genistein. *Cancer Metastasis Rev.* **2010**, *29*, 465–482. [CrossRef] [PubMed]

32. Jaiswal, N.; Akhtar, J.; Singh, S.P.; Ahsan, F.; Badruddeen. An overview on genistein and its various formulations. *Drug Res.* **2018**, *69*, 305–313. [CrossRef] [PubMed]

33. Rahmani, F.; Karimi, E.; Oskoueian, E. Synthesis and characterisation of chitosan-encapsulated genistein: Its anti-proliferative and anti-angiogenic activities. *J. Microencapsul.* **2020**, *37*, 305–313. [CrossRef] [PubMed]

34. Castro, C.; Zuluaga, R.; Álvarez, C.; Putaux, J.-L.; Caro, G.; Rojas, O.J.; Mondragon, I.; Gañán, P. Bacterial cellulose produced by a new acid-resistant strain of Gluconacetobacter genus. *Carbohydr. Polym.* **2012**, *89*, 1033–1037. [CrossRef] [PubMed]

35. Shi, Z.; Zhang, Y.; Phillips, G.O.; Yang, G. Utilization of bacterial cellulose in food. *Food Hydrocoll.* **2013**, *35*, 539–545. [CrossRef]

36. Meneguin, A.B.; Barud, H.D.S.; Sábio, R.M.; de Sousa, P.Z.; Manieri, K.F.; de Freitas, L.A.P.; Pacheco, G.; Alonso, J.D.; Chorilli, M. Spray-dried bacterial cellulose nanofibers: A new generation of pharmaceutical excipient intended for intestinal drug delivery. *Carbohydr. Polym.* **2020**, *249*, 116838. [CrossRef] [PubMed]

37. Kolakovic, R.; Laaksonen, T.; Peltonen, L.; Laukkanen, A.; Hirvonen, J. Spray-dried nanofibrillar cellulose microparticles for sustained drug release. *Int. J. Pharm.* **2012**, *430*, 47–55. [CrossRef]

38. Khorasani, A.C.; Shojaosadati, S.A. Bacterial nanocellulose-pectin bionanocomposites as prebiotics against drying and gastrointestinal condition. *Int. J. Biol. Macromol.* **2016**, *83*, 9–18. [CrossRef]

39. Peng, Y.; Han, Y.; Gardner, D. Spray-drying cellulose nanofibrils effect of drying process parameters on particle morphology and size distribution. *Soc. Wood Sci. Technol.* **2012**, *44*, 448–461.

40. Adepu, S.; Khandelwal, M. Ex-situ modification of bacterial cellulose for immediate and sustained drug release with insights into release mechanism. *Carbohydr. Polym.* **2020**, *249*, 116816. [CrossRef] [PubMed]

41. Geraili, A.; Xing, M.; Mequanint, K. Design and fabrication of drug-delivery systems toward adjustable release profiles for personalized treatment. *View* **2021**, *20200126*, 1–24. [CrossRef]

42. Mukund, V.; Mukund, D.; Sharma, V.; Mannarapu, M.; Alam, A. Genistein: Its role in metabolic diseases and cancer. *Crit. Rev. Oncol. Hematol.* **2017**, *119*, 13–22. [CrossRef] [PubMed]

43. Chen, X.; Wu, Y.; Gu, J.; Liang, P.; Shen, M.; Xi, J.; Qin, J. Anti-invasive effect and pharmacological mechanism of genistein against colorectal cancer. *BioFactors* **2020**, *46*, 620–628. [CrossRef] [PubMed]

44. Wong, R.S.Y. Apoptosis in cancer: From pathogenesis to treatment. *J. Exp. Clin. Cancer Res.* **2011**, *30*, 87. [CrossRef] [PubMed]

45. Liu, B.; Yan, X.; Hou, Z.; Zhang, L.; Zhang, D. Impact of Bupivacaine on malignant proliferation, apoptosis and autophagy of human colorectal cancer SW480 cells through regulating NF-κB signaling path. *Bioengineered* **2021**, *12*, 2723–2733. [CrossRef]

46. Coffey, R.N.T.; Watson, R.W.G.; Fitzpatrick, J.M. Signaling for the caspases: Their role in prostate cell apoptosis. *J. Urol.* **2001**, *165*, 5–14. [CrossRef]

47. Gobeil, S.; Boucher, C.C.; Nadeau, D.; Poirier, G.G. Characterization of the necrotic cleavage of poly(ADP-ribose) polymerase (PARP-1): Implication of lysosomal proteases. *Cell Death Differ.* **2001**, *8*, 588–594. [CrossRef]

48. Luo, H.; Liang, H.; Chen, J.; Xu, Y.; Chen, Y.; Xu, L.; Yun, L.; Liu, J.; Yang, H.; Liu, L.; et al. Hydroquinone induces TK6 cell growth arrest and apoptosis through PARP-1/p53 regulatory pathway. *Environ. Toxicol.* **2017**, *32*, 2163–2171. [CrossRef]

49. Whitacre, N.A.B.C.M.; Zborowska, E.; Willson, J.K. Detection of poly(ADP-ribose) polymerase cleavage in response to treatment with topoisomerase I inhibitors: A potential surrogate end point to assess treatment effectiveness. *Clin. Cancer Res.* **1999**, *5*, 665–672.

50. Bergamaschi, D.; Samuels, Y.; O'Neil, N.; Trigiante, G.; Crook, T.; Hsieh, J.-K.; O'Connor, D.J.; Zhong, S.; Campargue, I.; Tomlinson, M.L.; et al. iASPP oncoprotein is a key inhibitor of p53 conserved from worm to human. *Nat. Genet.* **2003**, *33*, 162–167. [CrossRef]

51. Kalpage, H.A.; Wan, J.; Morse, P.T.; Zurek, M.P.; Turner, A.A.; Khobeir, A.; Yazdi, N.; Hakim, L.; Liu, J.; Vaishnav, A.; et al. Cytochrome c phosphorylation: Control of mitochondrial electron transport chain flux and apoptosis. *Int. J. Biochem. Cell Biol.* **2020**, *121*, 105704. [CrossRef]

52. Marchenko, N.D.; Moll, U.M. Mitochondrial death functions of p53. *Mol. Cell. Oncol.* **2014**, *1*, e955995. [CrossRef] [PubMed]

53. Redza-Dutordoir, M.; Averill-Bates, D.A. Activation of apoptosis signalling pathways by reactive oxygen species. *Biochim. Biophys. Acta (BBA)—Mol. Cell Res.* **2016**, *1863*, 2977–2992. [CrossRef] [PubMed]

54. Ding, Y.; Wang, H.; Niu, J.; Luo, M.; Gou, Y.; Miao, L.; Zou, Z.; Cheng, Y. Induction of ROS overload by alantolactone prompts oxidative DNA damage and apoptosis in colorectal cancer cells. *Int. J. Mol. Sci.* **2016**, *17*, 558. [CrossRef] [PubMed]

55. Circu, M.L.; Aw, T.Y. Reactive oxygen species, cellular redox systems, and apoptosis. *Free Radic. Biol. Med.* **2010**, *48*, 749–762. [CrossRef] [PubMed]

56. Liu, C.; Zhang, K.; Shen, H.; Yao, X.; Sun, Q.; Chen, G. Necroptosis: A novel manner of cell death, associated with stroke. *Int. J. Mol. Med.* **2018**, *41*, 624–630. [CrossRef]

57. Linkermann, A.; Green, D.R. Necroptosis. *N. Engl. J. Med.* **2014**, *370*, 455. [CrossRef]
58. Feoktistova, M.; Leverkus, M. Programmed necrosis and necroptosis signalling. *FEBS J.* **2014**, *282*, 19–31. [CrossRef]
59. Czajka-Francuz, P.; Cisoń-Jurek, S.; Czajka, A.; Kozaczka, M.; Wojnar, J.; Chudek, J.; Francuz, T. Systemic Interleukins' profile in early and advanced colorectal cancer. *Int. J. Mol. Sci.* **2021**, *23*, 124. [CrossRef]
60. Whittington, R.; Faulds, D. Interleukin-2. A review of its pharmacological properties and therapeutic use in patients with cancer. *Drugs* **1993**, *46*, 446–514. [CrossRef]
61. Douillard, J.Y.; Bennouna, J.; Vavasseur, F.; Deporte-Fety, R.; Thomare, P.; Giacalone, F.; Meflah, K. Phase I trial of interleukin-2 and high-dose arginine butyrate in metastatic colorectal cancer. *Cancer Immunol. Immunother.* **2000**, *49*, 56–61. [CrossRef]
62. Ben-Sasson, S.Z.; Hu-Li, J.; Quiel, J.; Cauchetaux, S.; Ratner, M.; Shapira, I.; Dinarello, C.A.; Paul, W.E. IL-1 acts directly on CD4 T cells to enhance their antigen-driven expansion and differentiation. *Proc. Natl. Acad. Sci. USA* **2009**, *106*, 7119–7124. [CrossRef] [PubMed]
63. Schenten, D.; Nish, S.A.; Yu, S.; Yan, X.; Lee, H.K.; Brodsky, I.; Pasman, L.; Yordy, B.; Wunderlich, F.T.; Brüning, J.C.; et al. Signaling through the adaptor molecule MyD88 in CD4$^+$ T cells is required to overcome suppression by regulatory T cells. *Immunity* **2014**, *40*, 78–90. [CrossRef] [PubMed]
64. Mantovani, A.; Dinarello, C.A.; Molgora, M.; Garlanda, C. Interleukin-1 and related cytokines in the regulation of inflammation and immunity. *Immunity* **2019**, *50*, 778–795. [CrossRef] [PubMed]
65. Huang, Q.; Jacquelot, N.; Preaudet, A.; Hediyeh-Zadeh, S.; Souza-Fonseca-Guimaraes, F.; McKenzie, A.N.J.; Hansbro, P.M.; Davis, M.J.; Mielke, L.A.; Putoczki, T.L.; et al. Type 2 Innate Lymphoid Cells Protect against Colorectal Cancer Progression and Predict Improved Patient Survival. *Cancers* **2021**, *13*, 559. [CrossRef]
66. Hong, I.-S. Stimulatory versus suppressive effects of GM-CSF on tumor progression in multiple cancer types. *Exp. Mol. Med.* **2016**, *48*, e242. [CrossRef] [PubMed]
67. Li, Y.-P.; Du, X.-R.; Zhang, R.; Yang, Q. Interleukin-18 promotes the antitumor ability of natural killer cells in colorectal cancer via the miR-574-3p/TGF-β1 axis. *Bioengineered* **2021**, *12*, 763–778. [CrossRef]
68. Feng, X.; Zhang, Z.; Sun, P.; Song, G.; Wang, L.; Sun, Z.; Yuan, N.; Wang, Q.; Lun, L. Interleukin-18 is a prognostic marker and plays a tumor suppressive role in colon cancer. *Dis. Markers* **2020**, *2020*, 6439614. [CrossRef]
69. Lyu, S.-X.; Ye, L.; Wang, O.; Huang, G.; Yang, F.; Liu, Y.; Dong, S. IL-27 rs153109 polymorphism increases the risk of colorectal cancer in Chinese Han population. *OncoTargets Ther.* **2015**, *8*, 1493–1497. [CrossRef]
70. Song, X.; Traub, B.; Shi, J.; Kornmann, M. Possible roles of interleukin-4 and -13 and their receptors in gastric and colon cancer. *Int. J. Mol. Sci.* **2021**, *22*, 727.
71. Scheller, J.; Chalaris, A.; Schmidt-Arras, D.; Rose-John, S. The pro- and anti-inflammatory properties of the cytokine interleukin-6. *Biochim. Biophys. Acta Mol. Cell Res.* **2011**, *1813*, 878–888. [CrossRef]
72. Turano, M.; Cammarota, F.; Duraturo, F.; Izzo, P.; De Rosa, M. A potential role of IL-6/IL-6R in the development and management of colon cancer. *Membranes* **2021**, *11*, 312. [CrossRef] [PubMed]
73. Crucitti, A.; Corbi, M.; Mc Tomaiuolo, P.; Fanali, C.; Mazzari, A.; Lucchetti, D.; Migaldi, M.; Sgambato, A. Laparoscopic surgery for colorectal cancer is not associated with an increase in the circulating levels of several inflammation-related factors. *Cancer Biol. Ther.* **2015**, *16*, 671–677. [CrossRef] [PubMed]
74. Kwon, K.A.; Kim, S.H.; Oh, S.Y.; Lee, S.; Han, J.-Y.; Kim, K.H.; Goh, R.Y.; Choi, H.J.; Park, K.J.; Roh, M.S.; et al. Clinical significance of preoperative serum vascular endothelial growth factor, interleukin-6, and C-reactive protein level in colorectal cancer. *BMC Cancer* **2010**, *10*, 203. [CrossRef] [PubMed]
75. Yamaguchi, M.; Okamura, S.; Yamaji, T.; Iwasaki, M.; Tsugane, S.; Shetty, V.; Koizumi, T. Plasma cytokine levels and the presence of colorectal cancer. *PLoS ONE* **2019**, *14*, e0213602. [CrossRef] [PubMed]
76. Johdi, N.A.; Mazlan, L.; Sagap, I.; Jamal, R. Profiling of cytokines, chemokines and other soluble proteins as a potential biomarker in colorectal cancer and polyps. *Cytokine* **2017**, *99*, 35–42. [CrossRef] [PubMed]
77. Ghaderi, A.; Abtahi, S.; Davani, F.; Mojtahedi, Z.; Hosseini, S.V.; Bananzadeh, A. Dual association of serum interleukin-10 levels with colorectal cancer. *J. Cancer Res. Ther.* **2017**, *13*, 252–256. [CrossRef] [PubMed]
78. Oliveira, S.B.; Monteiro, I.M. Diagnosis and management of inflammatory bowel disease in children. *BMJ* **2017**, *357*, j2083. [CrossRef] [PubMed]
79. Chen, T.; Guo, J.; Cai, Z.; Li, B.; Sun, L.; Shen, Y.; Wang, S.; Wang, Z.; Wang, Z.; Wang, Y.; et al. Th9 cell differentiation and its dual effects in tumor development. *Front. Immunol.* **2020**, *11*, 1026. [CrossRef] [PubMed]
80. Czajka-Francuz, P.; Francuz, T.; Cisoń-Jurek, S.; Czajka, A.; Fajkis, M.; Szymczak, B.; Kozaczka, M.; Malinowski, K.P.; Zasada, W.; Wojnar, J.; et al. Serum cytokine profile as a potential prognostic tool in colorectal cancer patients—One center study. *Rep. Pract. Oncol. Radiother.* **2020**, *25*, 867–875. [CrossRef] [PubMed]
81. Mansoori, B.; Mohammadi, A.; Davudian, S.; Shirjang, S.; Baradaran, B. The different mechanisms of cancer drug resistance: A brief review. *Adv. Pharm. Bull.* **2017**, *7*, 339. [CrossRef] [PubMed]
82. Redmond, K.M.; Wilson, T.R.; Johnston, P.G.; Longley, D.B. Resistance mechanisms to cancer chemotherapy. *Front. Biosci.* **2008**, *13*, 5138–5154. [CrossRef] [PubMed]
83. Nathanson, D.A.; Gini, B.; Mottahedeh, J.; Visnyei, K.; Koga, T.; Gomez, G.; Eskin, A.; Hwang, K.; Wang, J.; Masui, K.; et al. Targeted Therapy Resistance Mediated by Dynamic Regulation of Extrachromosomal Mutant EGFR DNA. *Science* **2014**, *343*, 72–76. [CrossRef] [PubMed]

MDPI

St. Alban-Anlage 66

4052 Basel

Switzerland

www.mdpi.com

Molecules Editorial Office

E-mail: molecules@mdpi.com

www.mdpi.com/journal/molecules